15-75

Landform systems

Victoria Bishop

Head of Geography, Lutterworth Upper School and Community College

Robert Prosser

Lecturer, CURS, University of Birmingham

Collins Educational

An imprint of HarperCollins*Publishers*

Contents

Skills matrix
(distribution of numbered tasks)

Chapter	Understanding of text newspaper extracts and classification	Graphical/mapping methods and annotated diagrams	Analysis of data from tables, graphs and diagrams	Analysis of photographs	Analysis of maps	Statistical analysis/methods	Values enquiry	Project work based on library/fieldwork/research/ data collection	Writing: essays, reports
1	1, 2, 11, 12, 13, 14	3, 7, 9	2, 3, 4, 5, 6, 8, 10	11, 12					1, 2, 12
2	5, 6, 7, 11, 14, 15, 16, 17, 18, 25	4, 5, 9b, 15	1, 2, 4, 8, 9, 10, 11, 12, 16, 19, 20, 21, 22, 23, 24, 25	7, 13		3, 12			9, 10, 11, 23, 25
3	3, 6, 7, 8, 9, 10, 11, 12, 13, 14, 17, 18	1, 7	2, 4, 5, 15, 16	1					3, 4, 5, 6, 9, 10, 11, 12, 13, 14, 16, 17, 18
4	7, 8, 11, 12, 14	7, 11	1, 2, 3, 4, 6, 9, 10	5	13, 15, 16, 17, 18				3, 4, 8, 9, 10, 11, 12, 14, 15, 18
5	12, 14, 33	2, 8, 10, 13, 15, 16, 17, 20, 21, 36, 38	2, 3, 4, 5, 6, 7, 9, 11, 12, 16, 17, 18, 19, 23, 25, 26, 27, 28, 29, 30a, 31, 32, 34	10, 15	1	25, 29, 30a			5, 9, 13, 14, 22, 24, 27, 30b, 33, 35, 37
6	1, 4, 5, 6, 16		3, 8, 10, 11	2, 12, 13, 14	18		15, 17	18	1, 4, 5, 6, 7, 8, 9, 12, 16, 19
7	1, 9, 10, 11, 13, 14, 15, 21	5, 12a, 12b, 14, 15, 16	6, 8, 13a, 14, 18, 19, 20	17, 18	2, 3	12d, 16a, 16c, 16d		4	1, 7, 8, 11, 12c, 17
8	3, 13, 18, 20, 21, 23, 24, 25, 26, 28	14, 27	2, 3, 4, 5, 6, 7, 8, 9, 10, 11, 15, 17, 19, 22, 29, 31, 35		12	8, 9, 10			1, 3, 5a, 13, 16, 17, 23, 30, 32, 33, 34, 36, 37
9	1, 5, 7, 8, 14, 16, 19	7, 10, 20, 25	2, 9, 13, 17, 22		3, 4, 6, 10, 11, 12, 23	2, 4, 6, 24			1, 15, 18, 21

To the student

This is a book about geomorphology – the study of the forms and shapes which make up the earth's surface. A landform is an individual physical feature on this surface. There is an enormous diversity of landforms, from cliffs to canyons, from deltas to dunes, from pingos to pediments. Any specific natural landscape is made up of a collection of individual landforms. Each landscape is unique; there are, however, certain commonalities, and so we talk of desert landscapes, glaciated landscapes etc., in which we expect to find distinctive landforms.

Such general classifications are based upon the principal landforming agents in a particular environment. These agents are wind, water, ice and tectonic forces. The first three provide the basis for the structure of this book, and there are separate chapters devoted to the work of the sea, rivers, glaciers, wind and water in arid environments. Beware, however, for one of the most important understandings you will gain is that in any landscape, more than one agent is, or has been, at work. So glaciated landscapes are not only the result of the work of ice; coastal landforms are the result of more than the work of the sea, and so on.

As the title suggests, the book adopts the systems approach. This approach is introduced in Chapter 1, showing that any landform, from an individual sand-dune to a stretch of coast, may be studied in a similar way. Processes involving energy and material (inputs) are at work on the landform. Some of the energy and materials are stored in the landform, while some are lost (outputs). So we learn to ask some straightforward questions about any landform – what is it? where is it? what caused it? why is it there? how is it changing? It is important, therefore, that you understand what a system is and how it can be used as a general model or idea for studying landforms. As a result, you should read Chapter 1 carefully first. After this, the topics may be covered in any sequence, to suit your syllabus.

Throughout the book we have illustrated key terms and processes with case studies. Learning these terms, and the functioning of the diverse processes through the various themes, will increase your understanding of how landscapes evolve. The suggested activities should improve your skills in completing assignments and answering examination questions. These skills will prove useful in other areas of your geography studies. Try to apply the understandings of the processes to new situations as they arise, e.g. in current events, and to use your knowledge to consider the impacts of human interactions with the natural environment.

1 Landform systems

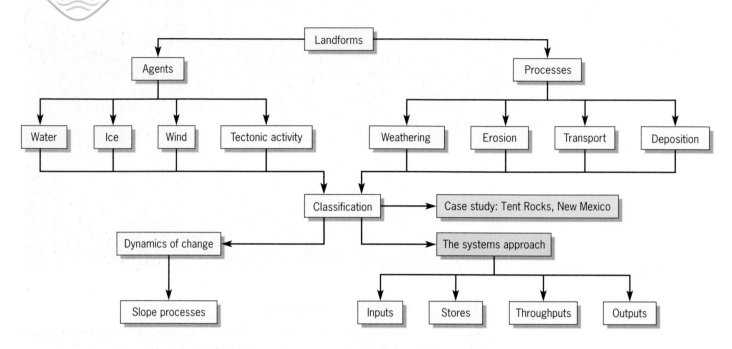

Landforms

Agents

Processes

Water | Ice | Wind | Tectonic activity

Weathering | Erosion | Transport | Deposition

Classification

Case study: Tent Rocks, New Mexico

Dynamics of change

The systems approach

Slope processes

Inputs | Stores | Throughputs | Outputs

1.1 Introduction

Take a trip into the countryside, or walk along a beach or cliff-top. As you look around, one part of your enjoyment is likely to be the shapes you see in the landscape: a sharp hill, a deep river valley, a vertical cliff, and so on. Each individual feature is a *landform*, and any landscape you gaze upon is made up of a unique assembly of landforms. The study of landforms is called geomorphology, i.e. 'geo' meaning earth, and 'morphology' meaning structure and shape.

The aim of this book is to improve your understanding of these shapes in the landscape, by considering questions such as those set alongside Figure 1.1. These understandings will help you to consider how environments should be managed and used. You may also find that you get more enjoyment from 'the great outdoors' if you understand what is going on around you!

?

1 With reference to Figure 1.1, explain the difference between a landscape and landforms.

Key questions

• What is it? • What shape is it? • What is it made of? • How did it get like this? • Why is it in this location? • How long has it taken to get like this? • Is it still forming and changing? • How can we tell what is happening? • What human influences are affecting it? • What will happen to it in the future?

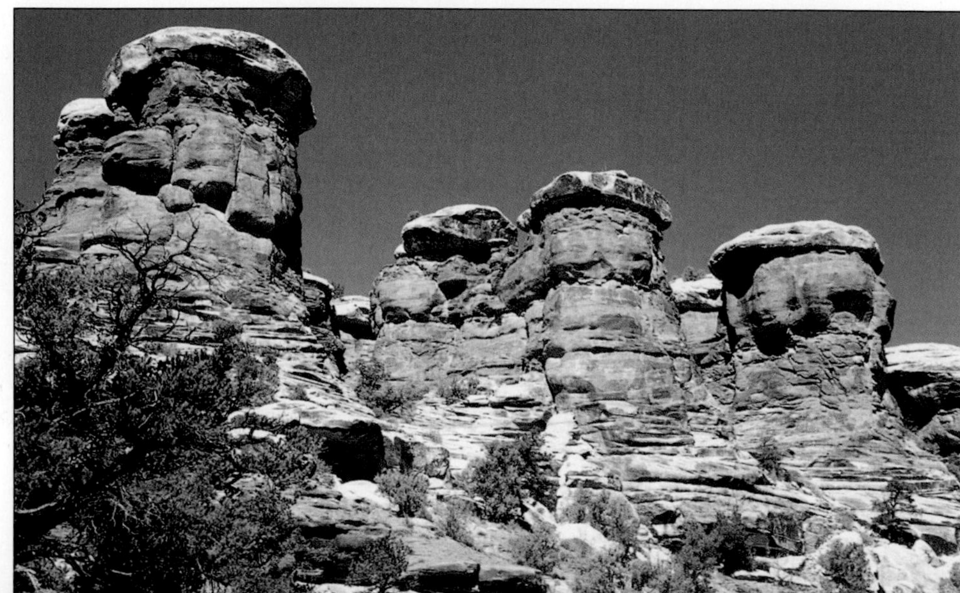

Figure 1.1 Spectacular landforms in Elephant Canyon, Canyonlands National Park, Utah, USA

Figure 1.2 Mount Rainier, Washington State, USA. This huge volcano, the result of tectonic activity, is being modified by the work of ice, water, wind and temperature fluctuations. Several glaciers are carving into the lava and ash beds, transporting and depositing much of the weathered and eroded material down the mountain.

1.2 Agents and processes

Landforms are the result of the work of four major **agents**, often functioning in combination: water (rivers and seas); ice; wind; tectonic activity. This book focuses upon the work of the first three. **Tectonic activity**, e.g. earthquakes and volcanic eruptions, produces such features as fault troughs, mountain ranges and volcanoes which are then worked on by the other agents (Fig. 1.2).

This work of landform formation is achieved by a variety of processes of *weathering, erosion, transportation, deposition*. Thus the butte landform of Figure 1.3 has been formed by processes associated with wind and water, assisted by temperature fluctuations and gravity to transport and deposit the weathered material, over thousands of years. It is useful at this point to distinguish between weathering and erosion.

Figure 1.3 A sandstone butte, Canyonlands National Park, Utah, USA. This isolated tower is surrounded by a 'skirt' of debris weathered and eroded from the solid rock above.

Weathering: The disintegration or decomposition of rock on site (*in situ*) at or near the earth's surface by physical breakup or chemical change.

Erosion: The wearing away and removal of surface and near-surface materials by a range of processes due to the action of agents such as waves, running water (e.g. streams and rivers), wind and moving ice (e.g. glaciers). The key difference is that erosion includes the element of transportation (removal), whereas weathering refers to the processes at work on site (*in situ*) only. The wearing down of the earth's surface by both weathering and erosion is called *denudation*.

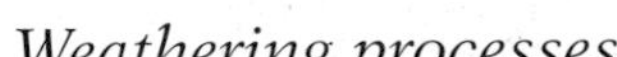

Weathering processes

Rocks are broken down by three sets of *weathering processes* – mechanical, chemical and biological. *Mechanical weathering* (Table 1.1) involves the breakup (disintegration) of material by physical processes, e.g. impacts, pressure release, or temperature changes (see, for example, the debris around the base of the butte in Figure 1.3). *Chemical weathering* (Table 1.2) or decomposition is achieved by water acting as a dilute acidic solution, which causes chemical reactions within the mineral constituents of a rock.

Table 1.1 Mechanical weathering processes

Process	Comments
Pressure release	As overlying rocks are eroded the deeper rocks come near to the earth's surface. The reducing pressure load causes expansion of the rock mass. Joints develop in the rock, often parallel to the ground surface. This process is best developed in granite. At a smaller scale, the arrangement of the minerals in the rock can provide lines of weakness for the rock to break up by *exfoliation* (thin sheets breaking off) or *granular disintegration* (individual grains or crystals break up).
Insolation weathering	Results from temperature changes e.g. diurnal changes in hot deserts, causing thermal expansion and contraction of the outer layers of the rock surfaces. The different minerals in the rock will have different rates of thermal change. Fires have a similar though less regular effect.
Frost weathering	The breakdown of rock as a result of the stresses caused by the freezing of water. Studies suggest that for this process to be active the freezing must be rapid, temperatures must reach at least –5°C, and this must occur frequently. This process will be most important in mountain and periglacial environments rather than glacial environments where temperatures remain constantly low rather than changing above and below the freezing point.
Salt weathering	The precipitation of salts in gaps and voids, and the expansion of salt crystals through hydration or heating. Most active in arid environments with high evaporation rates and in coastal environments.
Wetting and drying	Simple wetting and drying of rock surfaces causes expansion and contraction of rock minerals. Clay minerals are especially susceptible as they can absorb relatively large amounts of water. This process is important in the coastal zone.

Table 1.2 Chemical weathering processes

Name	Process	Comments
Solution	Water acts as a solvent.	Solubility of rock minerals depends upon temperature and pH of the water. The mineral halite is very soluble.
Hydration	Rock minerals absorb water into their crystal structure.	The crystal structure of the mineral is weakened allowing other weathering processes to operate, e.g. iron oxide is changed to hydrated iron oxide.
Hydrolysis	Water acts to replace metal cations in the mineral. The metal cations e.g. K^+, Na^+ and Ca^+ are replaced by H^+ ions.	Feldspar minerals are changed into clay minerals, e.g. kaolinite and montmorillonite.
Carbonation	Carbon dioxide in water reacts with the mineral calcite to produce bicarbonate ions.	Important in the weathering of limestones.The factors affecting the rate of carbonation are complex. Temperature and solubility of atmospheric and soil carbon dioxide are important. At high temperatures chemical reactions are quicker, but carbon dioxide is more soluble at low temperatures.
Oxidation	Oxygen dissolves in water causing the mineral to lose an electron.	The minerals increase their positive charge. This is most common with iron, titanium and manganese. We are all familiar with this process when iron rusts.
Reduction	Mineral ions gain an electron.	Mineral structure can collapse or become more susceptible to other weathering processes.
Cation exchange	One cation is exchanged for another of a different element in a rock mineral structure.	Commonly affects clay minerals.
Chelation	Biochemical weathering process. Metal cations in the mineral are mobilised.	Organic processes produce the chelating agents, e.g. secretions from lichen, or from the decomposition of humus.

The acidic component may be supplied from the atmosphere, e.g. CO_2 (carbon dioxide) to combine with H_2O (water) to create dilute H_2CO_3 (carbonic acid); SO_2 (sulphur dioxide), to create H_2SO_4 (sulphuric acid), or a variety of organic acids from plant roots and decomposing organic matter, such as peat. This last source activates an important sub-type of chemical weathering, called *biochemical weathering*.

The third type of weathering is *biological*. This occurs due to the action of plants and animals, e.g. where rock is broken down by the penetration of plant roots, and is therefore a form of mechanical disintegration. However, water can percolate along the root paths and, through the addition of organic acids from the plant roots, biochemical weathering becomes active.

As weathering breaks down rock into smaller particles (mechanical weathering) or new minerals which are more stable in the environment of the earth's surface (chemical and biochemical weathering), the weathered material forms a *weathering mantle* or **regolith** on the unweathered solid rock beneath. The boundary between the weathered and unweathered bedrock is called the *weathering front*. The regolith may eventually form soil.

Figure 1.4 Dramatic karst landforms developed on thick-bedded and well-jointed limestones along the River Li, near Guilin, south-central China

Rates of weathering

The rates of weathering at any location will depend upon the interrelationships between the rock type, climate, topography and time.

Rock type

Rock type determines the rock minerals which are decomposed by weathering processes and the zones and lines of weakness in the rock (Table 1.3). Rocks are formed under conditions of heat (igneous) and pressure on and compression of sediments (sedimentary) or both (deep-seated igneous and metamorphic rocks). They are therefore unstable under the relatively low temperature and pressure conditions of the earth's surface. How susceptible a rock type is to chemical weathering will depend upon the mineral composition of the rock. Minerals such as olivine, augite, hornblende and mica are the most susceptible to chemical weathering, whereas feldspars and quartz are the least susceptible. However, it is important to realise that it only needs the breakdown of one mineral in the rock to weaken its whole structure. For example, sandstones are usually composed of quartz grains which are very resistant to weathering. However, the cement of the rock may be composed of another less resistant mineral which as it weathers will cause the quartz grains to be detached and the whole rock to break down. The quartz grains will form the regolith.

Two common rock types susceptible to chemical and biochemical weathering are limestones and granites. Limestones, which contain much calcium carbonate ($CaCO_3$), are vigorously attacked by dilute H_2CO_3 to produce distinctive **karst** landscapes (Fig. 1.4). Granites are crystalline igneous rocks consisting mainly of quartz, micas and feldspars (Table 1.3).

Table 1.3 The major rocks, formation and susceptibility to weathering

Rock type	Formation	Main minerals	Weathering processes and properties
Granite	A light-coloured, coarse-grained (>5mm) igneous rock formed by slow cooling deep in the earth's crust.	Quartz, feldspars and micas	Quartz is relatively resistant to mechanical and chemical weathering. Micas are most susceptible minerals. Both micas and feldspars weather to form clay minerals. Granite covers approximately 15% of the earth's surface and has an average porosity of 1%.
Basalt	A fine-grained (<1mm), dark-coloured igneous rock formed by rapid cooling at or near the earth's surface, e.g. in lava flows.	Feldspars, pyroxenes and olivines	Pyroxenes and olivines susceptible to chemical weathering processes. Cooling and unloading joints aid weathering processes. Covers 3% of the earth's surface. Low porosity of 1% on average.
Gabbro	A coarse-grained, dark-coloured igneous rock formed by slow cooling as with granite. Similar chemical composition to basalt.	Feldspars, pyroxenes and olivines	Mineral susceptibility as with basalt.
Mudstones	These are a group of fine-grained, sedimentary rocks, including shales and clays. The clay minerals result from the chemical weathering of previously existing rocks.	Clay minerals, e.g. kaolinite and montmorillonite	Susceptible to wetting and drying. Further chemical weathering will produce new clay minerals. These rocks are common and cover approximately 52% of the earth's surface. Porosity varies between 18–45%.
Sandstone	A medium-grained sedimentary rock. Quartz is a common mineral forming sandstones since it is slow to weather and is detached from the original rock as sand particles. Transportation, deposition and consolidation, of these grains produces the sandstone rock.	Quartz	Quartz is relatively resistant to weathering, but the cement binding the quartz grains together may be more easily weathered resulting in the breakdown of the rock structure. Bedding planes, joints and cross-bedding structures will allow weathering processes to penetrate the rock. Covers 15% of the earth's surface. Porosity averages 18%.
Limestone	Formed by deposition of organic material or remains, e.g. shells and bone fragments. Chalk is a type of limestone.	Calcite	Most susceptible rock to the process of carbonation. Covers 7% of the earth's surface. Average porosity 10%.
Metamorphic rocks	Formed by change to previously existing rocks due to heat and/or pressure effects of tectonic or igneous activity.	Variable, but commonly feldspars, micas, hornblende, quartz and garnet	Micas and hornblende are particularly susceptible to chemical weathering. The minerals are often in layers, which allows water penetration into the body of the rock.

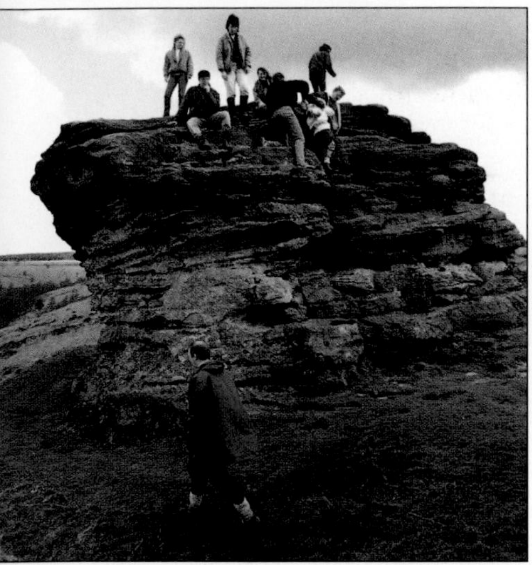

Figure 1.5 Granite tor, Dartmoor. The form of the tor is controlled by the joint system, along which chemical and mechanical weathering are most active. (Note the power of students as weathering agents!)

It is the rotting or decomposition of the micas and then the feldspars which helps to weaken the rock fabric. The quartz may not chemically weather at all in many environments. One end-product of the chemical weathering of granite is the economically valuable china clay (kaolinite).

Rock porosity and structural weakness

Rock porosity will determine how much water can enter the body of the rock for chemical weathering to occur. In addition, structural weaknesses such as joints and bedding planes are zones and lines of weakness which allow water to enter the rock and increase rates of weathering. For example, the pattern of the limestone 'towers' shown in Figure 1.4 is controlled by the bedding plane and joint network of the limestones. The tors of Dartmoor (Fig. 1.5) reflect the vigorous chemical and mechanical weathering along the joint system of the granite.

Climate

Climate is a crucial factor in the nature and rates of the weathering processes operating (Fig. 1.6 and Table 1.4). As temperature increases, so do the rates of chemical reactions and therefore chemical weathering. However, water is vital for these processes as well. Overall, maximum rates of chemical weathering occur in the humid tropics where there are constantly high temperatures and moisture availability. If either variable is reduced, then so is the rate of weathering. As well as rates of weathering, different climatic conditions influence the type of weathering processes operating (Table 1.4).

Local conditions

Weathering rates vary over short distances. Topography and slope position influence the rate of water movement through the regolith to the unweathered bedrock (the weathering front). Aspect can also influence rates of weathering, since slopes facing into the sun may have different temperature and moisture conditions from those facing away from the sun.

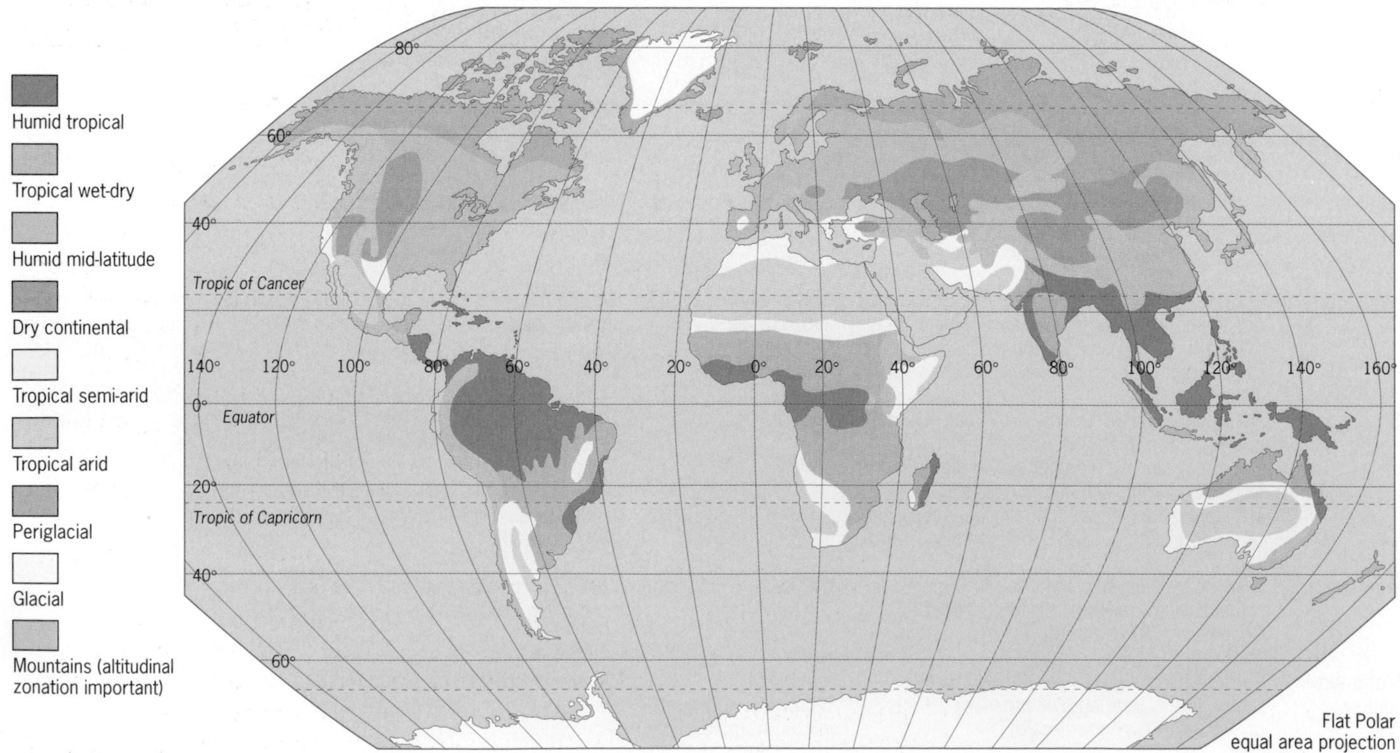

Figure 1.6 Global distribution of morphoclimatic zones

Table 1.4 The earth's major morphoclimatic zones (*Source:* Summerfield, 1991)

Morphoclimatic zone	Mean annual temperature (°C)	Mean annual precipitation (mm)	Relative importance of geomorphic processes
Humid tropical	20–30	>1500	High potential rates of chemical weathering; mechanical weathering limited; active, highly episodic mass movement; moderate to low rates of stream corrasion but locally high rates of dissolved and suspended load transport.
Tropical wet–dry	20–30	600–1500	Chemical weathering active during wet season; rates of mechanical weathering low to moderate; mass movement fairly active; fluvial action high during wet season with overland and channel flow; wind action generally minimal but locally moderate in dry season.
Tropical semi-arid	10–30	300–600	Chemical weathering rates moderate to low; mechanical weathering locally active especially on drier and cooler margins; mass movement locally active but sporadic; fluvial action rates high but episodic; wind action moderate to high.
Tropical arid	10–30	0–300	Mechanical weathering rates high (especially salt weathering); chemical weathering minimal; mass movement minimal; rates of fluvial activity generally very low but sporadically high; wind action at a maximum.
Humid mid-latitude	0–20	400–1800	Chemical weathering rates moderate, increasing to high at lower latitudes; mechanical weathering activity moderate with frost action important at higher latitudes; mass movement activity moderate to high; moderate rates of fluvial processes; wind action confined to coasts.
Dry continental	0–10	100–400	Chemical weathering rates low to moderate; mechanical weathering, especially frost action, seasonally active; mass movement moderate and episodic; fluvial processes active in wet season; wind action locally moderate.
Periglacial	<0	100–1000	Mechanical weathering very active with frost action at a maximum; chemical weathering rates low to moderate; mass movement very active; fluvial processes seasonally active; wind action rates locally high.
Glacial	<0	0–1000	Mechanical weathering rates (especially frost action) high; chemical weathering rates low; mass movement rates low except locally; fluvial action confined to seasonal melt; glacial action at a maximum; wind action significant.
Azonal mountain zone	Highly variable	Highly variable	Rates of all processes vary significantly with altitude; mechanical and glacial action become significant at high elevations.

Weathering and slope processes

The rate of weathering may be influenced by the rate of removal (transportation) of regolith debris. For example, in lowland areas of the humid tropics, there is often a thick regolith produced by rapid chemical weathering processes. However, when the regolith is deep, there is a slower rate of water movement down to the weathering front. Thus, the rate of chemical weathering is reduced, even in an environment where rates of weathering are potentially very high. On mountain slopes in the same region, the situation is different. Deep layers of regolith do not build up because the rapid mass movement processes (see below) remove regolith debris as it forms. The regolith remains thin and rates of weathering high.

This link between rates of regolith production and rates of removal by mass movement introduces the idea of *weathering-limited slopes* and *transport-limited slopes*. Weathering-limited slopes occur where the erosional processes are faster than the rate of weathering to supply new material. There is little or no regolith cover, and the form of the slope will be controlled by the rock types and structures forming the slope and their relative resistance to weathering. Transport-limited slopes are more common globally and occur where the rates of weathering are greater than the erosional processes. The slopes have a regolith cover with a depth which reflects the balance between weathering inputs and erosion outputs.

Classification of landforms

The most obvious way to classify landforms is by the primary agent of origin, i.e. whether they result from the work of rivers, the sea, ice or wind. This approach has been used as the basis for the structure of this book. However, as examples throughout the book illustrate, more than one agent may have played its part in the evolution of a particular landform.

A second approach is to classify a landform according to the dominant process, e.g. mechanical weathering (Table 1.4). Once again, we must be careful, as more than one process may have been involved.

In any landscape, we need to consider time as a factor. Over time, climatic change may have changed the principal agent or process operating or changed their relative importance. For example, many of the landforms in Britain are the result of the work of ice (see Chapters 7 and 8) during the Pleistocene Period. These are being modified by present-day processes which are dominated by fluvial (river) activity.

An alternative process-based general classification is to assign a landform according to whether it is the result of erosional or depositional processes, e.g. a sea cliff is essentially an erosional landform, while a sand-spit is formed by depositional processes.

A third approach is to classify a landform according to its stage of development. This introduces three fundamental understandings you will find useful in all chapters. First, that any individual landform evolves over time; second, that examples of the various stages of development of this landform type can be found in different places at a particular moment in time; third, that the timescales involved in the 'life cycle' vary widely according to landform. Thus, as we travel along a river valley, we may see features such as meanders, terraces, waterfalls and floodplains at all stages of development (see the Case Study – Tent Rocks, New Mexico).

Some features will be slow-forming and long-lasting, while others will form quickly, be short-lived or constantly changing. The butte of Figure 1.3 is a fully developed feature. It began as part of a spur, became isolated, and will eventually disintegrate into a mound of debris which itself will ultimately be eroded away. This progression takes many thousands of years. The eroded material will provide material for a depositional landform elsewhere.

Landform development over space and time: Tent Rocks, New Mexico, USA

The high plateaus of New Mexico (altitudes above 2000m) are built of great thicknesses of sedimentary rocks, interspersed with beds of volcanic lavas and ashes. As the land has slowly been uplifted over millions of years, so rivers have incised deep canyons such as that of the Rio Grande (Fig. 1.7). South-west of Santa Fe, a minor tributary of the Rio Grande has cut Cochiti Canyon through thick beds of volcanic ash. Within these ash beds lie several hard lava flows (Fig. 1.8).

Along the flanks of Cochiti Canyon we can see all stages in the development of some very distinctive landforms. At one location, at one time, we can learn how a landform evolves over thousands of years. Figure 1.8 shows vividly why these features are known as Tent Rocks: numerous pointed hills looking like conical tents. The dark bands and blocks are the lavas set within the thick, pale beds of ash.

The diagram of Figure 1.9 divides the landscape into four components, each representing a stage in the evolution of the landforms. At Stage 1, the plateau surface is intact, with a continuous capping of a resistant lava bed. Joint systems are developing from tensions caused by the slow regional uplift. At Stage 2, weathering has opened up the main joints, exposing the less resistant ash beds. Separate flat-topped buttes, capped by the lava bed, develop. As weathering proceeds, the more resistant lavas act as

Figure 1.7 Deep canyons carved by the Rio Grande, USA

Figure 1.8 Tent Rocks, Cochiti Canyon, New Mexico, USA. The thick beds of pale, coarse volcanic ash are interspersed with darker lava beds.

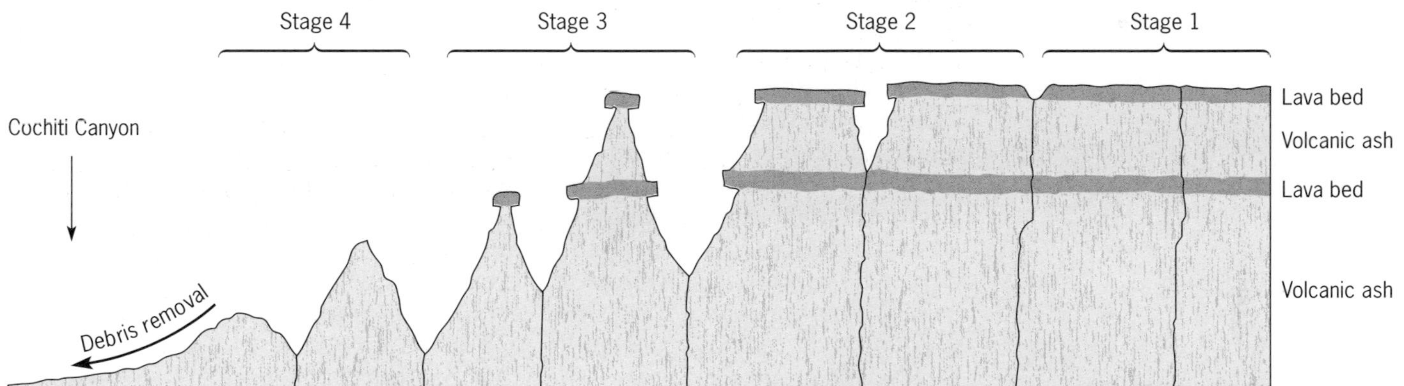

Figure 1.9 Landform evolution at Tent Rocks

perched caps, protecting the ash beds below. The location of each 'tent' is controlled by joint systems (stages 1 and 2 on Fig. 1.9). By Stage 3, the 'tent' shapes are well developed. As the gaps between the hills widen, so the lava beds are increasingly exposed to the weathering processes of wind, rain, snow and freeze–thaw. Weathering along the joint network penetrates progressively deeper as the gaps above open up. The resistant lavas remain as perched caps, but eventually topple. The steep slopes ensure the rapid removal of any loosened material. Once the lava cap is removed (Fig. 1.10) weathering and erosion of the conical 'tents' accelerates (Stage 4).

Figure 1.10 Tent Rocks. The final stage of evolution (stage 4 on Fig. 1.9). The protective lava caps have toppled and the ash 'tents' are eroded vigorously by heavy summer rain storms. The water runs as sheets across the steep surfaces, removing loosened surface particles.

?

2 Describe and explain briefly how the landforms of today's Stage 2 in Figure 1.9 will develop in the future.

1.3 The dynamics of landforms: a general model

One of the most important understandings about how landscapes and individual landforms evolve is that *changes are irregular*. For example, on a field trip, for safety and comfort reasons, you are likely to be taking measurements on a beach or along a stream, in fairly quiet conditions – and little seems to be happening. Return to the same locations during a violent storm or flood, however, and landforms may change radically as you watch, e.g. cliffs or river banks collapse. (Look again at the cover photograph of this book.)

The work done in a landscape is related to the amount, character, location and duration of energy available. The energy available in the wind, water or ice at a given location varies. In times of increasing energy, erosional and transportational processes dominate. As available energy decreases, so depositional processes increase. In general, a landscape or landform tends to exist for relatively long periods with slow rates of change. These periods of relative stability are interrupted by briefer episodes of abrupt changes in environmental conditions, e.g. surges of energy applied through the wind, water or ice. As you use the materials in this book, you will learn that the timescales of these events vary widely, e.g. gale-force winds every few months, flood discharges along a river every few years, accelerated glacier flow measured over decades or even centuries.

During the periods of relative stability, processes will still be at work. The rate of change may be slow, but few landforms are ever in a static state. They exist in overall balance with the weathering agents and processes, but adjust constantly to fluctuations in the energy available. This is called *dynamic equilibrium*: a condition in which a landscape or landform fluctuates about a mean value which is itself altering over time. For example, a beach adjusts to the daily and monthly tidal rhythms (Chapter 5); a sand-dune adjusts to seasonal shifts in predominant wind direction (Chapter 9).

Such gradual changes may lead eventually to instability, and so to a sudden modification of the landform. For example, slow but prolonged weathering along the joints in the sandstone butte of Figure 1.3 will finally cause a section of the tower to collapse. However, most radical changes to landforms occur during the relatively brief disruptive episodes. During these episodes, the balanced relationship between the landform and the forces working upon it, i.e. the condition of dynamic equilibrium, breaks down – for example, a disastrous flood cuts a new river channel, or a storm may throw up a pebble beach bar well above the normal limits of wave action (Fig. 1.11). These are high-energy events, but reductions in available energy

Figure 1.11 A typical storm beach

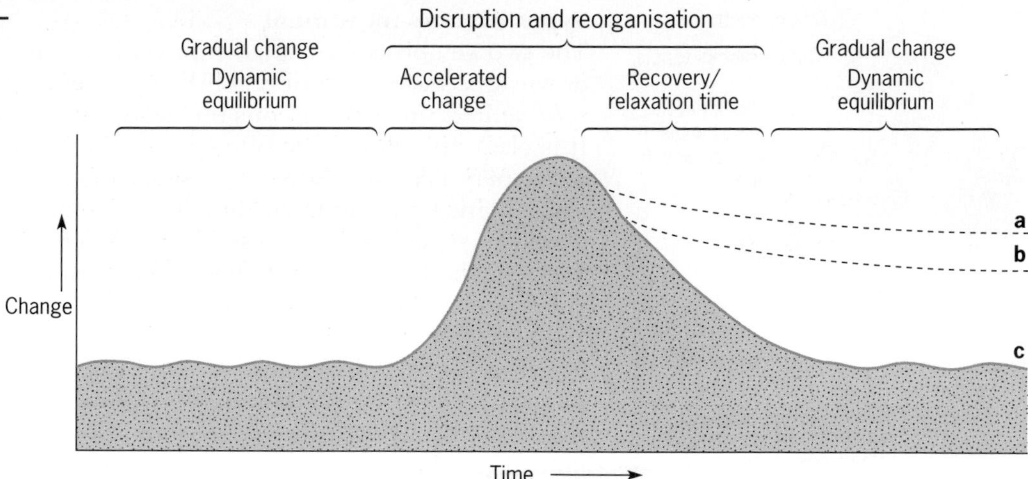

Figure 1.12 Landform development – a dynamic model

can also lead to rapid landform change, e.g. several unusually warm years (increased inputs of solar energy) may cause increased ice melting and reduced glacier movement, leading to increased deposition of debris (Chapter 7).

As the disruptive episode ends, the landform takes on a new shape. The dynamic model in Figure 1.12 suggests either that the change will be permanent, with processes working to establish a new dynamic equilibrium (*a* and *b*), or that the processes will work to return the landform to its original condition (*c*). Thus, a stream may stay in its new channel or may eventually return to its original course; the storm beach may be long-lasting, or may be modified fairly quickly.

It is important to bear in mind that the model of Figure 1.12 is very generalised. It functions somewhat differently for each of the agents of landform formation, as examples throughout this book illustrate.

1.4 Mass movement

As almost all land surfaces have some slope, or gradient, **gravity** is an important factor influencing the movement of weathered material (Fig. 1.13). We call this general process *mass movement*: 'the detachment and downslope transport of soil and rock material under the influence of gravity' (Chorley et al., 1984). The term includes the downslope movement of both unweathered rock and regolith. Where only regolith debris is involved, the correct term is *mass wasting*.

Speed of movement

Mass movement occurs at various speeds, some of which are extremely slow. Across gently rolling grassy hills, in a moist temperate climate such as that of Britain, a particle may move downslope at less than 1mm a year. The slope is stable, in dynamic equilibrium, with the input of weathered material to the upper slope approximately equal to the output from the slope foot.

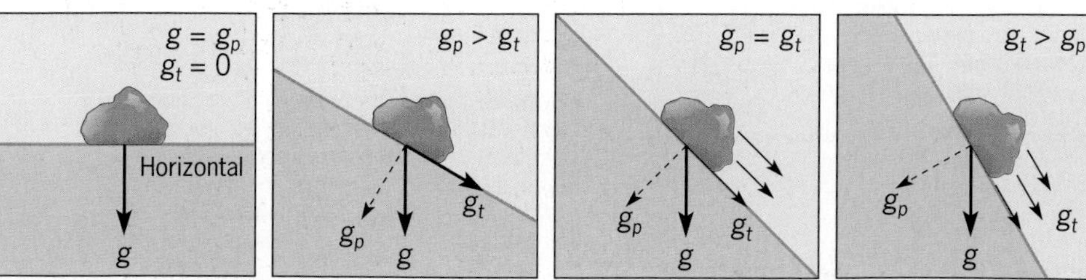

Figure 1.13 The effects of gravity on a rock lying on a hillslope

Types of slow movement

The two key processes of slow movement are *creep*: 'the imperceptibly slow downslope movement of regolith' (Skinner and Porter, 1987); and *solifluction*: the slow viscous downslope movement of waterlogged regolith. It is clear, therefore, that often gravity does not work alone. Much mass movement is assisted by the presence of water and air within the materials. For instance, in cold environments, seasonal freezing of water in the regolith produces *frost-heaving*, caused by the 9 per cent expansion of water when it freezes. The heaving and subsequent subsidence on melting moves particles slowly downslope (see Chapter 8).

Types of rapid movement

At the other extreme, sudden slope failure may cause huge rock and debris masses to roar downslope at perhaps 100 km/hr, causing fundamental landscape changes in a matter of minutes. This diversity of mass movement fits our general model of Figure 1.12: in any given landscape, slow processes dominate for long periods but are interrupted by briefer episodes of rapid activity. The relative importance of slow and rapid processes varies over time and space. For instance, in the mountainous upper Rhine basin of Switzerland, and in central Taiwan, rapid failures account for about 85 per cent of the total annual movement. Across the steep but forested slopes of the Adelbert Range in Papua New Guinea, the figure is around 70 per cent, while on slopes of 15°–45° in southern Sweden, it is 50 per cent.

Rapid mass movements involve four types of motion: falling, sliding, flowing and slumping (Fig. 1.14). Which type is dominant depends upon such factors as slope angle, rock type and structure, depth and nature of the regolith, vegetation cover, occurrence of water, and tectonic activity. For example, rock falls are most likely where strong, well-bedded and jointed steep rock faces are exposed to active mechanical weathering. Slow movement may lead eventually to sudden failure (Fig. 1.15).

A sudden increase in the presence of water can overload slope materials and reduce friction on even quite gentle gradients, and cause sliding and flowage. Slumping too, which involves rotational slippage along moderate to steep slopes, is often activated in saturated conditions, especially if the slope foot is being steepened, for instance, by stream or wave erosion, e.g. Barton Clay cliffs, Hampshire (Chapter 5).

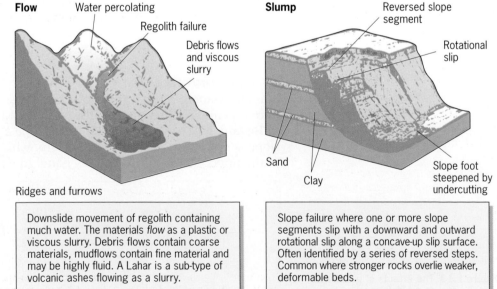

Fall

Joints opened

Large blocks topple

Block debris lies at angle of repose at cliff foot

Detached blocks of bedrock topple from a cliff face or steep slope. Common where mechanical weathering, e.g. freeze–thaw action is active on exposed rock surfaces and joint systems.

Avalanche

Rockfall

Impact smashes rocks and traps air

Debris rushes downslope

1–10km

Rockfall material is pulverised by impact at the cliff/slope foot and continues to travel downslope, often at great speed. It is thought that this velocity is sustained by a cushion of air trapped beneath the mass, cf. a hovercraft.

Slide

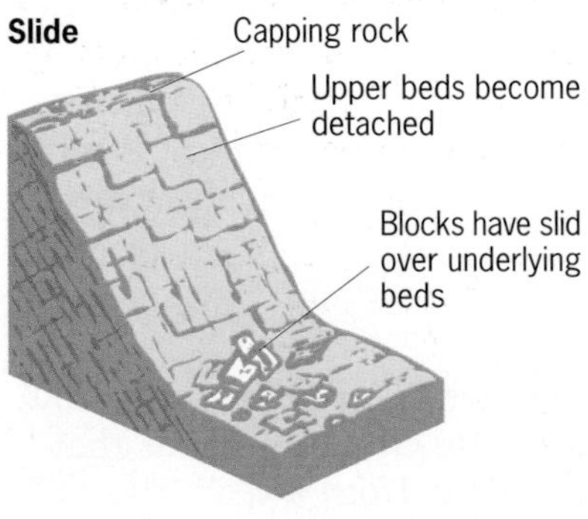

Rockslide Blocks become detached from underlying rocks and slide over them, moving as a solid material.
Debris slide The sliding of weathered regolith debris over bedrock.

Flow

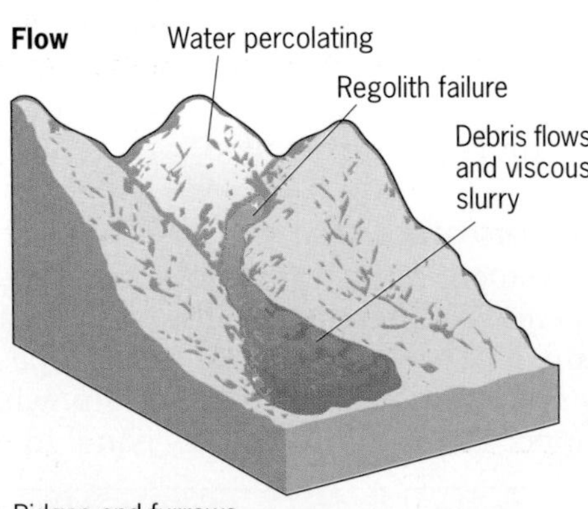

Downslide movement of regolith containing much water. The materials *flow* as a plastic or viscous slurry. Debris flows contain coarse materials, mudflows contain fine material and may be highly fluid. A Lahar is a sub-type of volcanic ashes flowing as a slurry.

Slump

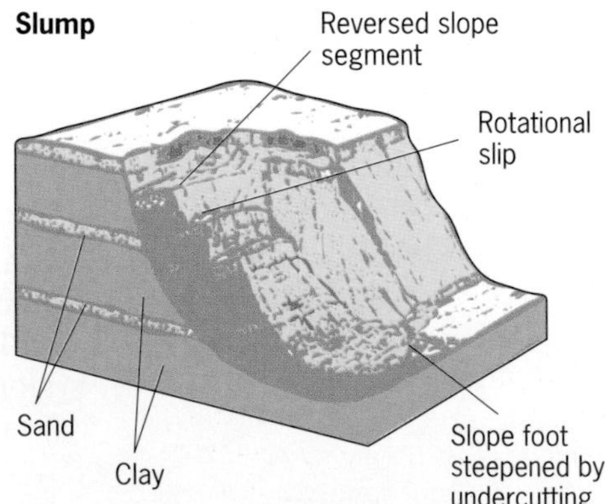

Slope failure where one or more slope segments slip with a downward and outward rotational slip along a concave-up slip surface. Often identified by a series of reversed steps. Common where stronger rocks overlie weaker, deformable beds.

Figure 1.14 Types of mass movement

Figure 1.15 Pueblo Bonito is the largest of a series of multi-roomed community houses built in Chaco Canyon, New Mexico, USA, before AD 1000, by the Anasazi ('The Ancient Ones') people. In 1941, a huge sandstone block (90 m long; 30 m high; 9 m thick) toppled from the cliff face, almost destroying this magnificent ruin. Chaco Canyon receives only 210 mm of precipitation a year, and so mass movement is generally slow. Geologists estimate that the block began to separate from the cliff face around 2300 years ago. Almost 1000 years ago a gap of 0.3 m had opened up, and the local people wedged logs beneath the block because they felt threatened. By the 1930s, the logs had rotted away and the block was moving more quickly, especially during the freeze–thaw period of winter: 1 cm in 1937; 5 cm in 1939; 6 cm in 1940. In January 1941 the block toppled.

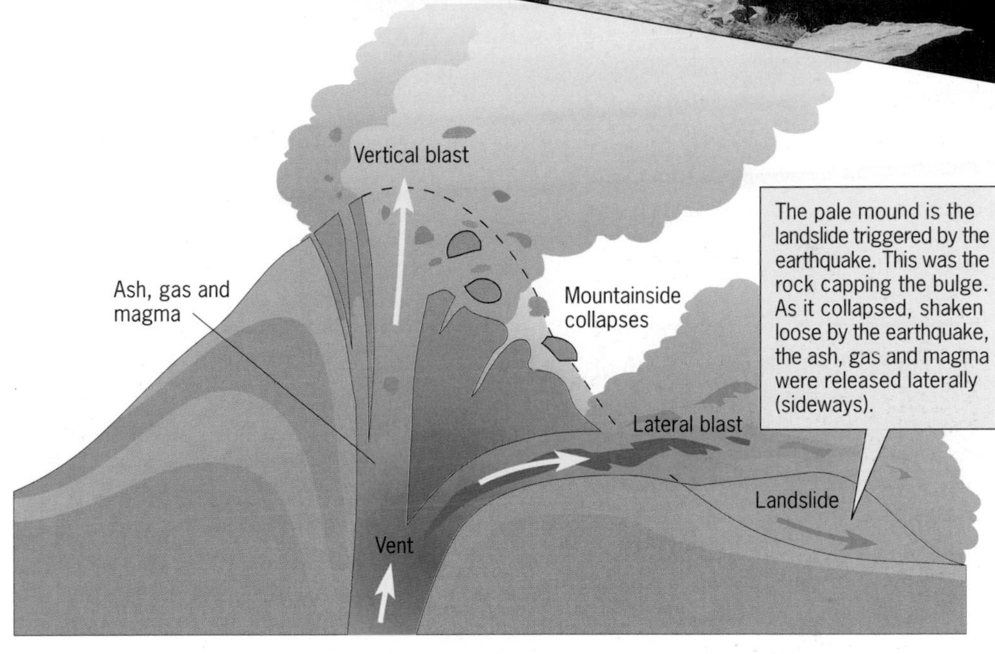

Figure 1.16 Slope failure on Mount St Helens, Washington State, USA. Early in 1980, rising gas and magma forced the north side of the mountain to bulge. On 18 May, an earthquake caused this bulge to collapse, producing a massive landslide and releasing an explosive eruption of gas, ash and lava.

?

3 Study Figure 1.17. Draw a sketch of the Monsoon Himalayas and Western Europe curves. Label the periods of constant form (steady state), the landslide event (rapid change) and landform adjustment. Use Figure 1.12 to help.

4 Compare the frequency of landsliding events for these two areas.

5 Which environment is showing the most rapid landform evolution? Suggest reasons for the differences.

6 The two New Zealand curves illustrate the influence of human activity. The pasture is artificial grassland on formerly forested areas. Compare the two curves in terms of the timescale of landslide events and the rate of landform change. Suggest why these differences occur in an area with similar climatic and relief characteristics.

Where a slope or landform is near to a critical threshold of instability, earth tremors can trigger sudden failure. This is why violent mass movements are common along the zones of crustal plate margins, especially in mountainous environments, e.g. the Andes, Alaska, Japan, Indonesia (Fig. 1.16). Human activity such as construction work or heavy traffic is also capable of generating vibrations sufficient to trigger slope failure.

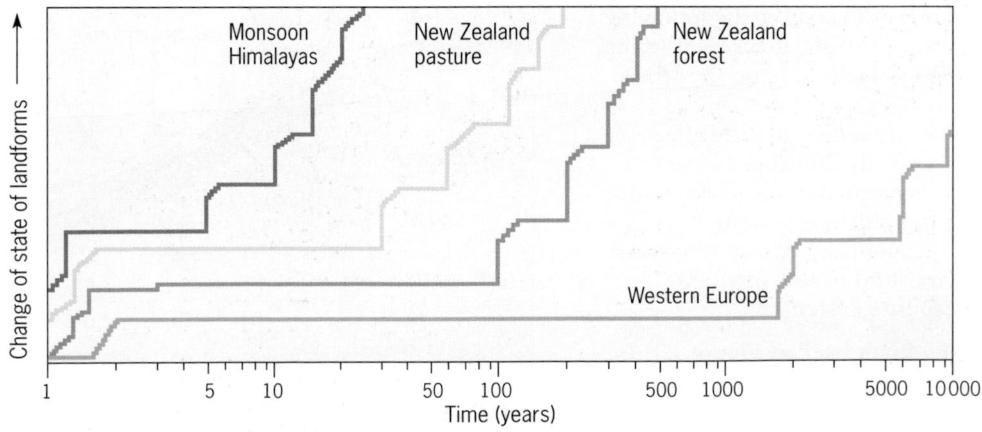

Figure 1.17 Curves showing the magnitude and frequency of landform change by landsliding for selected environments

The systems approach to landforms

Each agent of landform formation carries out its work in different ways, yet we can use the same general model or concept to develop our understandings about all of them. It is called the *systems approach* (Fig. 1.18). In its most basic form, an environmental system consists of inputs of material and energy which travel along one or more pathways (throughputs) to a store and then travel from the store as outputs. For instance, a slope is often studied as a system, with the regolith as the store. Inputs of materials (e.g. water, rock particles) enter, pass through, and finally leave the slope as outputs (Fig. 1.19). Energy, too, is input, stored and exits.

Figure 1.18 A basic systems diagram

Negative feedback = Mechanisms and processes working to retain balance or equilibrium.

Positive feedback = Mechanisms and processes which create progressive change ('the runaway mechanism').

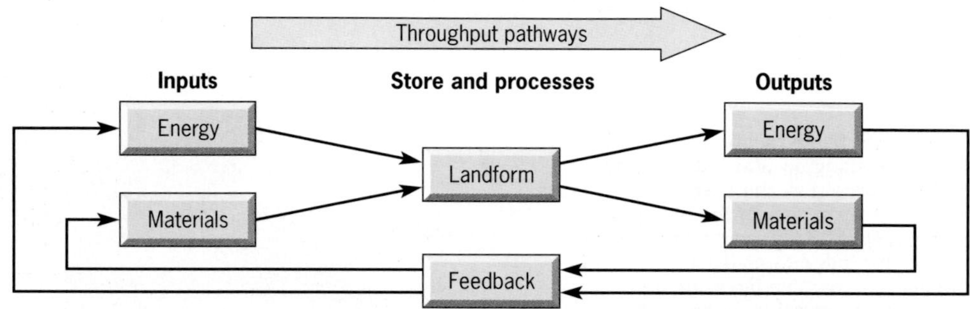

At another scale, the landforming agent itself becomes the system. For example, a glacier or a river becomes the store by which inputs of energy and material are stored and moved, and through which the processes of erosion–transportation–deposition are achieved. In this case, individual glacial or fluvial landforms become subsystems of the glacier or river system. Figure 1.19 shows how systems are linked. The hillslope system is linked to the river, glacial or coastal system acting to remove the material from the hillslope base (basal removal).

Environments and even individual landforms can be complex, but we can build more detailed structures from the basic system. Thus, the throughputs of energy and materials may travel along several pathways, into more than one store, and emerge as outputs taking several routes (Fig. 1.21). We must remember that, although systems diagrams may look complicated, they are all built upon the basic logic of Figure 1.18.

7 Think of the butte landform of Figure 1.3 as the store of a system. Draw a systems diagram to show the inputs to and outputs from this store (NB Include solar energy).

8 Study Figure 1.19. What are the inputs, outputs and throughputs of this system?

9 Draw the hillslope system as an inputs/outputs diagram.

10 Figure 1.20 shows the hillslope as a process-response system. This means that the form or morphology of the hillslope is a function of the processes operating on it.

a Follow the arrows on the diagram to answer the following:
- What factors affect weathering?
- What factors influence debris transfers?
- What factors directly affect the form of the hillslope?

b What is the role of vegetation in the hillslope system?

c Explain why basal removal is important to this process–response system.

11 Look back at Figure 1.1. Is this a weathering-limited or a transport-limited slope? Explain your answer.

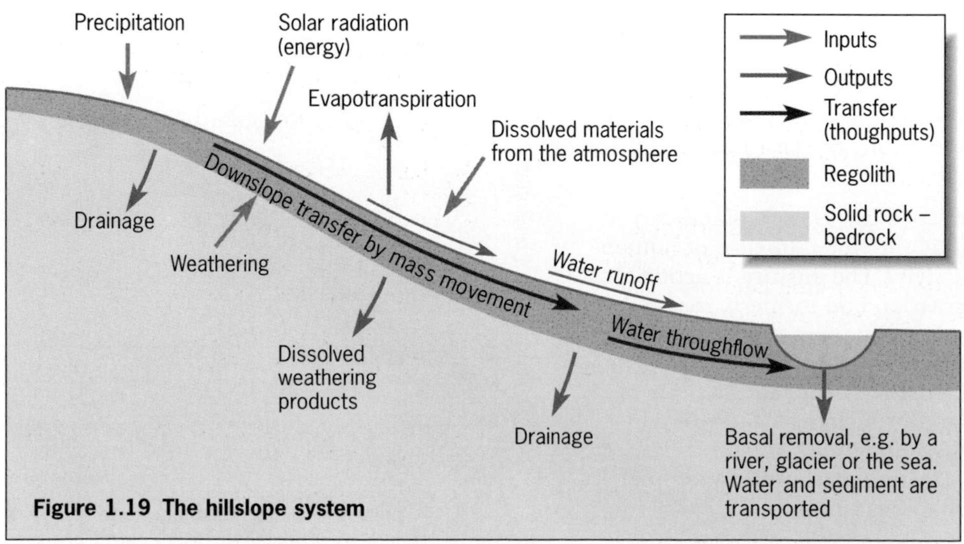

Figure 1.19 The hillslope system

Climate

Vegetation

Weathering

– rates of weathering and nature of the regolith

Debris inputs

Debris transfers

(mass movement)

Debris outputs

Geology

Rock type and structure

Hillslope form

Weathering-limited slope – inputs from weathering are the controlling factor, e.g. hard rocks or slow weathering environment. Slope form will be determined by rock type and structure. Transport-limited slope – weathering inputs exceed outputs by basal erosion, e.g. an inactive river. As material builds up on the slope the bedrock is increasingly protected from weathering. Negative feedback occurs and the inputs of debris to the slope are reduced.

Basal erosion

The amount of removal of material from the slope foot by rivers, glaciers or the sea

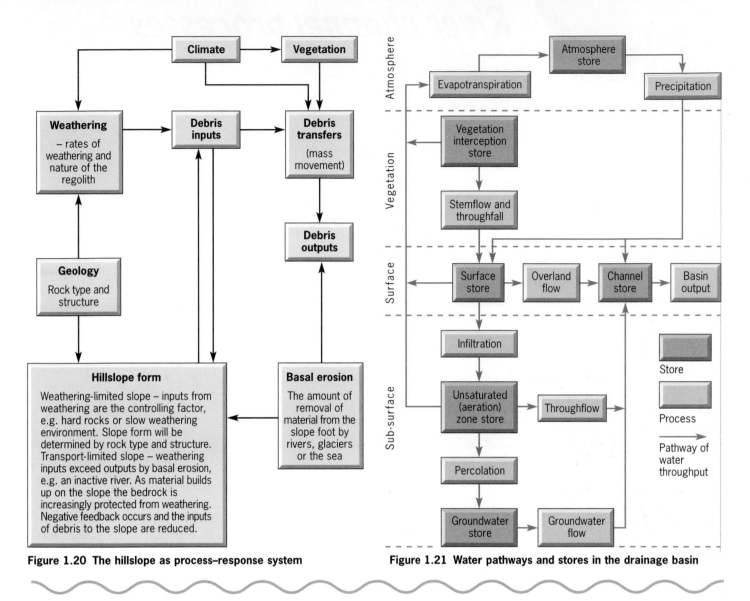

Figure 1.20 The hillslope as process–response system

Atmosphere

Atmosphere store

Evapotranspiration

Precipitation

Vegetation

Vegetation interception store

Stemflow and throughfall

Surface

Surface store

Overland flow

Channel store

Basin output

Sub-surface

Infiltration

Unsaturated (aeration) zone store

Throughflow

Percolation

Groundwater store

Groundwater flow

Store

Process

Pathway of water throughput

Figure 1.21 Water pathways and stores in the drainage basin

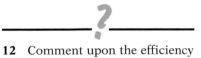

12 Comment upon the efficiency of basal removal processes in Figure 1.3. Suggest why it may be difficult to classify this slope as weathering-limited or transport-limited.

13 It has often been said that 'the present is the key to the past' in understanding landform development. Suggest what this statement means with reference to some specific landscapes.

14 Assess the extent to which landscapes change and landform development is gradual or occurs suddenly.

Summary

- Landform formation is achieved by a variety of processes of weathering, erosion, transportation and deposition. The main agents of landform development are water, ice and wind. Tectonic activity produces the world's macro-scale landform features which are denuded (weathering and erosion combined) by the other agents.
- Weathering processes can be divided into three groups – mechanical, chemical and biochemical – which interact to break down rock into regolith. These processes vary in importance in different environmental conditions and with the nature of the rock.
- Landforms can be classified by their agent of origin, whether they are erosional or depositional, or the stage in their development.
- Landforms exist in a state of overall balance or dynamic equilibrium with the weathering agents and processes operating. Most change to landforms occurs during relatively brief disruptive episodes followed by a new dynamic equilibrium.
- Mass movements transfer weathered material downslope by a variety of processes and at various rates.
- Landforms can be viewed as systems with inputs, stores, throughputs and outputs.

2 River channel processes

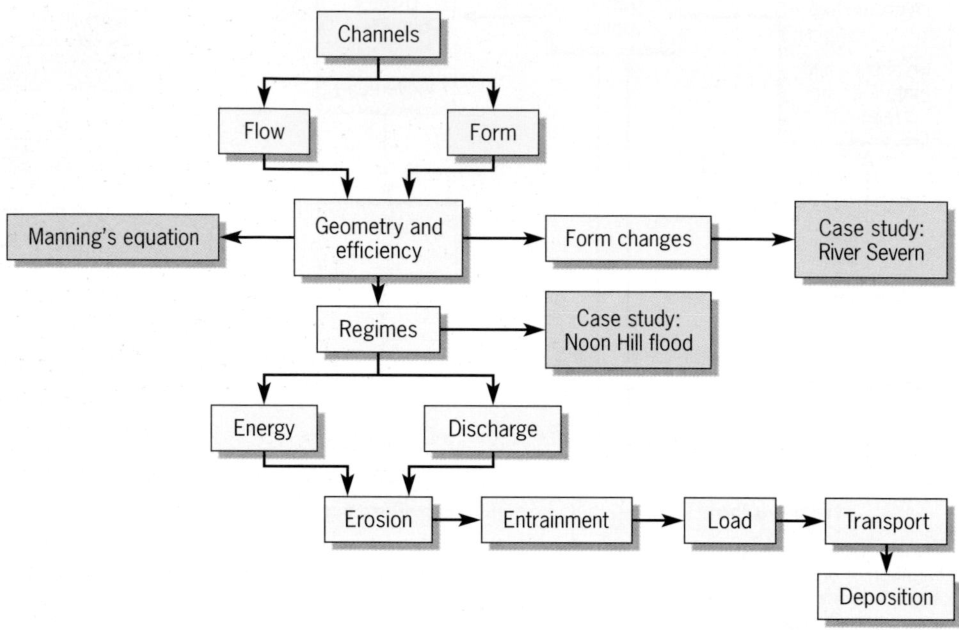

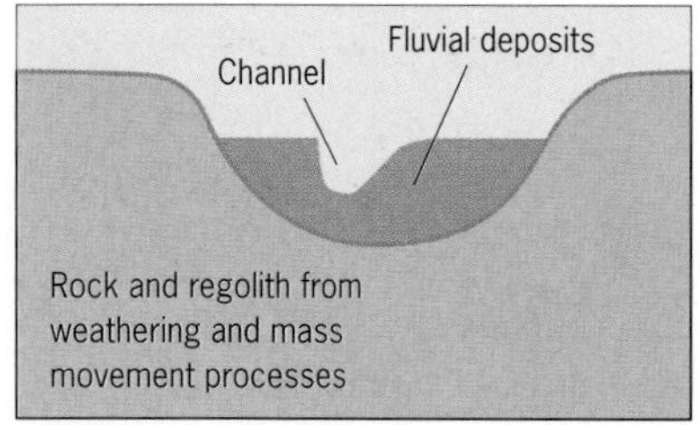

Valley slope and topography

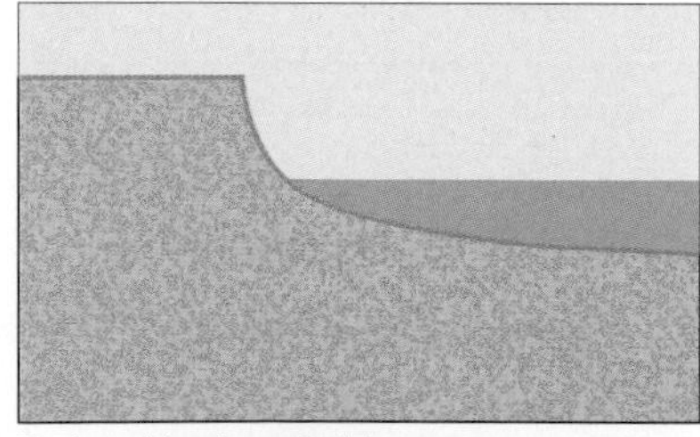

Bed and bank materials

Figure 2.1 Boundary characteristics of a channel (*After:* Thornes, 1992)

2.1 Introduction

Water is a primary agent of landform formation, and rivers are a major component in the achievement of this work. We call the results **fluvial landforms**. In Chapter 1 we saw that the fundamental system for studying fluvial processes is the **drainage basin** (Fig. 1.21), through which water and sediment move and are stored. In this chapter we shall consider the water and sediment once they have arrived in the stream channel. The channel is the solid boundary – bed and banks – within which the movement and storage takes place (Fig. 2.1)

One of the key themes of this chapter is the three-way relationship between the channel, the water and the sediment. Keep in mind, too, that the channel is a three-dimensional phenomenon consisting of cross-section, long profile and plan (Fig. 2.2), and that all these dimensions adjust constantly.

Energy is required to move the **driving variable** of water downstream along the channel. **Energy**, simply defined, is the ability to do work, and **force** is the application of this energy. (It is important to remember that energy is a complex phenomenon, as any physics textbook illustrates. For the purposes of this chapter, energy is used as a general term, unless a

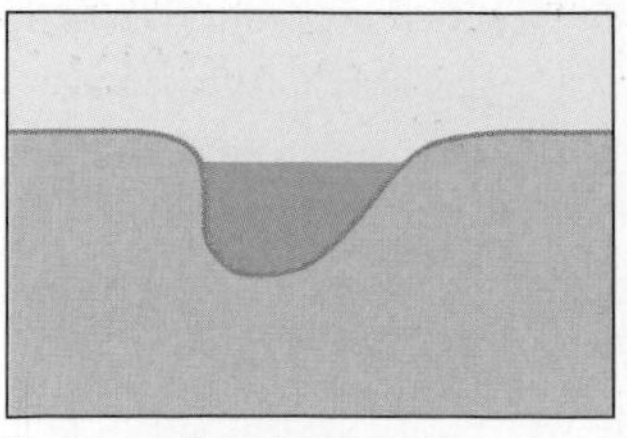

Cross-sectional geometry (width, depth, maximum depth)

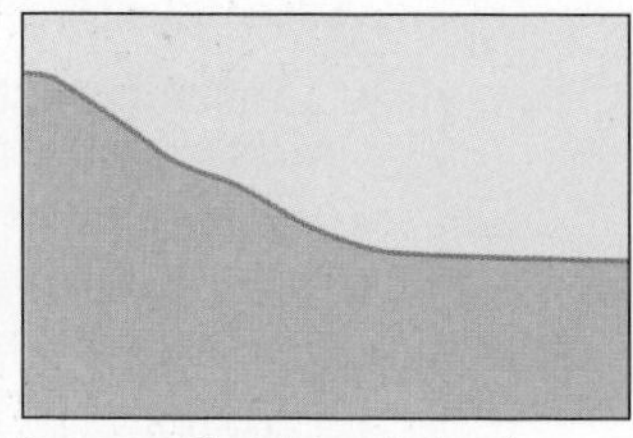

Long profile (channel slope)

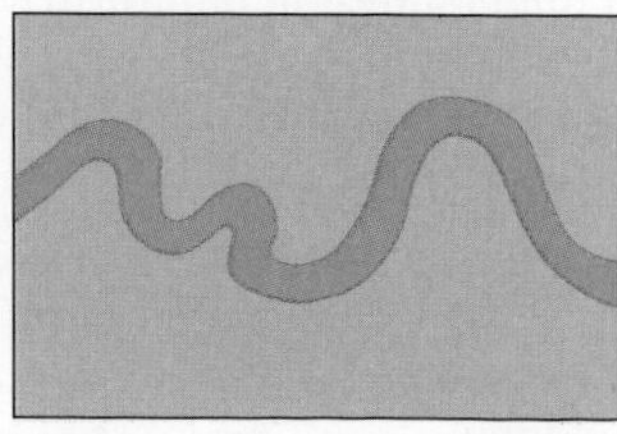

Planform

Figure 2.2 Dimensions of channel form (*After:* Thornes, 1992)

Figure 2.3 Why study channels? The purpose of taking stream measurements is to identify reasons why a stream channel changes shape and why its ability to transfer water and sediment varies. We do this not only to help explain the landforms created by rivers, but also to help water managers, planners and farmers to improve their ability to predict and forecast how a river is likely to behave, e.g. when it might flood and how. This applied research is aimed, too, at improving water management and the quality of environmental impact assessments for proposed channel and land use changes.

distinct form is specified, e.g. kinetic or mechanical energy.) Therefore, the erosion and transportation of sediment can take place only when sufficient energy to apply the force is available to do the work. For this reason we shall first examine the motion of water and its relation to channel shape, before adding the driving variable of sediment.

2.2 Channel variables and flow characteristics

A stream channel is an open funnel whose function is the transfer of water and sediment downstream. At one level it is a component of the river basin system (Fig. 1.21), yet it can also be seen as an **open system** in its own right, with inputs, stores, throughputs and outputs (Fig. 2.4). The channel of a permanent stream is constantly adjusting to changes in inputs of water and sediment. The purpose of the adjustments is to achieve a form which enables the channel to transfer these water and sediment inputs. This constantly adjusting balance we call the state of dynamic equilibrium.

?

1 Referring to Figure 2.4, list how human activity may affect inputs, stores and outputs.

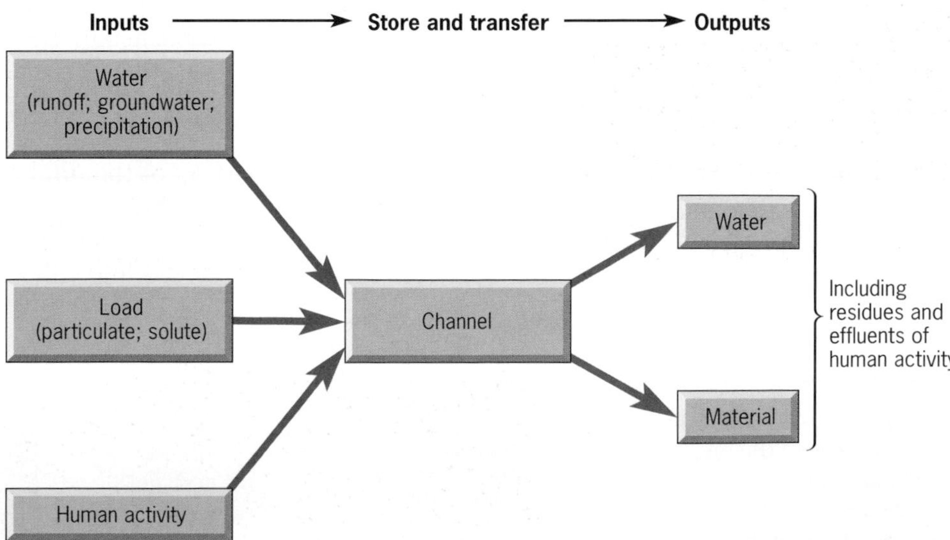

Figure 2.4 The stream channel as an open system

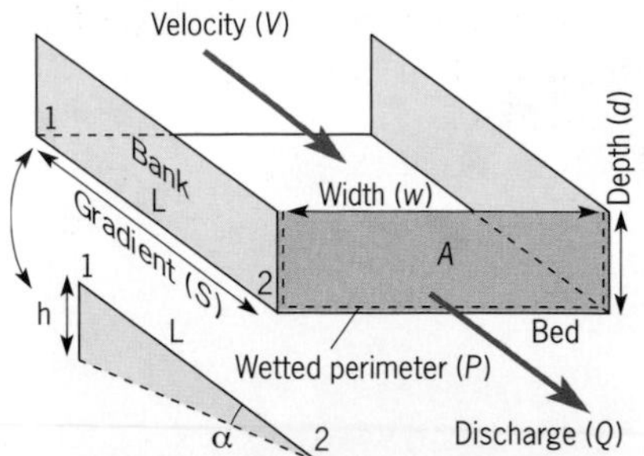

Velocity (*V*): Speed of water flow, i.e. distance travelled per unit of time, e.g. metres per second (ms^{-1}).

Discharge (*Q*): Volume of water which passes through cross-section *A* per unit times. Usually measured in cubic metres per second (cumecs).

Channel width (*w*): Distance across the channel, usually measured at the water surface.

Channel depth (*d*): Height from water to stream bed. As in natural channels the bed is uneven, the mean of several measurements (*d*) is used.

Cross-sectional area (*A*): Calculated as the product of width and mean depth: $A = wd$

Wetted perimeter (*P*): The length of the channel margin (banks and bed) around the cross-section in contact with the water. In the case of the model of the diagram: $P = 2d + w$; but in a 'real' stream the relationship would not be so straight-forward.

Note: When **bankfull channel capacity** is to be measured *A* is calculated from *w* measured from bank-top to bank-top, and *d* by measurements vertically from the line joining these two points.

Gradient (*S*): Change in channel bed elevation per unit length of channel. Calculated as:

$S = h/h \cos \alpha$, or $\tan \alpha$

L = unit length of channel (1–2)
h = height change between 1 and 2

Figure 2.5 Basic channel variables

Channel geometry

Several variables, which together make up the channel geometry (Fig. 2.5), are involved in the constant adjustment. For instance, follow Figure 2.5 carefully: if *Q* increases, the channel might adjust by increasing *d*, thereby increasing *S*, and so increasing *V*. If *d* and *w* both increase, then *A* increases and so enlarges the channel capacity. This illustrates how a channel tries to maintain an equilibrium (balance) between form and the throughput of discharge and load by **negative feedback** mechanisms (Fig. 1.18). When a channel is overwhelmed by a rapid input surge of water and/or solids, e.g. catastrophic flood or landslide, **positive feedback** may take over to create a fundamentally different channel.

In our study of stream channels, therefore, an understanding of water hydraulics is important. The key questions become:

1 How does water behave?

2 What factors influence this behaviour?

3 How does this behaviour influence landforms and, in turn, human activities?

Water hydraulics: energy and flow

Water is a fluid and, unlike a solid, is unable to resist shear stresses. Thus it deforms, i.e. changes shape, according to the strength and direction of the forces applied to it. This property enables it to flow under its own weight within the solid channel perimeter. What we observe as **streamflow** is the outcome of the balance between the tractive or driving force (i.e. gravity) and the resisting force (i.e. friction with the bed and banks).

Types of flow

The simplest form of flow is *laminar*, where layers of water slip past each other (Fig. 2.6a). Water speed increases with height above the channel bed, as frictional resistance generated by the bed materials decreases. At periods of higher flow and/or where the stream bed is rough, the flow may become *turbulent*. There is increasing mixing of the water, and a more even distribution of shear through the channel, causing eddies (Fig. 2.6b). The turbulence causes energy loss due to viscosity friction within the water, but over 90 per cent of energy is applied in overcoming the frictional or drag effect of the bed and banks against the water. The remaining energy, approximately 5 per cent, is available as kinetic energy to move the sediment load.

2.3 Channel efficiency

That only a small proportion of the total energy in a stream is available for the processes of erosion and transportation is an important understanding. Any factors which increase this available energy will influence the amount of geomorphological 'work' the stream can achieve. For instance, velocity is one

a Laminar flow

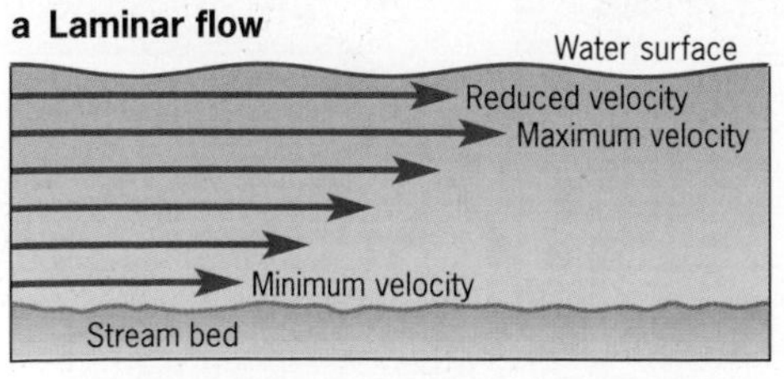

b Turbulent flow

Water surface
Maximum velocity zone
Minimum velocity zone
Stream bed

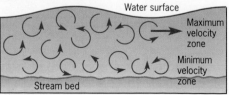

Figure 2.6 Streamflow types – extreme forms

?

2 Using the variables of Figure 2.5, suggest how the channel might react to
a a decrease in discharge *Q*;
b an increase in gradient *S*.

influential factor: at a given steady speed, material above a certain size and density will fall to the channel bed. As stream velocity declines, less dense particles, as well as the heavier materials, are deposited and there is increased deposition per metre of channel bed. The lower the proportion of total energy applied in overcoming viscosity friction and bed/bank drag, the more *efficient* is the channel form.

Hydraulic radius

If we examine Figure 2.7 carefully, we can see that velocity increases as distance from the channel bed increases. The compact channel is more efficient than the wide, shallow channel, because in the compact channel a higher proportion of the total water volume is not in contact with the channel bed, even though both channels may have the same cross-sectional area. The key variable here is the *hydraulic radius* (HR). We define the hydraulic radius as the proportion of the water in a channel cross-section which is in contact with the channel margin. This proportion, and therefore the hydraulic radius, is influenced by the length of the *wetted perimeter*, the length of bed and banks in contact with the water. This is shown by the length a–b on the three cross-sections of Figure 2.8.

The formula we use to calculate the hydraulic radius becomes:

$$\text{Hydraulic radius (HR)} = \frac{\text{Cross-sectional area}}{\text{Wetted perimeter}}$$

The higher the hydraulic radius, the more efficient the channel.

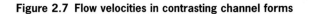

3 Calculate the approximate hydraulic radius of each of the Keskadale Beck cross-sections in Figure 2.8. At which location is the channel most efficient?

4 Use Figures 2.7 and 2.8 to explain the idea that a channel is most efficient at its bankfull stage (labelled diagrams will help).

5 Use annotated diagrams and notes to support the following statement: As discharge increases downstream, so cross-sectional area increases. As a result a river tends to be progressively more efficient in the downstream direction.

6 If a channel is more efficient downstream, what implications are there for energy loss through friction, water velocity and the ability to transport sediment?

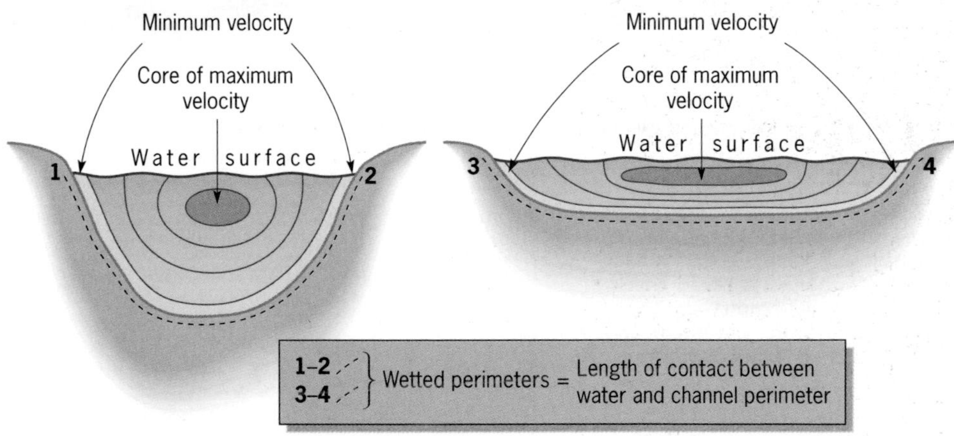

Figure 2.7 Flow velocities in contrasting channel forms

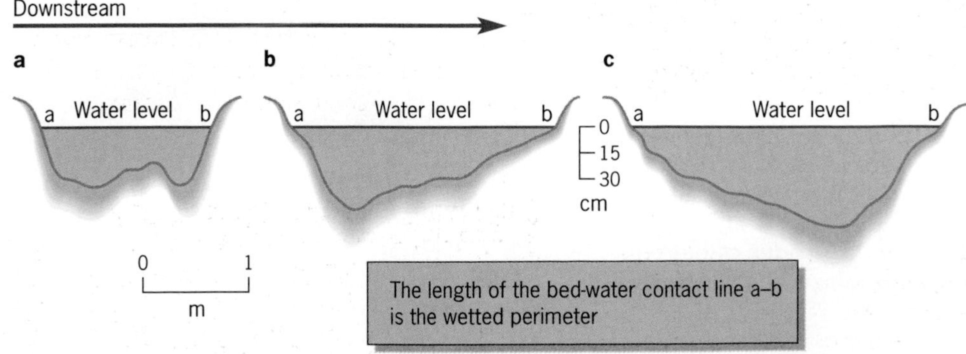

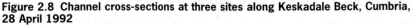

Figure 2.8 Channel cross-sections at three sites along Keskadale Beck, Cumbria, 28 April 1992

Figure 2.9 This small mountain stream has low volume, a steep gradient, and appears to be flowing quickly, but this is deceptive. The boulder bed is rough, which gives a long wetted perimeter, low hydraulic radius and relatively low velocity. The clarity of the water shows that little sediment is being transferred.

Figure 2.10 Meltwater on Child's Glacier, Alasaka. This small channel on the glacier surface is efficient. The gradient is gentle and the stream volume small, but the velocity is high because the ice forms a smooth, dish-shaped channel bed.

Figure 2.11 The lower Mississippi at Natchez. The channel is more than one kilometre wide, often 30 m deep, and the current swirls along vigorously. Gradient is slight, volume is huge, the channel is broad and deep, with fine bed materials. The opaque colour of the water indicates the enormous sediment load being carried.

7 Use the concepts of wetted perimeter, bed roughness and hydraulic radius to explain
a why the River Mississippi (Fig. 2.11) is flowing quickly;
b what the opaque colour tells you about the available energy; and
c why the river can be said to have an 'efficient' channel form.

Bed roughness

Equally important with channel form in determining efficiency is *bed roughness*. The rougher the bed, the greater the wetted perimeter, and therefore a higher proportion of the total water volume is in contact with or close to the bed. In turn, the river must use a higher proportion of its total energy in overcoming friction. This reduces its velocity. Bed roughness is thus an important variable in the calculation of velocity. For a given volume of water, the rougher the bed, the less efficient the channel.

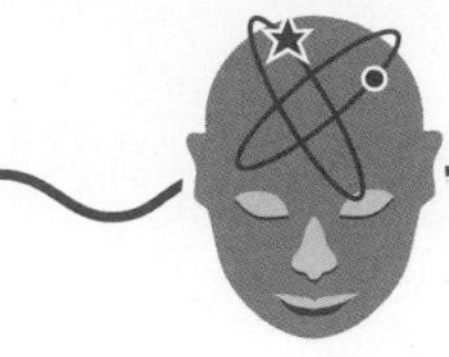

Manning's equation

The most popular calculation to estimate velocity using the channel form and bed roughness variables is the Manning equation, often known as *Manning's 'n'*:

$$\text{Velocity } (V) = \frac{R^{\frac{2}{3}} S^{\frac{1}{2}}}{n}$$

Where:

n = The Manning roughness coefficient (based on variables such as size and shape of bed material; wetted perimeter; vegetation in and fringing the channel; discharge)

R = hydraulic radius

s = channel gradient

The Manning roughness coefficient ranges between 0.03 in straight, clean channels to 0.15 for densely vegetated channels. Table 2.1 gives some typical values for natural channels.

Table 2.1 Typical Manning roughness coefficient values

Channel type	Coefficient range
Small mountain stream, pebble and boulder bed	0.040–0.070
Small, clean, straight lowland stream	0.025–0.033
Small weedy stream with deep pools	0.075–0.150
Floodplain stream in pastureland	0.025–0.035
Floodplain stream in heavy woodland	0.100–0.150
Large streams (width > 33 m)	0.025–0.060

2.4 Changes in channel form from source to mouth

If we were able to follow a stream from its source to its mouth, taking measurements of channel form and discharge at intervals as we moved downstream, we would find that all the channel and flow components change. If we then compare our data with the results from other rivers we may find that some patterns are common to most streams. That is, although each stream is unique in detail, we can make certain general statements about stream channel, discharge and energy relationships.

Stream gradient

One basic generalisation is that stream gradient decreases downstream. We can express this most simply in terms of the **long profile** (Fig. 2.12). Over long periods a stream will work towards attaining the long profile which balances energy with load: there is sufficient discharge and gradient to transport the sediment input gradually downstream. When a stream achieves this equilibrium, we say it has attained *grade* and has a *graded profile*.

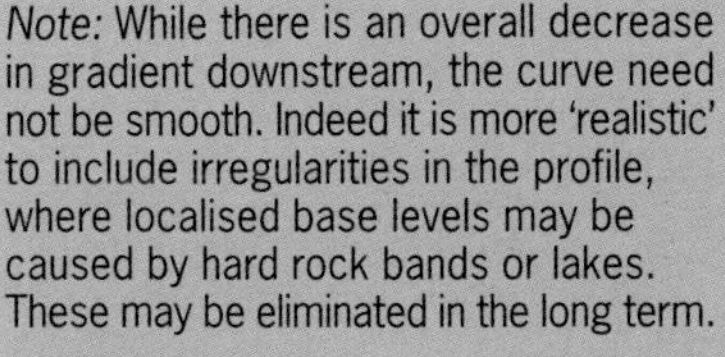

Note: While there is an overall decrease in gradient downstream, the curve need not be smooth. Indeed it is more 'realistic' to include irregularities in the profile, where localised base levels may be caused by hard rock bands or lakes. These may be eliminated in the long term.

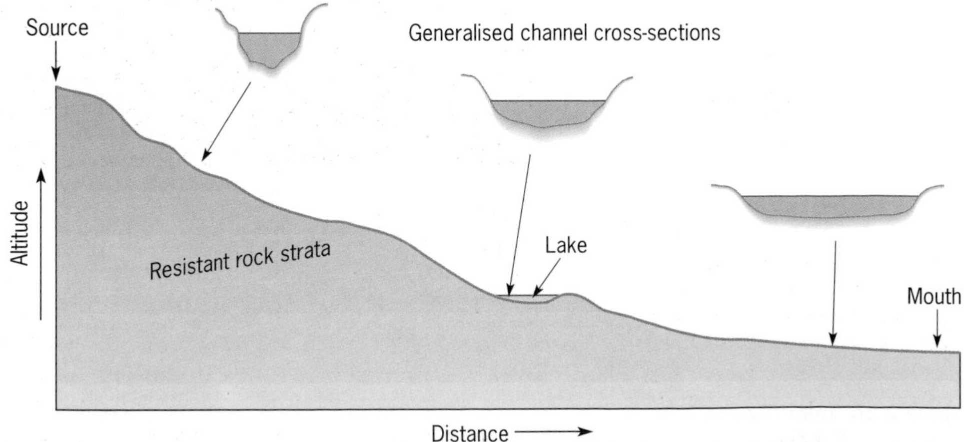

Figure 2.12 A graded long profile. There is an overall decrease in gradient downstream, but the curve need not be smooth. Localised base levels may be caused by hard rock bands or lakes, although the model suggests that in the very long term the stream will work to remove these irregularities.

Figure 2.13 **Channel and discharge relationships down a stream course**

8 Figure 2.13 summarises the general channel and discharge trends along the stream course by identifying the relationship between discharge and a second variable. Remember that in a humid climate discharge will increase downstream as water is added by tributary rivers and from the drainage basin. For each graph (except graph A which is done for you), state in one sentence the general relationship identified. For example, the relationship in graph A may be expressed as: 'Channel width increases downstream especially during periods of high flow.'

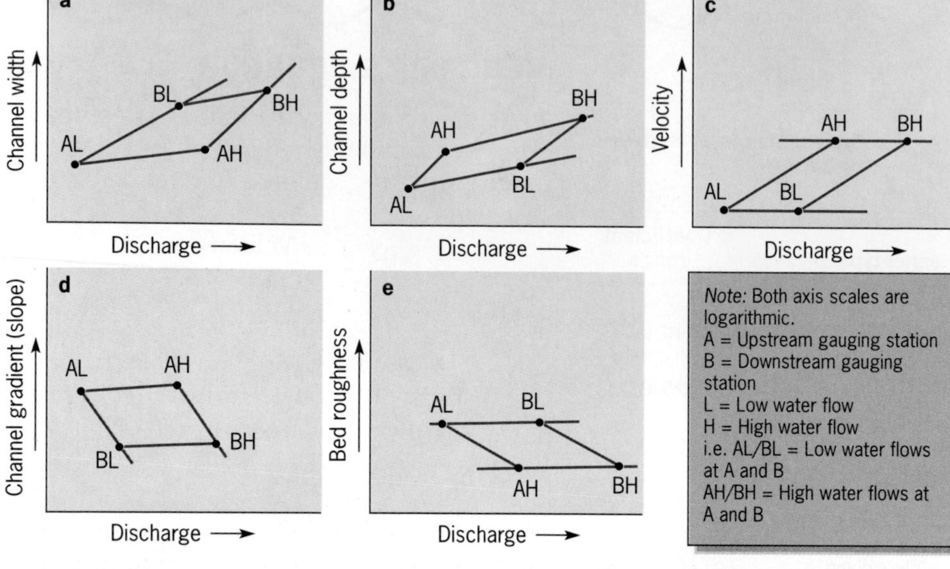

Remember, stream discharge and load are constantly changing, and therefore the achievement of grade will be a temporary state. Indeed, some hydrologists believe the graded long profile is useful only as a general or theoretical idea, and that it is better to think of streams working towards a state of dynamic equilibrium.

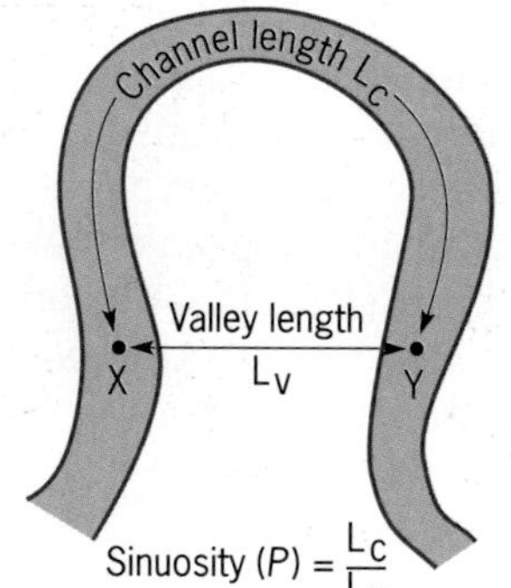

Figure 2.14 The sinuosity ratio

2.5 Channel and flow sequences along a river

All natural rivers have a tendency to develop a winding course over time, because a component of the force is directed laterally (sideways). This property of a river is called its **sinuosity**, and is measured in a standard way, allowing comparison with other rivers or with other stretches of the same river (Fig. 2.14).

Riffle-and-pool patterns

If we plot the long profile of the bed and water depth along a stretch of river where sinuosity is well developed, we are likely to observe a regular sequence (Fig. 2.15). Along straighter stretches, where sediment banks have built up on the stream bed, we find shallow flows called **riffles**. Around the bends, as the core of maximum velocity concentrates along the **thalweg** (the line of maximum flow velocity), **pools** of deeper water develop. There is also another type of flow, a helical or helicoidal flow, which is dealt with in Chapter 3 (p. 46).

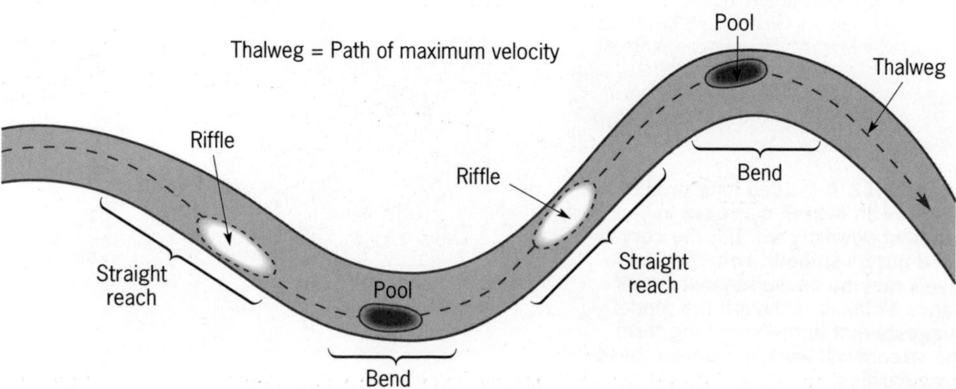

Figure 2.15 The riffle-and-pool pattern along a sinuous channel

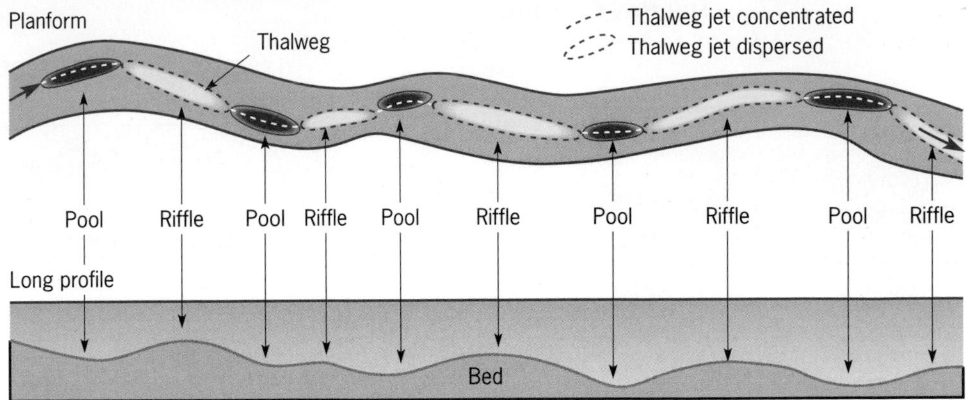

Figure 2.16 The riffle-and-pool pattern along a channel with low sinuosity

The riffle-and-pool pattern may also evolve along less sinuous stretches (Fig. 2.16). In such situations the thalweg wanders across the channel and also varies in its concentration. Where this core flow is a concentrated jet, channel deepening occurs to give a pool. Where the flow is dispersed, and water slows down, deposition takes place, creating a riffle. Once this concentration–dispersion rhythm is established, it is self-sustaining under normal flow conditions.

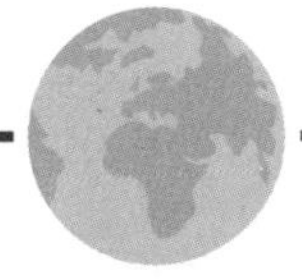

Riffle-and-pool patterns along the River Severn near Shrewsbury, England

The River Severn above and below Shrewsbury exhibits a sinuous planform within a well-developed floodplain (Fig. 2.17). Thus we would expect to find a riffle-and-pool sequence along the channel which follows the general model of Figure 2.15. Surveys of two 1-km stretches of channel at Montford and Leighton reveal that a sequence does occur, but is more complex than the model would predict.

The upstream stretch at Montford has a lower sinuosity ratio than the Leighton stretch. However, both channels show a riffle-and-pool sequence (Figs 2.19, 2.22). At Montford the riffle-and-pool pattern has a mean spacing of 300 m. The channel gradient at Leighton is higher than at Montford, and along both stretches the channel is slightly wider at the riffles than the pools. Figures 2.18–2.23 set out the characteristics of the two stretches and allow us to compare them with the model of Figure 2.15.

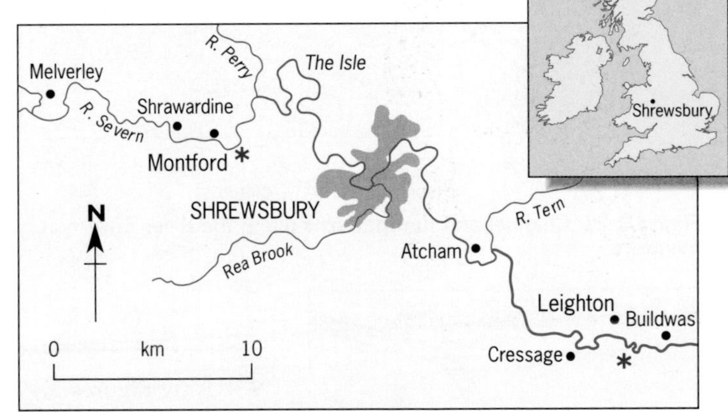

Figure 2.17 The Montford and Leighton sites

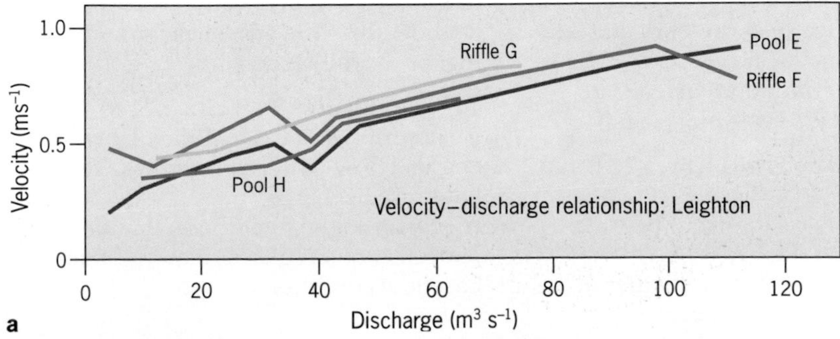

Figure 2.18 Channel and flow patterns along the River Severn at Leighton

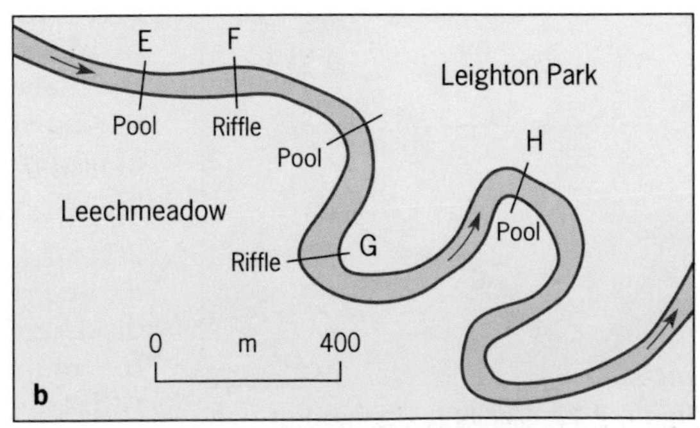

Figure 2.19 Channel planform at Leighton

The River Severn

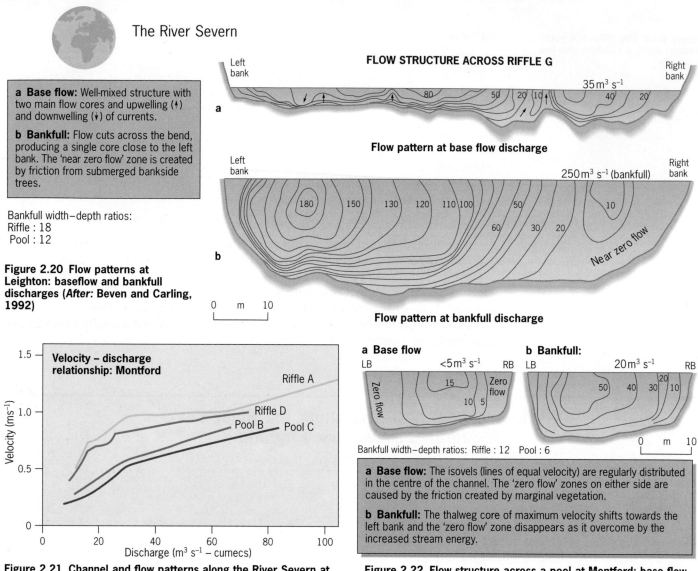

a Base flow: Well-mixed structure with two main flow cores and upwelling (↑) and downwelling (↓) of currents.

b Bankfull: Flow cuts across the bend, producing a single core close to the left bank. The 'near zero flow' zone is created by friction from submerged bankside trees.

Bankfull width–depth ratios:
Riffle : 18
Pool : 12

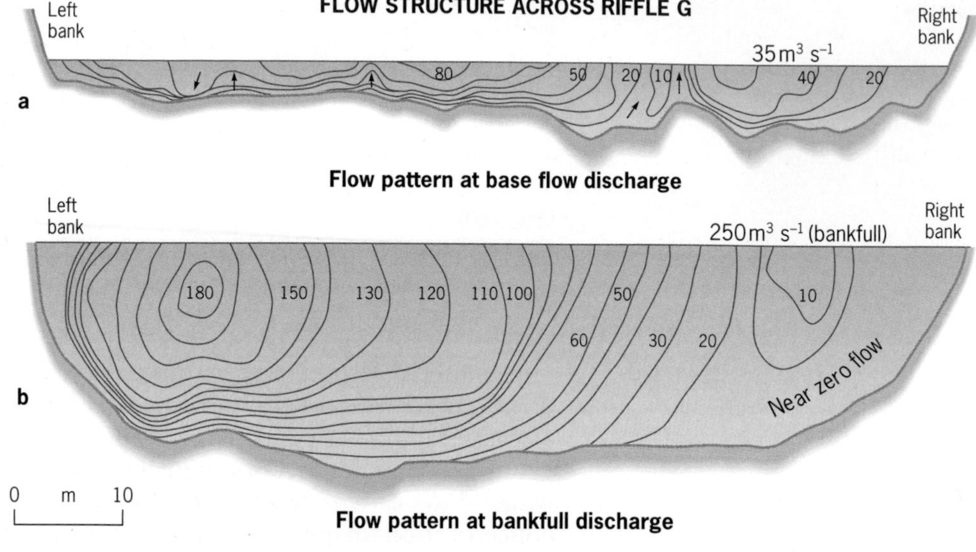

Figure 2.20 Flow patterns at Leighton: baseflow and bankfull discharges (*After:* Beven and Carling, 1992)

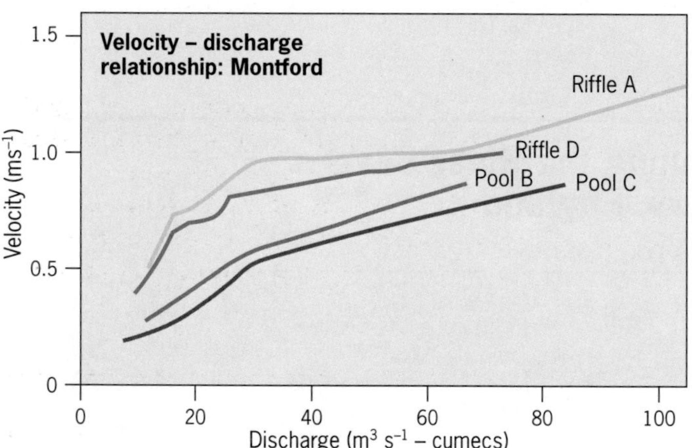

Figure 2.21 Channel and flow patterns along the River Severn at Montford

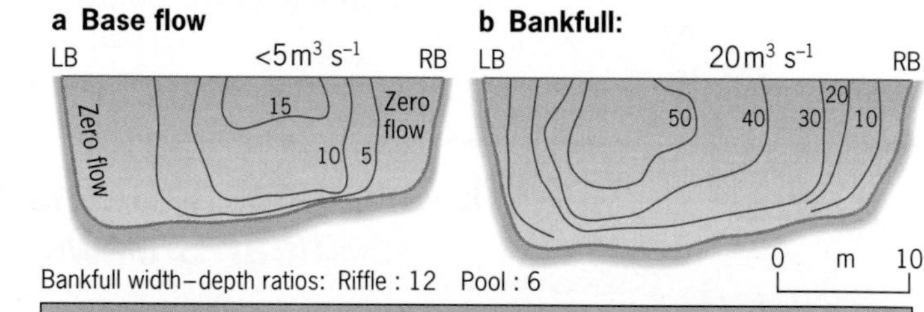

a Base flow: The isovels (lines of equal velocity) are regularly distributed in the centre of the channel. The 'zero flow' zones on either side are caused by the friction created by marginal vegetation.

b Bankfull: The thalweg core of maximum velocity shifts towards the left bank and the 'zero flow' zone disappears as it overcome by the increased stream energy.

Figure 2.22 Flow structure across a pool at Montford: base flow and bankfull

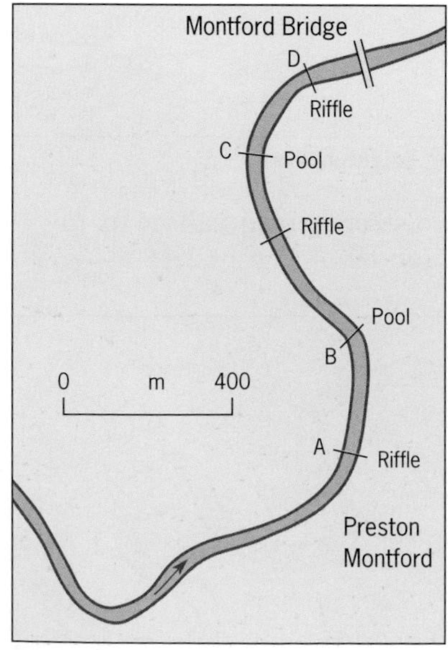

Figure 2.23 Channel planform at Montford

?

9 Using Figures 2.19 and 2.23,

a describe the locations of the riffle-and-pool patterns along the two stretches;

b attempt to explain these patterns in terms of the model of Figures 2.15 and 2.16. (You may find it useful to copy the planforms of the two stretches and add the thalweg line as in Figure 2.16, then use explanatory annotations at appropriate points on your plans.)

10 Use the graphs of Figures 2.18 and 2.21 to

a describe the relationship between discharge and velocity;

b compare streamflow velocities at riffles and pools. Remember, the velocities plotted on the graphs are the mean speeds recorded throughout the channel water body, *not* the maximum velocity along the thalweg core flow. Thus there may be a core of highest velocity at pools which enables channel deepening, but across the water body as a whole the riffle stretches may sustain a higher overall velocity.

11 Use Figures 2.20 and 2.22 to

a describe and explain the changes in channel flow patterns from base flow to bankfull discharge;

b examine the hypothesis that flow patterns are simpler at bankfull than at baseflow conditions.

Meanders

When sinuosity attains an extreme form, i.e. a sinuosity ratio greater than 1.5, the river is said to have developed a *meander* planform. Meanders show progressively exaggerated lateral movement and a highly developed riffle-and-pool pattern. (Meanders and braided channels are discussed in detail in Chapter 3.)

12 Using the formula P = Lc/Lv, calculate the sinuosity ratios for the Montford and Leighton stretches of the River Severn (Figs 2.19, 2.23). Do either of these stretches have a meandering planform?

2.6 Fluctuations in flow and energy

All rivers have distinctive rhythms or **regimes** of discharge and runoff. Even in equatorial environments with year-round moisture budget surpluses, rivers exhibit marked seasonal regimes. Every river has its own rhythmic pattern or 'fingerprint', recorded by its hydrograph, although there will be significant variations from year to year. It is these rhythms, plus shorter-term, irregular fluctuations in discharge, which allow us to divide the stream hydrograph into *baseflow* and *quickflow* components (Figs 2.24, 2.25).

Energy–flow relationships

During periods of base flow, the river is in a relatively low energy phase and little geomorphological 'work' is being done. If you do your fieldwork at this time, the weather may be pleasant but the stream will give you little indication of its potential and you may conclude that 'nothing much is happening'. Channel adjustments are likely to be the result of deposition (Fig. 2.25) as the discharge and energy decline down the falling limb of the storm hydrograph (Fig. 2.24).

?

13 Study Figure 2.25 carefully. Make a list of the features of this stream channel which suggest that it is at baseflow condition and give reasons for the items on your list.

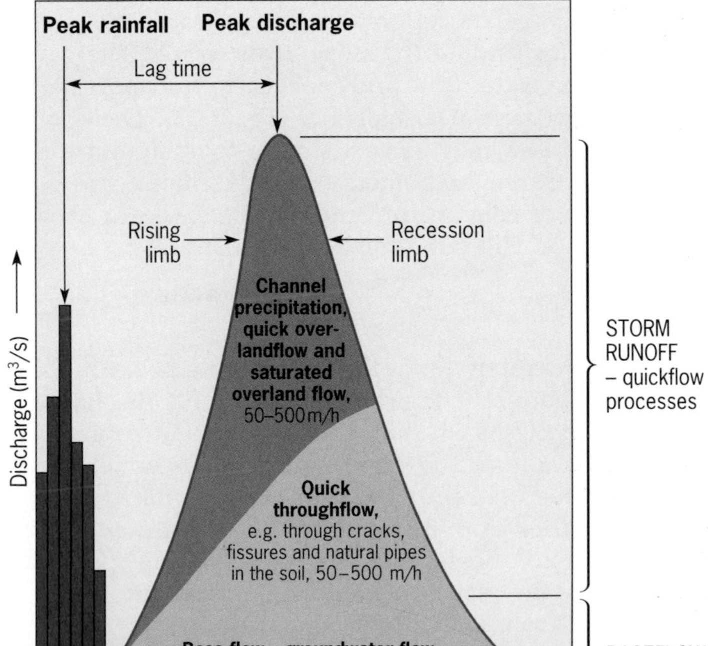

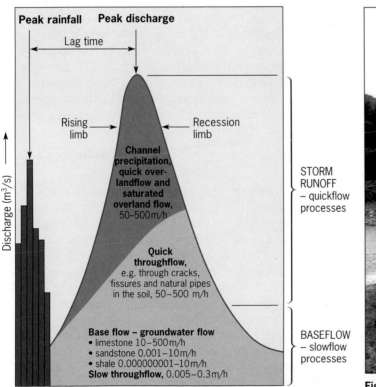

Figure 2.24 The storm hydrograph: terminology and processes

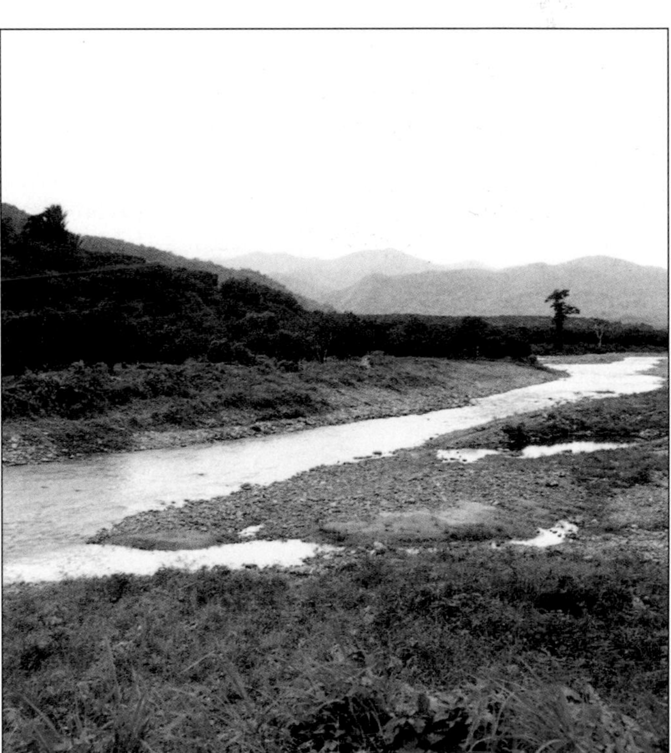

Figure 2.25 A stream at base flow, Belize, Central America, late in the dry season

When rivers do their work

From our understandings of stream regimes, we can propose three related hypotheses:

1 High-energy conditions in a channel occur with sudden or sustained inputs of water, such as those following rainfall events and snow or ice melt.

2 Rivers carry out most of their erosion and transportation during these spasmodic high-energy episodes.

3 Most changes in channel gradient, shape and planform will take place at times of extremes of available energy. (NB The term 'energy' is used in this section to include both kinetic energy and the energy transfer, i.e. applied force, which determine the 'workrate' of the stream.)

These hypotheses are compatible with the broader geomorphological principle: that landforms and landscapes experience relatively long periods of stability or minor fluctuations, interspersed with short episodes of accelerated change generated by surges or pulses of material and energy inputs (see Fig. 1.12).

Increasing energy

In river channels, this water input surge is indicated by the rising limb of a storm hydrograph. (Sediment inputs are discussed later in section 2.7.) Thus, as the quickflow processes deliver water to the channel, volume and energy increase. Some of this additional energy is applied to increase the channel capacity and efficiency. For example, the channel bed is scoured to increase depth and perhaps increase the gradient. Banks are eroded to increase the channel width. These processes increase the velocity and the throughput of water (Fig. 2.26). Stream channels tend to be at their most efficient during bankfull conditions (section 2.3).

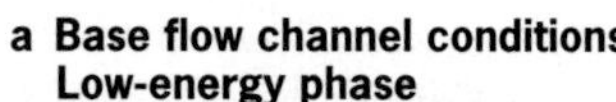

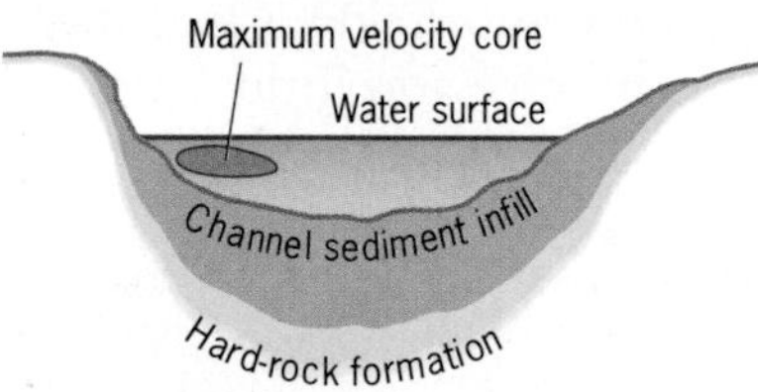

b **Bankfull channel conditions High-energy phase**

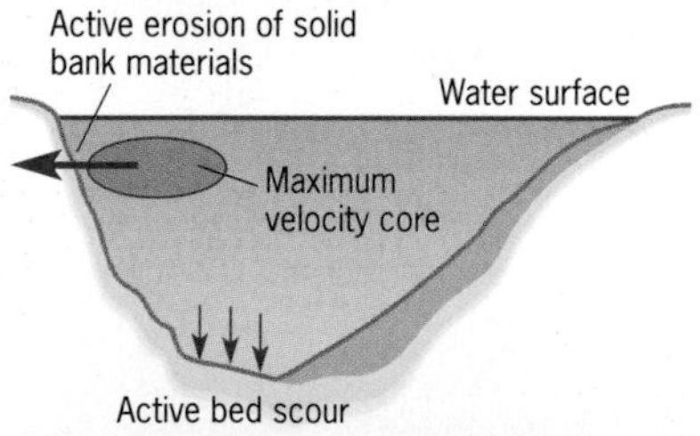

Figure 2.26 Relationships between discharge, channel form and processes

Decreasing energy

As the input surge subsides, discharge and velocity decline towards the baseflow levels, following the falling limb of the storm hydrograph. The channel adjusts to the diminishing water flow and capacity to transport load. Deposition makes the channel narrower and shallower (Fig. 2.26). The new equilibrium between channel and flow may be very similar to or significantly different from the conditions before the high-energy episode (lines *a* and *b* on Fig. 1.12). This **recovery time** or **relaxation time** may be weeks or even years, as the case study of the Noon Hill flood (opposite) illustrates.

Flow-form relationships

It is not easy to make a simple statement about the relationship between streamflow and the form of the channel. The problem here is that discharge is not 'steady', but highly variable. Research shows that channel form in humid environments seems to be adjusted to storm runoff events which occur every 3–5 years. This is called the **recurrence interval**. In more arid regions the recurrence interval varies from 30 to 100 years. It is these medium-magnitude events which are most important in controlling channel form, while extreme events, with long recurrence intervals, or return periods, may achieve abrupt shifts in channel form and location.

During such extreme events, **critical thresholds** in the functioning of the channel system are crossed. The existing channel can no longer cope, and positive feedback takes over. For instance, in a severe flood, there is a temporary period of chaos before the system reorganises itself, perhaps in a fundamentally new form and location.

An extreme event: the Noon Hill flood, northern Pennines, England

The event

On 17 July 1983, Noon Hill, a small headwater catchment of the Tees–Wear watershed, experienced an extreme storm (Fig. 2.27). The highly localised cloudburst, covering only 3.5 km^2, yielded 105 mm of rain at Ireshope Plains between 1430 and 1700 hours. Two-thirds fell between 1445 and 1600 hours. The Noon Hill storm is estimated as having a recurrence interval of 20–30 years.

The Noon Hill environment

The moorland has a peat blanket cover, with mainly heather vegetation. The upper stream channels draining this peat cover have cobble and boulder beds, are less than 3 m wide and are confined by steep valley walls. Downstream, the valleys widen to allow some floodplain development.

The flood surge

Langdon Beck, West Grain and Ireshope Burn (see Fig. 2.27), which had to cope with the storm inputs, are small moorland becks with 'flashy' hydrographs, i.e. are adjusted to sudden upland downpours, yet sustain baseflow from the peat store during dry spells. During the 17 July storm they went from baseflow to peak discharge in less than 15 minutes, and peak flows lasted less than three hours (Fig. 2.28). Runoff reached 8 cumecs per square kilometre, and a floodwave surged down-channel at almost 5 km/hr. Figure 2.29 indicates the dimensions of this floodwave in comparison with the normal bankfull conditions. In Langdon Beck the flood became a slurry as slope failure dumped 30 000 tonnes of peat into the channel. As an eyewitness said, it was 'a wall of mud with the appearance and consistency of chocolate sauce' (Carling, 1986).

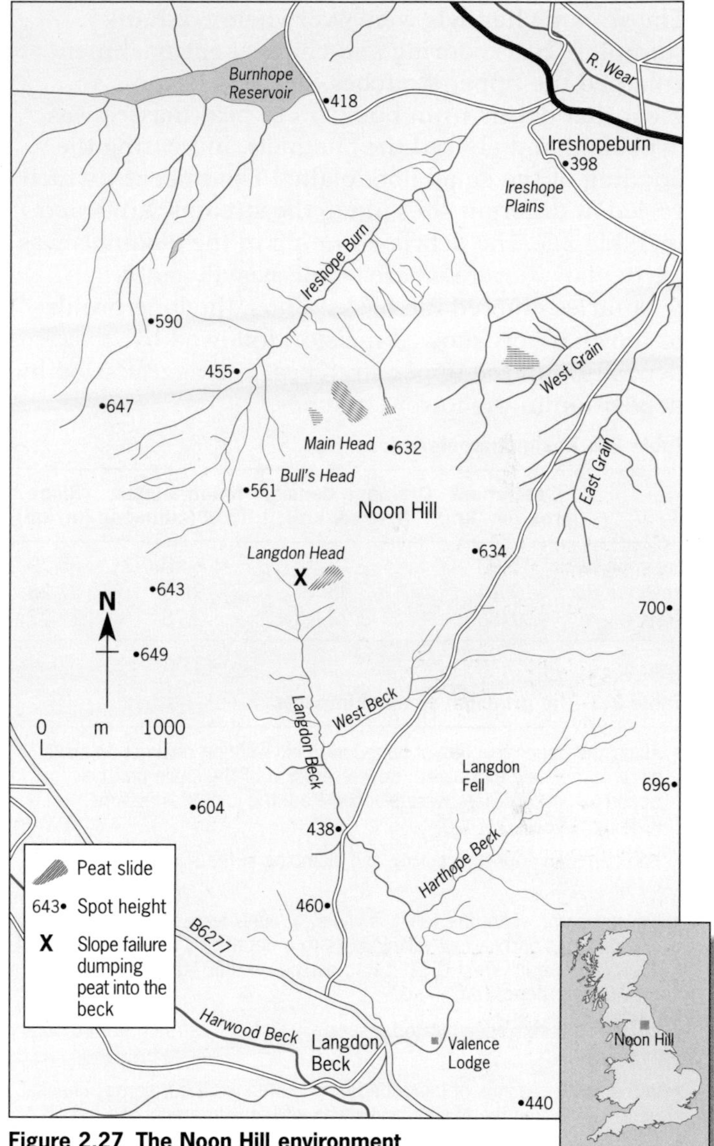

Figure 2.27 The Noon Hill environment

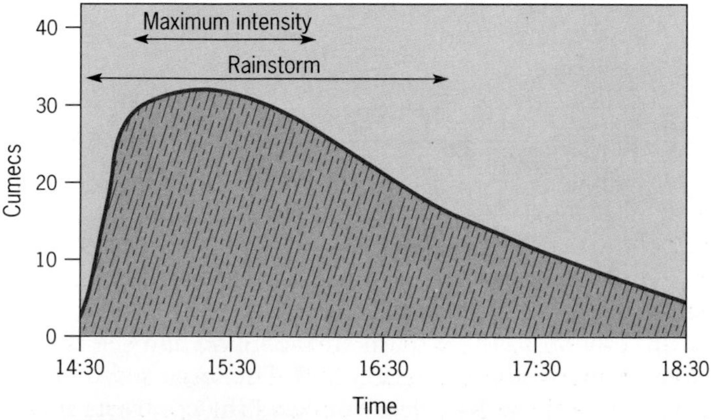

Figure 2.28 Hydrograph for Langdon Beck, 17 July 1983

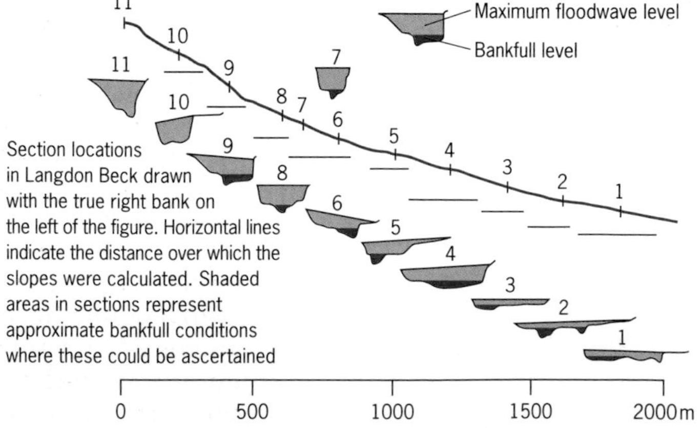

Figure 2.29 The Langdon Beck flood wave (*After:* Carling, 1986)

The Noon Hill flood

Channel impacts

The stream channels were overwhelmed. Bank slumping, bed scouring and channel entrenchment enlarged the upper stretches, the load became so great that debris from boulders to peat masses was deposited and choked the channels, increasing the flooding of the small floodplains. The impacts, which varied at different sites along the streams, are listed in Table 2.3. The debris dumping in the channels was particularly severe where small, heavily laden tributaries entered the becks. Thus, the long profiles of the channels showed incision followed by deposition in the upper stretches, and aggradation by deposition in the lower stretches.

Table 2.2 Basin characteristics

	Catchment area (sq. km)	Drainage density (km/sq. km)	Mean annual flood (cumecs)	Slope (m/km)
Langdon Beck	5.93	2.76	6.60	35.70
Ireshope Burn	7.14	3.95	9.73	37.28
West Grain	1.86	3.92	2.78	89.97

Table 2.3 The principal channel impacts

- Along the upper reaches of Langdon Beck 'existing channel deposits were completely evacuated. Severe erosion of the shale bedrock occurred and boulders were plucked from the jointed limestone' (Carling, 1986).
- Bank erosion caused cracking and slumping in the surrounding peat cover.
- Downstream, where the valley is wider, 'gravels were deposited, infilling the channels and narrow valley floors to a depth of 1 m' (Carling, 1986). Even in the small West Grain it was estimated that 10 000 tonnes of gravel were deposited.
- Boulder mounds accumulated, especially just downstream from tributary junctions.
- In the lower reaches of Langdon Beck 'chutes were cut across channel bends and often the main channel was infilled with gravel and the post-flood discharge diverted down the chutes' (Carling, 1986).
- 'Chute bars, large point bars and gravel splays one pebble thick were widely developed on the inside of channel bends' (Carling, 1986).
- The infilling of meander bends and the incision of chutes resulted in a straightening of the stream course.

Channel adjustments

Despite the severity of the impacts, by 1985 Langdon Beck had begun its work of readjustment. It was digging out its original channel and abandoning some of the storm chutes cut during the flood (chutes are short, straight sections of channel, incised by floodwaters, and often abandoned as normal flows return). Selective removal of finer materials was in progress. In the upper reaches, bank slumping caused by undercutting during the storm has since partially infilled the scoured channel. This acts as a supply of fine sediment for suspension load. Thus, the channels appear to be readjusting rapidly to gradients, cross-sections and planforms similar to those in place before the flood.

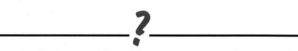

14 Divide the Noon Hill event into its major phases and describe the key features of each.

15 Describe the impact the storm had on the channels in
a the upper reaches and
b the lower reaches of the becks.
Draw 'before' and 'immediately after' long profiles of a beck and label the key changes.

16 What evidence is there that the relationship between discharge, debris supply and available energy changed progressively downstream during the flood event?

17 Use the information in the case study to support the following conclusions from research into the Noon Hill flood event:
a Extreme events move substantial volumes of debris into and down channels from one store to another. This increases the supply of sediment available in the channels to be worked on at times of more normal flows.
b Many small river basins in the Pennine uplands are resistant to permanent change, even after extreme and apparently catastrophic events.

18 Use the Noon Hill flood episode to illustrate what is meant by
a critical threshold;
b negative feedback;
c positive feedback.

2.7 Channels and sediment load

A stream obtains its load from two sources: approximately 90 per cent from weathering and mass movement of the slopes of the surrounding catchment basin, and the remainder from its own channel bed and banks. (A small amount of material falls directly on to the stream surface.) We can see the importance of the two sources by studying Figure 2.30. The large sediment load of the Amazon compared with the Rio Negro reflects the contrasts in their catchments in geology, soils, geomorphology and vegetation.

Figure 2.30 'The meeting of the waters' at Manaus. Look carefully and you will see that the waters on the right are dark and clear. This is the Rio Negro (Black River). On the left are the muddy brown waters of the Amazon.

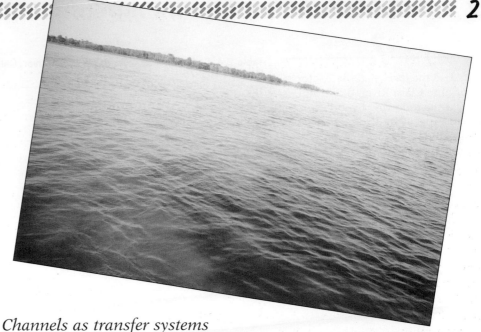

Channels as transfer systems

There are three components to the stream's work as a transfer system – erosion, transportation and deposition – and the relationship between these components depends upon the interactions between key variables (Fig. 2.31).

Capacity and competence

As we examine the relationships between a river and its solid load, it is useful to bear in mind two key properties:

1 The total load, measured by volume, mass and weight, a stream is capable of moving at any specific discharge or energy level. This is known as the stream *capacity*.

2 The dimensions of individual particles (the *calibre* of the materials). The maximum size/weight dimensions of individual particles which a stream can entrain is known as stream *competence*.

Clearly, the two are related: as available kinetic energy, i.e. the ability to apply forces to materials, increases we would expect both capacity and competence to increase, although not necessarily at the same rate.

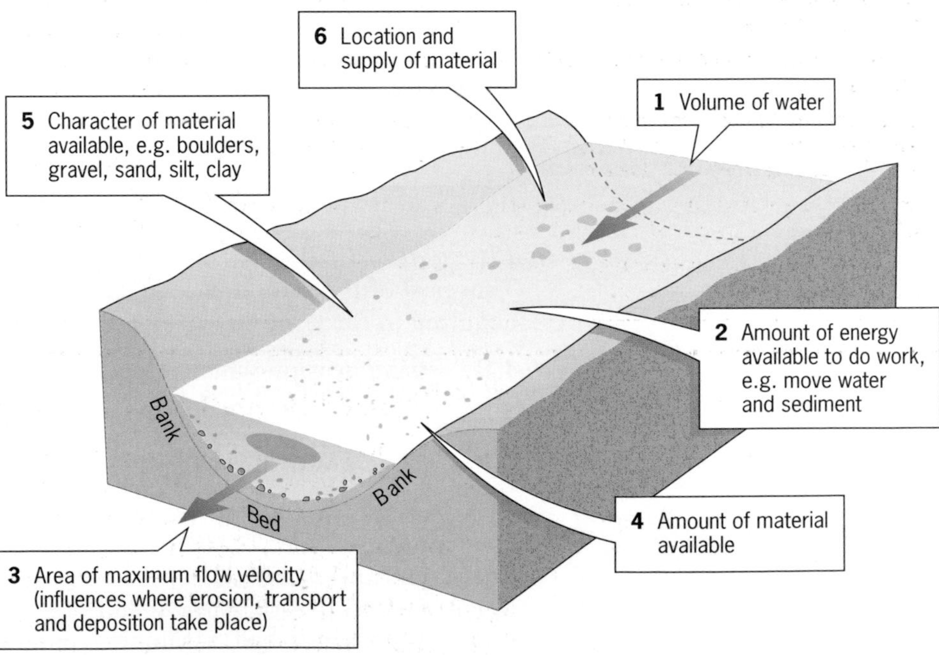

Figure 2.31 Variables influencing the work of a stream

?

19 Use the data of Table 2.4 for the Santa Clara River, California, to test the following hypothesis: 'Load capacity and competence of a stream increase as discharge increases.' (Select and justify the choice of appropriate graphical and statistical techniques to support your answer. NB You will need to use logarithmic graph paper.)

Table 2.4 Sediment statistics: Santa Clara River, California (Montalvo Station near river mouth) Period 1969–75

Sample	Discharge (m^3/sec)	Sediment discharge (tonnes per day)	Particle size: Clay per cent	Silt per cent	Sand and gravel per cent
1	13	232	73	27	0
2	4 420	736 000	26	34	40
3	1 860	756 000	30	33	37
4	1 645	534 000	33	41	26
5	1 830	321 000	36	35	29
6	923	92 000	35	37	28
7	40	323	72	27	1
8	870	105 000	33	41	26
9	350	103 000	26	38	36
10	6 250	1 196 000	23	37	40
11	15	172	57	40	3
12	1 350	120 000	37	40	23
13	1 980	253 000	55	41	4
14	3 200	513 000	25	34	41

Load components

We can divide the transported sediment into three categories: *bedload*, *suspension load* and *solute load*. Bedload consists generally of coarser fragments which move more slowly than the water. Suspension load consists of particles light enough to be carried along in the body of the water, at the same velocity as the water. The balance between the two categories varies, but bedload 'is commonly assumed to comprise less than about 10 per cent of the total load, increasing to as much as 55 per cent in favourable circumstances' (Chorley et al., 1984). Solute load consists of chemically weathered rock material which is carried by the river water.

2.8 Erosion and entrainment processes

A particle becomes part of a stream's load only when it is in motion. So it must first be dislodged from the channel bed or the banks, by erosion, and then set in motion by the process of **entrainment**. Once entrained, there must be sufficient energy available to keep it moving. Consequently, for an understanding of how a particle becomes part of a stream's load we need to answer two questions:

1 What starts a particle moving?

2 What keeps a particle moving?

Physical action

That part of the sediment load which comes from the work of the stream is acquired in three ways:

1 By *vertical* erosion which deepens the stream channel.

2 By *lateral* erosion of the banks which increases channel width.

3 By *headward* erosion which lengthens the stream channel.

This erosional work is carried out by a combination of several processes.

Corrasion and attrition

Much vertical erosion is the result of *corrasion* or *abrasion*, where bedload is dragged across the bed, scratching, smoothing and chipping away at the materials whether solid rock or debris. This process is most common at periods of high discharge in the upper reaches of streams, when angular fragments in the bedload work on the exposed bedrock.

A similar process, which also relies on physical impacts, takes place between the debris particles of the load itself: they collide with each other, causing rounding and gradual reduction in size. This is known as *attrition*. It helps to account for the tendency for the calibre of bedload particles to be smaller and more rounded progressively downstream. Such materials give a smoother, more efficient channel bed. Consequently, along lower reaches of a stream, channel efficiency compensates for reduced gradient to maintain load capacity, e.g. the large mass of fine particles of lowish density carried in suspension by rivers such as the lower Mississippi (Fig. 2.11). This capacity is assisted by local turbulence patterns in the streamflow.

Hydraulic action

A second set of erosion processes relies on the force of the water itself and is called *hydraulic action*. This is most effective in the middle and lower reaches of a river where the channel bed and bank materials are more likely to be unconsolidated sediments than solid bedrock. The force applied by the water is able to dislodge and entrain the fragments. When the energy is concentrated vertically, the channel will be deepened by *scour*, i.e. the removal of bed materials. As all rivers have some degree of sinuosity, a proportion of the force is applied laterally. This causes bank collapse and hence channel widening (Fig. 2.32).

Cavitation

An extreme form of this action is known as *cavitation*. This occurs when turbulent floodwater is hurled against a river bank: pressure release bursts air bubbles, which causes shock waves to pass into and weaken the bank sediments. Where the river is deeply incised, the repetition of this cycle of oversteepening, slumping and removal can produce impressive *bluffs* or *river cliffs*.

Headward erosion

Finally, a stream can be lengthened by headward erosion which is caused by two processes: the gradual process of *spring sapping*, which progressively cuts back into a slope; and the more rapid process of *gully formation* on a slope during a major storm. One or more gullies may then link up with the headwater channel and so extend the length of the stream.

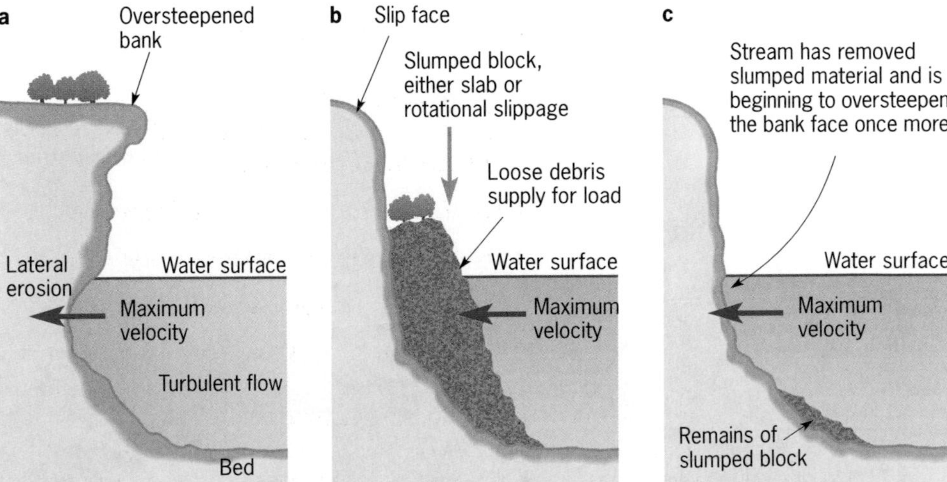

Figure 2.32 Hydraulic action and lateral erosion. Notice how the collapsed section of bank provides a fresh supply of loose material easily eroded and entrained by hydraulic action.

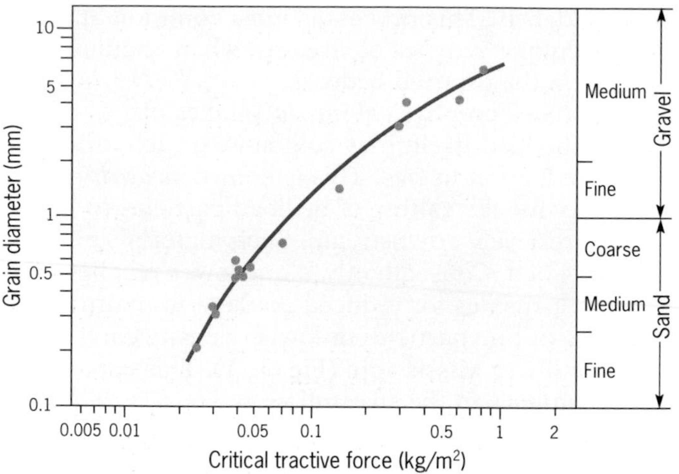

Figure 2.33 Erosion, entrainment and bedload: critical tractive force by particle size (*Source:* Chorley et al., 1984)

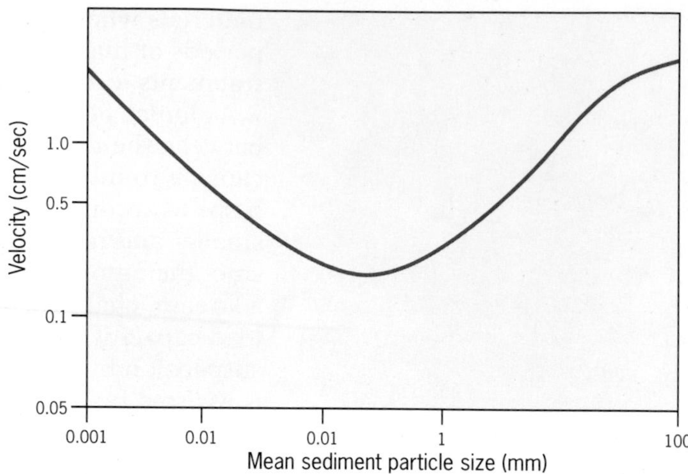

Figure 2.34 Relationships between particle size and velocity: the Hjulström curve

Critical forces

For sediment motion to take place, the upward and forward driving or tractive force must overcome the resisting force. As discharge and available energy increase, e.g. along the rising arm of a hydrograph, a **critical tractive force** is achieved, for particles of progressively greater mass to be moved. Thus, the critical tractive force is the level of force required to trigger motion in a particle of specific size (Fig. 2.33). Turbulent flow, where eddies produce an upward force and speed variations induce lift, is more effective than laminar flow.

Particle size–velocity relationships

The critical tractive force can be expressed in terms of stream velocity as the *critical erosion velocity*: the speed of flow at which particles of specific size, shape and density will be dislodged and moved (Fig. 2.34). The Hjulström curve shows that the particle size–velocity relationship is not simple. For instance, fine-grained clays and silts (diameter < 0.6 mm) have higher critical erosion velocities than medium-grained sands. This is explained by the smooth, cohesive bed formed by the tiny clay and silt particles, and the small but significant electrostatic bonds between them.

Figure 2.35 Critical erosion and deposition velocities

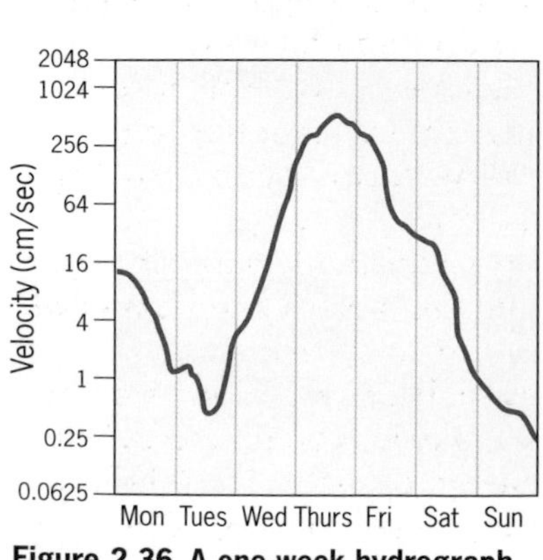

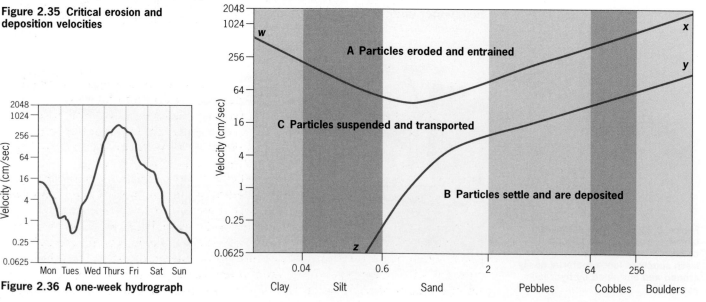

Figure 2.36 A one-week hydrograph

20 From Figures 2.34 and 2.35,
a give the critical erosion velocities for particles of 0.01 mm; 0.5 mm; 1 mm; 50 mm diameter;
b give names to each of these particle sizes.

21 At what stream velocities will sand grains of 0.8 mm diameter be • entrained • deposited?

22 At what stream velocities will small pebbles of 5 mm diameter be • entrained • deposited?

23 The graph of Figure 2.36 shows the mean flow velocity of a stream over a one-week period. By using Figure 2.35, describe and explain changes in sediment load during that week. (Include particle sizes, entrainment and deposition characteristics.)

Figure 2.35 allows us to examine this in more detail. The critical erosion velocity is given by line *w–x*. All material falling into zone A will be in motion. Conversely, as velocity and available energy decline (cf. the falling arm of the hydrograph) so material will be progressively deposited once more. This critical threshold is represented by line *y–z*. All material in zone B will be at rest.

The second complication is that higher velocities are required to entrain particles than to keep them in motion. For instance, a coarse sand grain with a diameter of 2 mm has a critical erosion velocity or tractive force of 1200 mm/s, but once entrained it will continue to move until the velocity falls to 100 mm/s. This phenomenon is indicated by zone C.

The gap between lines *w–x* and *y–z* highlights the third complicating feature: the distinctive attributes of fine particles, especially clay and silt. Once they are in motion, they will remain in suspension even though flow velocity almost ceases. Their *settling velocity* is very low, although their erosion and entrainment velocity is high.

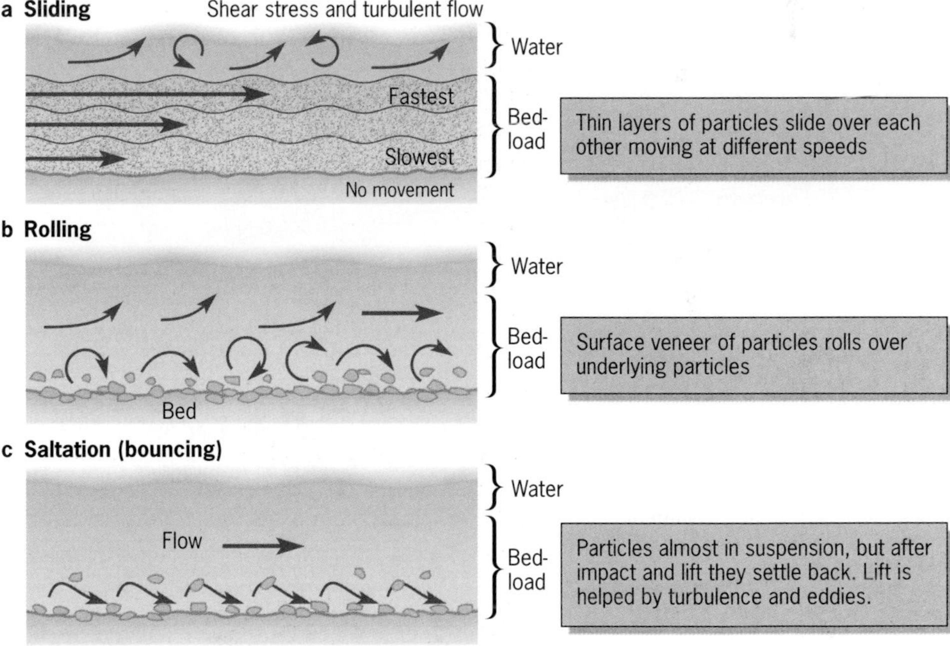

Figure 2.37 Bedload movement

Bedload movement

Bedload is that component of total load whose immersed weight is carried by repeated contact with the channel bed. The upward and forward movement of the particles occurs as a result of two forces: dynamic forces which are generated as particles collide, and forces imposed on the bed materials by the moving water. Movement takes three forms: sliding, rolling and bouncing (Fig. 2.37). Usually, all three will be found at work simultaneously. The balance between them changes constantly according to velocity, degree of turbulence, and the amount and character of bed material.

Most stream beds consist of materials of varying sizes. Erosion and entrainment are selective, dependent upon the available energy and velocity. This means that, over time, the finer materials are likely to be removed. The upper size limit of removal will depend on the energy available in high-flow episodes with a recurrence interval of a few years (see section 2.6). This selective removal of the finer materials may cause a **bed armour** to develop.

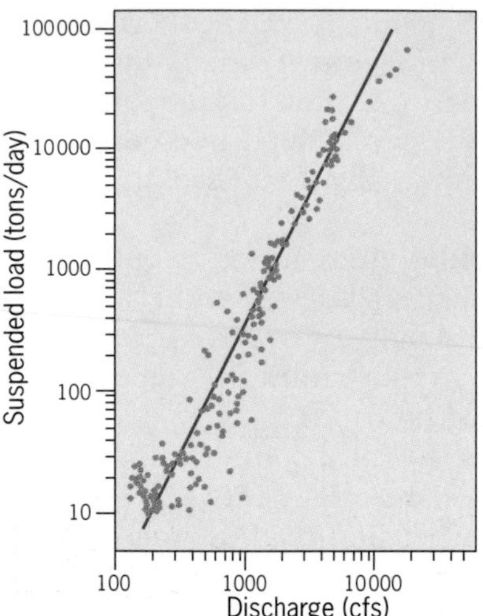

Figure 2.38 Suspension load/discharge relationships in two catchments: Brandywine Creek, Delaware, USA
(*After:* Chorley et al., 1985)

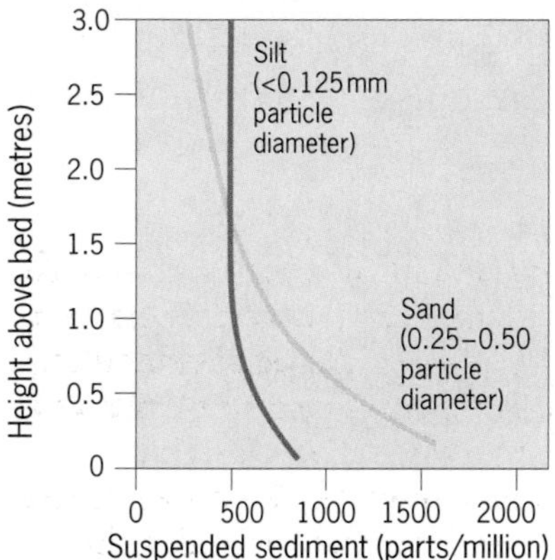

Figure 2.39 Suspended sediment variations with depth: Niobrara River, Nebraska, USA
(*After:* Chorley et al., 1985)

24 Use the graphs of Figures 2.33, 2.34 and 2.35 to illustrate
a the concept of a critical threshold for entrainment;
b the relationships between velocity and load;
c the relationships between discharge and load.

25 Use Figure 2.39 to describe and explain the vertical distribution of sediment in a stream. (Use specific figures from the graphs to add precision to your answer.)

The channel bed will be a surface made up of particles whose critical tractive force is above that which occurs in the medium-term flood event. For instance, a stream bed may consist of a layer of rounded pebbles laid out like the scales of a fish, aligned in an overlapping fashion and resistant to removal.

2.9 Suspension load

Suspension load is that part of total stream load whose weight is supported by the water. Turbulent flow is crucial for keeping sediment in suspension.

We can see from Figure 2.38 that as discharge increases so does suspension load. Remember, however, that once particles have been lifted into the body of the water, they will remain in suspension for a time even at low velocities (Fig. 2.35). This means that suspension load depends more on sediment supply than upon available force. None the less, generalisations can be dangerous. Sediment supply from the drainage basin may be limited in hilly catchments, and so suspension load may decline rapidly as discharge fades.

We find that suspension load declines towards the stream surface (Fig. 2.39). This change from bed to surface is less pronounced for fine particles: thus the coarser sand particles are concentrated mainly near the bed, while clay and silt particles are distributed more evenly through the body of water. This is despite the fact that the maximum turbulent velocity tends to occur towards the surface.

Summary

- A stream channel is an open funnel along which water and sediment are transferred.
- A channel consists of three dimensions: planform; long profile; cross-section.
- All natural stream channels have sinuous planforms.
- The form and functioning of a channel depend upon the constantly shifting balance between available energy, water and sediment.
- A channel works to sustain a state of dynamic equilibrium where its geometry is adjusted to the throughputs of water and sediments.
- At least 90 per cent of total energy in a stream is used in overcoming the friction of the channel surface, so channel form and efficiency have major effects upon the amount of energy available for erosion and transportation.
- There are direct but complex relationships between discharge, velocity, long profile, cross-section form and erosion/transportation/deposition characteristics along a channel.
- Most stream load is derived by vertical and lateral erosion from the channel bed and banks.
- There are three types of load: bed, suspension and solute.
- Most erosion and transportation take place during high-flow episodes with medium-term recurrence intervals, while irregular extreme events can trigger fundamental channel changes.
- All streams exhibit rhythmic flows (regimes) and hence distinctive sequences of erosion, transportation and deposition.

3 Fluvial landforms

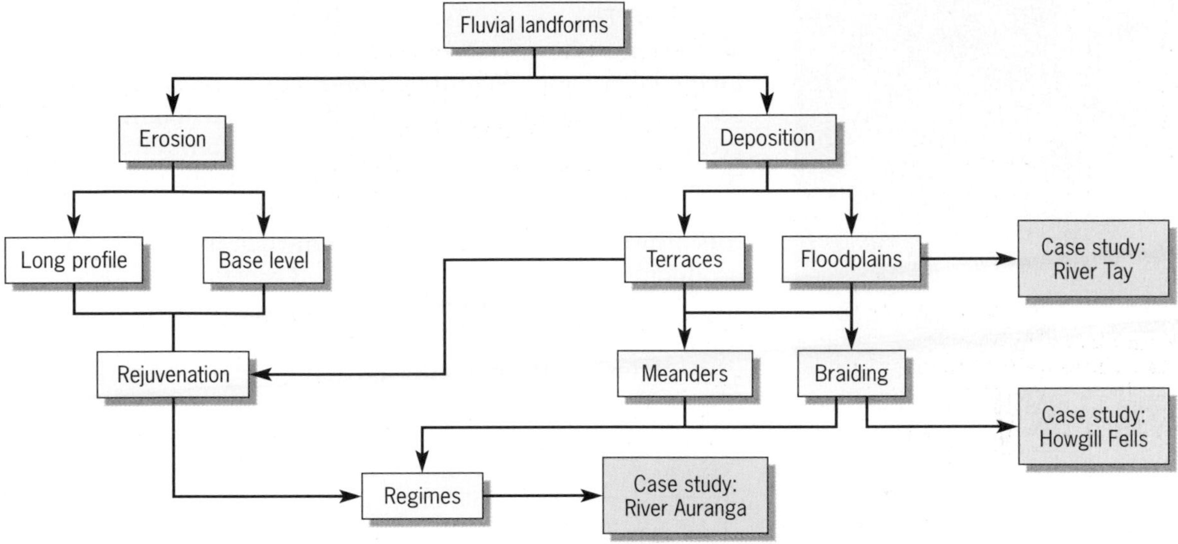

3.1 Introduction

Rivers influence landform formation at all scales and in all environments. At the drainage basin scale, the stream network sculpts the landscape, and at a local level, individual landforms are created. The results can be spectacular, as Figure 3.1 illustrates. Using the knowledge of how channels work, gained in Chapter 2, we can study this work of rivers in terms of the materials involved in the erosion–transportation–deposition sequence. Once again, the systems approach of inputs, stores, throughputs and outputs provides the framework (Fig. 1.21). For convenience, the chapter studies erosional and depositional landforms separately. Keep in mind, however, that *both* erosional *and* depositional processes are involved in the evolution of any given fluvial landform. Remember, too, that people are increasingly active in modifying the landforming capabilities of rivers, e.g. by dams which control discharge and load transfer.

Figure 3.1 Meander incision and vertical erosion: the 'goosenecks' of the San Juan River, USA

Figure 3.2 King's Canyon, South Fork, California, USA

3.2 Erosional landforms

Base level and the long profile

Energy in a stream can be applied as force both *vertically* and *laterally* to cause erosion, entrainment and transportation of material. The relative importance of these directional forces changes from the source to the mouth of a river. The key feature of the long profile is the **base level**. This is defined as 'the controlling theoretical level down to which ... a river can be reduced by fluvial erosion' (Whittow, 1984). For most rivers, base level is its mouth at sea level. In the headwater stretches, where height above base level is greatest, vertical erosion or downcutting dominates (Fig. 3.2). Towards the river mouth, closer to base level, an increasing proportion of force (the application of energy) is applied sideways, creating lateral erosion (Fig. 3.4).

Features such as lakes or resistant rock formations act as *local base levels*, producing a composite long profile. Over long periods of time the river will work to eliminate these irregularities, and so achieve a *graded profile* (see Chapter 2, p. 25). Meanwhile, however, above each local base level the vertical downcutting to lateral carving sequence is likely to be found (Fig. 3.3).

We can see, therefore, that if base level changes then the balance between vertical and lateral erosion will alter. In addition, as the amount and distribution of available energy change, so will the balance between erosion and deposition within the stream network alter.

Figure 3.3 The lower falls of the Yellowstone River, Wyoming, USA. The cap-rock of resistant lavas form the rim of the Yellowstone Plateau. It acts as a local base level. Upstream the river sways laterally across the plateau surface. Below the local base level, the high-energy river cuts deeply into less-resistant volcanic beds.

Figure 3.4 Floodplain of a tributary of the Rio Grande, Texas, USA

Valley incision

Valleys formed by streams well above base level, where erosional energy is applied to produce vertical downcutting (incision), are most commonly V-shaped (Fig. 3.2). Valley sides are rarely vertical. The diagrams of Figure 3.5 explain this characteristic. As the stream cuts downwards, so the slopes above are subjected to **subaerial** weathering and erosion processes, and gradually retreat. The uppermost valley slopes have been exposed longest, so their retreat is greatest. The angle of the slope is controlled by the balance between rates of vertical incision and lateral retreat, plus the rate of debris removal. This balance is influenced by such factors as time, stream energy, type and efficiency of weathering processes, underlying geology, slope aspect, vegetation and human activities (Fig. 3.6).

1 Produce a sketch contour map of a set of interlocking spurs, using the landscape of Figure 3.3 as a basis. Number your contour heights and label the stream and spurs.

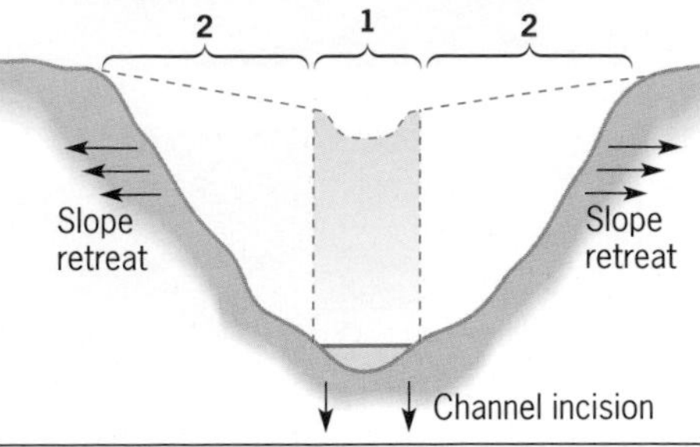

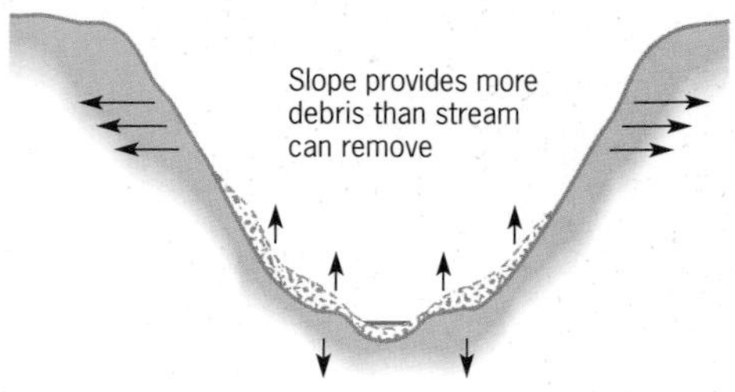

b Debris supply greater than available energy: 'V' shape becomes wider
Slope provides more debris than stream can remove
Valley floor and lower slopes aggraded. Stream removes part of debris supply by vertical use of energy. Only in flood discharge does bed scour permit vertical incision.

Figure 3.5 Stream work and valley form

Figure 3.6 Logging along steep valley sides of the Humboldt River in northern California has led to slope failure and rapid slope retreat

All stream channels have sinuous planforms, indicating that some of the available energy is applied laterally against the channel banks (see section 2.8). This undercuts the foot of the slope. In time the oversteepened slope will fail, resulting in mass slumping and slope retreat.

Only if the stream is capable of entraining and removing this eroded material which arrives in the channel will bed incision continue and the steepness of the 'V' shape be sustained (Fig. 3.7). The stream works towards a state of dynamic equilibrium where slope retreat, debris supply and downcutting retain a stable relationship over time. Where the downcutting is combined with a sinuous planform, a set of **interlocking spurs** may develop, as is beginning in the King's Canyon stream of Figure 3.2.

Rejuvenation

We use the term *rejuvenation* to describe the condition where a river sustains or regains sufficient energy for continued or renewed vertical downcutting (incision) in its efforts to achieve a graded long profile. A *gorge* or *canyon* is the extreme result of this battle (Fig. 3.8). There must be forces which, despite the work of the river, maintain or increase the height of the channel above base level. This can occur in two sets of circumstances: first, *tectonic uplift,* and second, *fall of base level.*

Figure 3.7 Keskadale, Cumbria. The floor of this glaciated valley has been filled with glacial and fluvioglacial deposits. The present-day stream has insufficient energy to entrain and remove all this debris. As a result, there is little valley incision.

Uplift

When uplift occurs across all or part of a drainage basin, the thalweg gradient (gradient of main channel flow) is maintained or increased over time. As the land rises, so the rivers cut down to restore grade in relation to base level. For example, across the mountain and plateau region of the

Figure 3.8 The gorge of the Red River, a tributary of the Rio Grande, New Mexico, USA

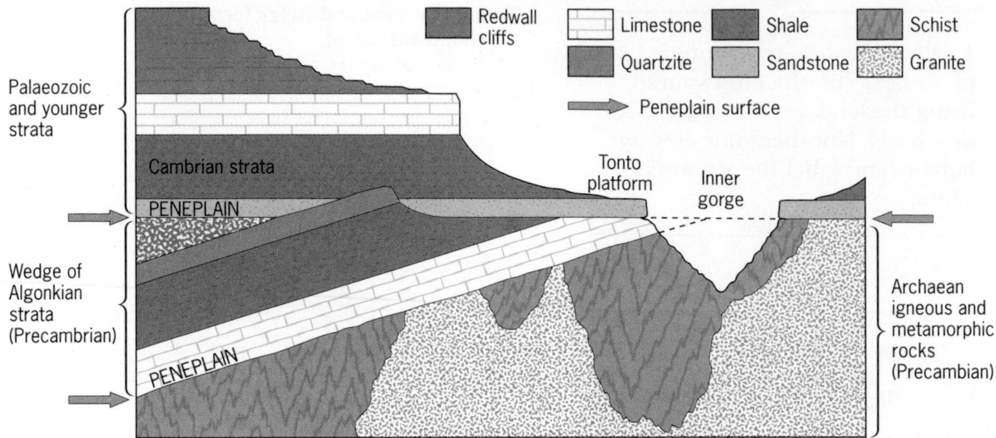

Figure 3.9 Lithological controls: the Grand Canyon, Arizona, USA (*Source:* Strahler, 1966)

western USA there has been prolonged, slow uplift. In response, rivers such as the Colorado, the Rio Grande and their tributaries have been able to counter this uplift by cutting their spectacular canyons (Fig. 3.8).

The canyon cross-sections illustrate the role of **lithology** as a control upon fluvial valley form (Fig. 3.9). The uplift may be intermittent, and so this rejuvenation may involve a *renewal* of energy within a drainage basin after a period of low-energy conditions. In such cases, the established stream network will begin to incise itself into the landscape, often retaining its planform courses. Where the pre-existing rivers had meandering courses, sets of **incised meanders** may evolve, as in the spectacular 'goosenecks' of the San Juan River in the USA (Fig. 3.1). The rejuvenation of meanders across floodplains produces quite different landforms, as the section on meanders later in this chapter makes clear (see section 3.3).

One thing that you should realise is that all these landforms can equally result from a lowering of the base level.

Lowering of base level

A fall in sea level lowers the base level, increasing the long profile gradient and triggering rejuvenation (Fig. 3.10). The incision commences at base level and moves progressively upstream. The upper limit of the zone of incision is the *knick point*: the location on the long profile below which the gradient increases and velocity accelerates. As Figure 3.10 illustrates, the long profile of a stream may reveal several knick points, each indicating a change of base level. Each knick point acts as a local base level. Because the knick points shift progressively upstream, the further upstream they occur the older they are. Thus, only the section of the long profile downstream from the lowest (youngest) knick point is graded to the current base level.

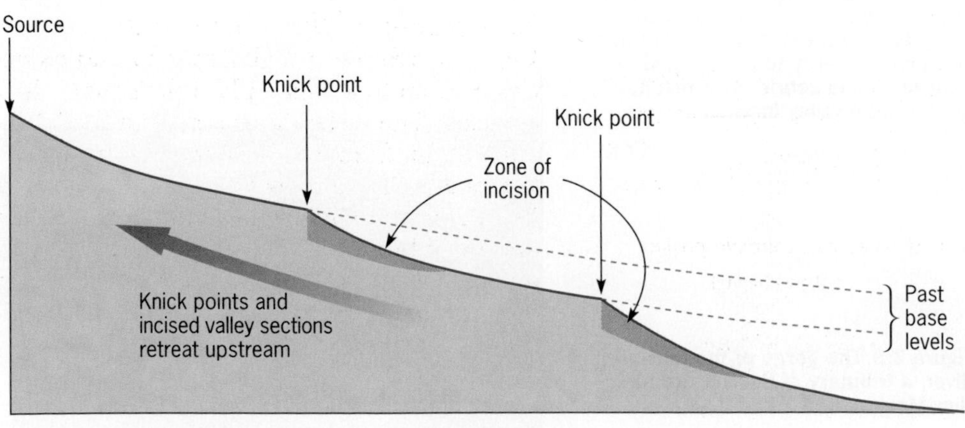

Figure 3.10 Rejuvenation and valley form: long profile with knick points

Geological and geomorphological controls
There are exceptions to the general idea that knick points increase in age upstream. Local base levels, and hence knick points, may be caused by resistant rock formations and lakes. These irregularities on the long profile have not moved upstream from the mouth of the river. Yet they *do* move upstream. For instance, the Yellowstone Falls of Figure 3.3, like all major waterfalls, are retreating headwards, leaving a lengthening gorge downstream.

Rejuvenation resulting in accelerated or revived downcutting may also be caused by sudden increases in discharge. **River capture** may enlarge the drainage basin, increasing the water and energy inputs. Downstream from the capture the channel is likely to be deepened and widened by the application of vertical and lateral forces, to cope with the increased flows. This in time enlarges the valley landform as slope retreat progresses.

River discharge can also be enlarged by **drainage reversal**, where a river changes course. For example, the River Severn originally flowed northwards from its headwater catchments in Wales to the Dee estuary. This was reversed late in the Pleistocene glaciation by the cutting of the Ironbridge Gorge. The head of this gorge still acts as a local base level.

Figure 3.11 An alluvial fan on the Colorado Plateau, USA

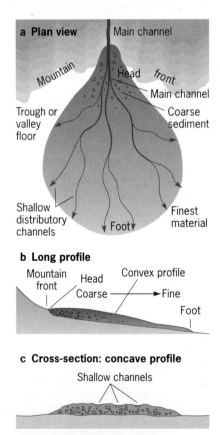

3.3 Depositional landforms

Deposition, like erosion, can occur in all sections of a river's course. In general, however, it tends to be progressively dominant downstream, as height above base level decreases, or as a local base level is approached. As rivers approach final base level, e.g. sea level, or local base levels, they tend to be less confined by valley slopes, and have more freedom to swing and move laterally. As they do so they not only erode laterally, but also deposit considerable bodies of sediment, i.e. both erosional and depositional processes are at work. Sedimentation landforms are likely to develop at three main locations:

1 *Topographical discontinuities*, i.e. where a significant decrease in gradient occurs, e.g. **alluvial fans** where streams emerge from mountains on to a plain.

2 *Valley infill*, where valley floors have been filled with alluvial deposits, e.g. floodplains and terraces.

3 *Water margins*, where streams flow into standing water and energy and tractive force are suddenly reduced, e.g. deltas (see Chapter 5).

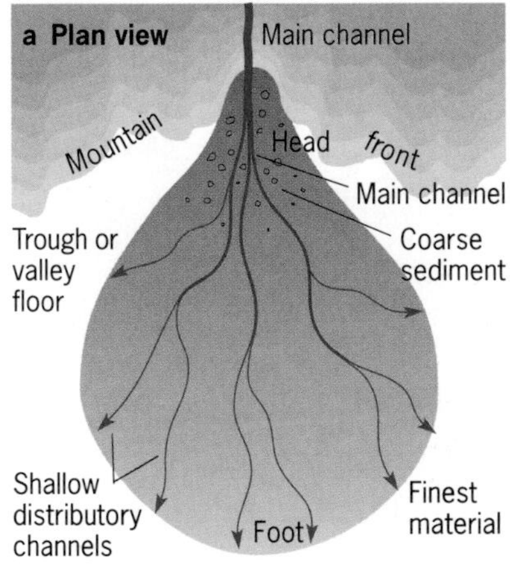

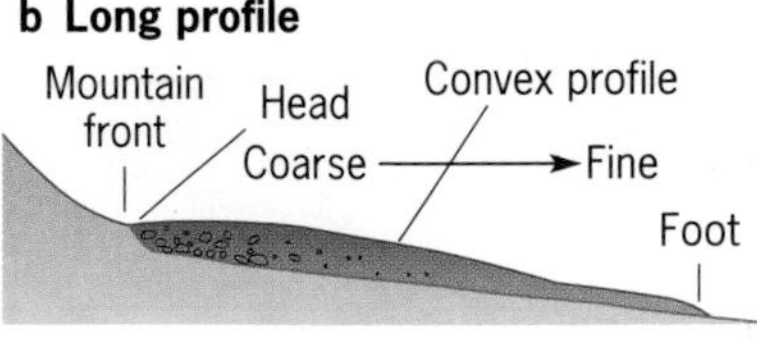

Figure 3.12 Features of a fan

Topographical discontinuities
Where a river with high capacity and competence, and ample debris supply, emerges on to a valley plain or trough, the available energy may be abruptly reduced. The valley floor acts as a local base level, and the result is an alluvial fan (Fig. 3.11). Such fans are most obvious in semi-arid or seasonally dry regions with incomplete vegetation cover, but they are found in all environments. They can reach 150 km in length and breadth and have depths of hundreds of metres, and so are major landforms. Each fan is unique and complex in detail, yet the principles are always the same: a stream finds itself with too much load, and so progressively deposits in and around its channel (Fig. 3.12).

Valley infill: floodplain development

As rivers approach base level, they apply an increasing proportion of their available energy to swing laterally, creating a broader valley (Fig. 3.4). Deposition tends to exceed erosion, creating valley floors which are built up, or **aggraded**, over time.

Floodplain variables

The extreme form of the aggradational valley floor is the **floodplain**. Floodplains are formed by a combination of lateral and vertical **accretion** of alluvial sediments. The key understanding is that the balance between these lateral and vertical components and processes is constantly changing. It varies spatially between rivers, along different stretches of the same river, and even along the same stretch over time.

The form and function of a floodplain are precisely what the name suggests: surfaces of low relief which accommodate and store excess waters when channels overflow their banks. The key controls on floodplain construction are (i) meander belt width, (ii) meander migration rate in relation to floodplain width, (iii) the frequency of floods, (iv) sediment supply, which is related to drainage basin size and character, e.g. rock type. These complex dynamics are dependent upon available stream energy and how it is applied.

Channel–floodplain relationships

A channel and its floodplain function as an open system. A large-scale example is shown in Figure 3.13 for the lower Mississippi between Cairo and the river mouth at the Gulf of Mexico. Notice that the sediment budget estimate is for 1880–1911, a period before human activity significantly altered the discharge and load characteristics. The floodplain is not only the

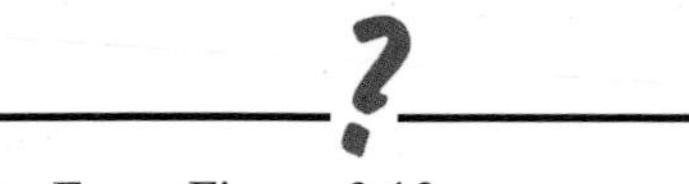

2 From Figure 3.13:
a make a table of inputs, stores and outputs for each of the channel sections;
b describe the way the channel/floodplain system works;
c in what ways do the upper and lower sections differ? Suggest reasons for these differences.

3 What changes are human activities likely to have had upon this discharge/load system?

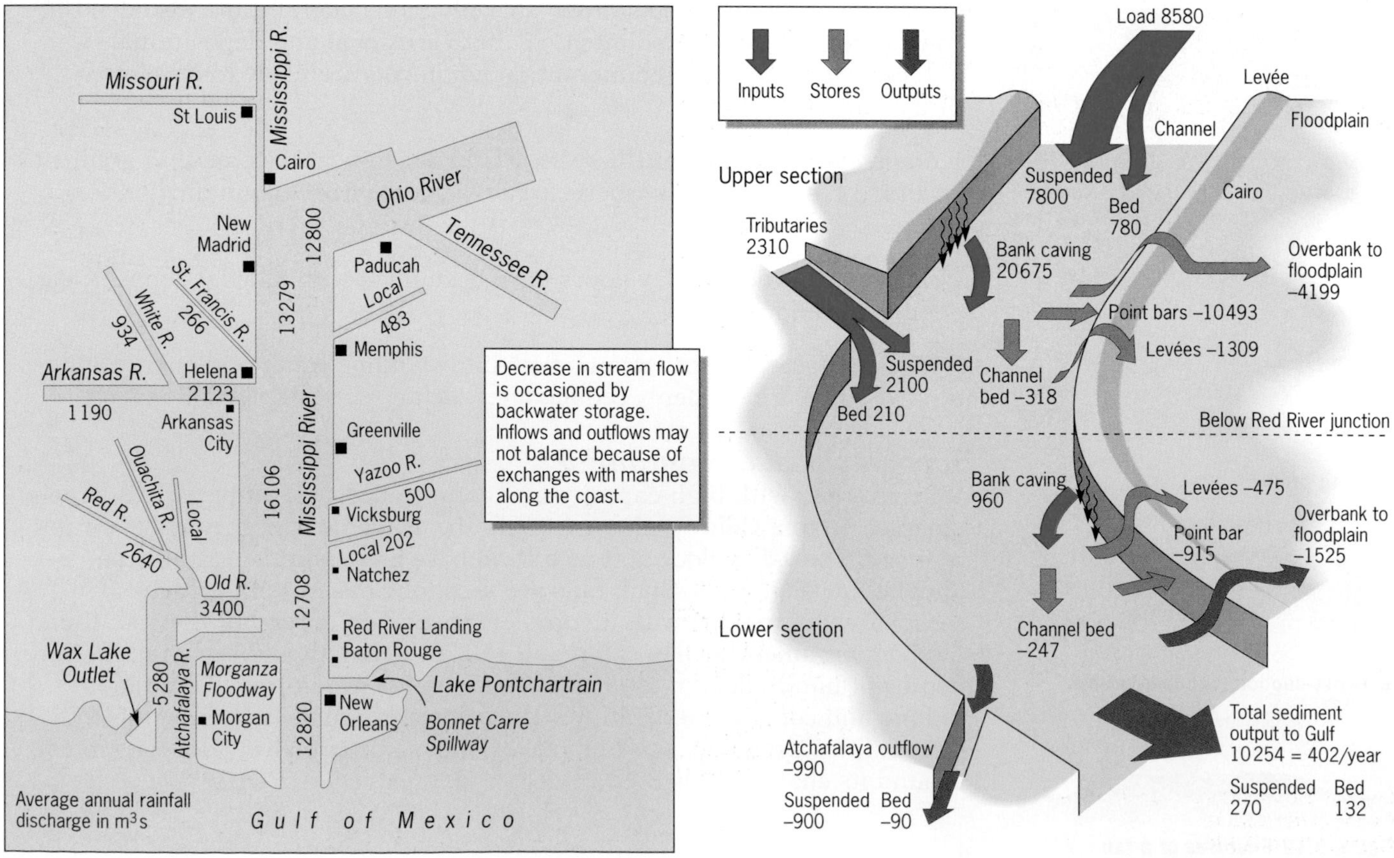

Figure 3.13 Right and left: Lower Mississippi sediment budget and average annual discharge, 1880–1911 (*After:* Kesel et al., 1992)

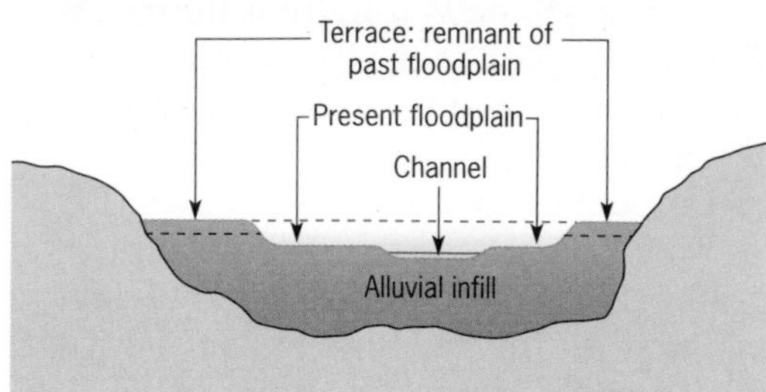

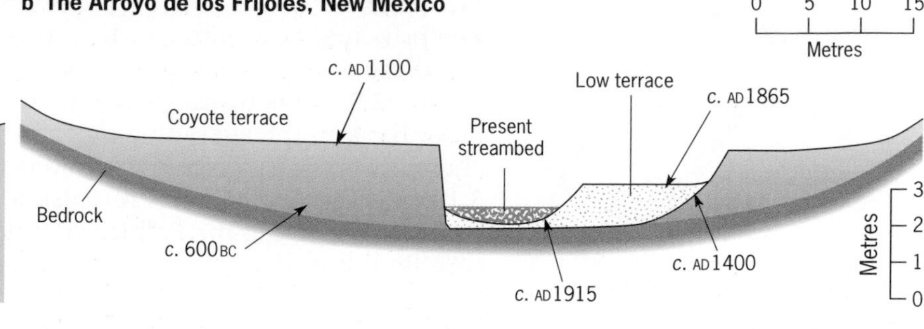

Figure 3.14 Terraces and floodplains

main store (point bars, levées, main floodplain) but also a major source of sediment input via bank caving, i.e. the erosion and collapse into the channel of earlier sediments stored in the floodplain.

Figure 3.15 River terraces, Pamplona, N.E. Spain. The river is cutting into its former floodplain, which now becomes a broad, flat river terrace.

River terraces

Environmental conditions fluctuate over time, changing the discharge, load and energy relationships. One result is that many floodplains and aggradational valley floors show evidence of periods of downcutting in their history. The most common visual evidence is the river terrace (Fig. 3.14). The floodplain of the lower River Severn has an excellent set of terraces. The incision into the existing floodplain may be triggered by a change in base level causing rejuvenation (sea level falls and/or land rises), or by an increase in discharge, or a reduction in load. These changes provide the stream with sufficient energy to apply increased vertical and lateral erosional forces (Fig. 3.15). A valley with a well-developed set of terraces illustrates how erosional and depositional processes combine to create landforms, and the assemblages of landforms we call landscapes.

Meanders and floodplains

Meanders are the key to understanding floodplains and the development of terraces. A meander is defined as 'a loop-like bend in a river, characterised by a river-cliff or bluff on the outside of the curve and a gently shelving slip-off slope [point bar] on the inner side of the bend' (Whittow, 1984). A river is said to have a meandering course when its planform has a **sinuosity ratio** >1.5. The dimensions of a meander set out in Figure 3.16 are significant, as studies of numerous meanders have shown that there are mathematical regularities between them. For instance, meander wavelength is commonly 7–10 times that of channel width; points of inflexion are commonly 5–7 channel widths apart.

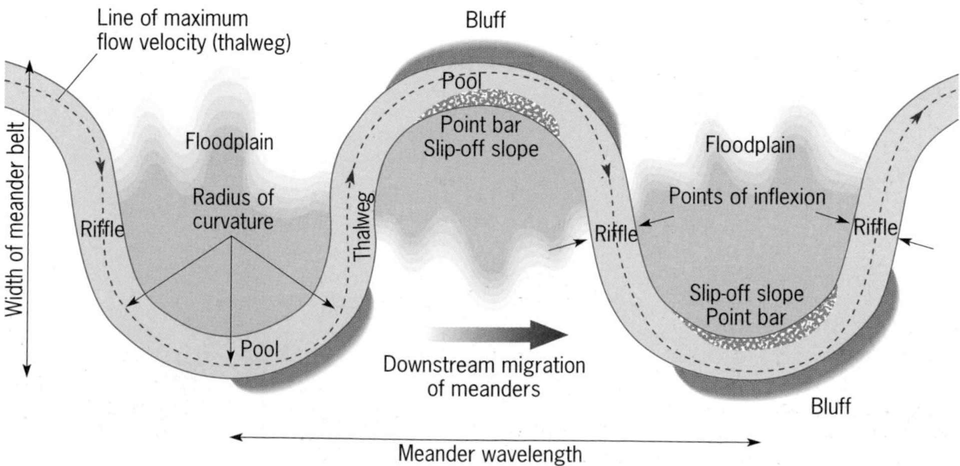

Figure 3.16 Dimensions and features of a meander reach (*After:* Collard, 1988)

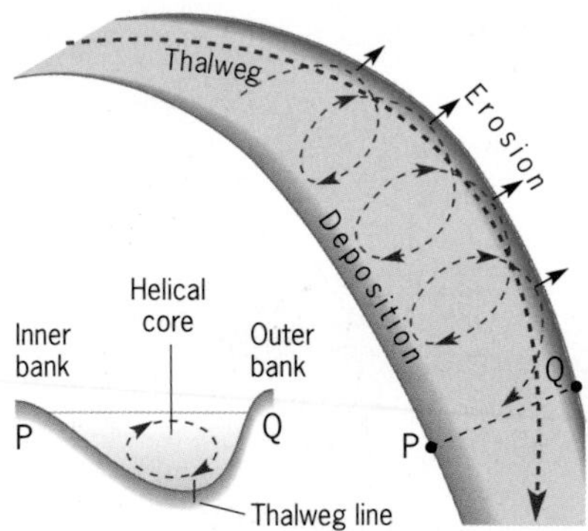

Figure 3.17 Helical flow: flow–channel relationships at a meander

Meander flows

The principal flow pattern which produces this shape is a **helical flow** or spiralling core of maximum velocity (Fig. 3.17). (For other flow patterns, see section 2.2.) This transfers sediment downstream, and from the deepened outer bank to the aggrading inner bank, producing the asymmetrical channel cross-section and the traditional model of meander hydraulics (Fig 3.18). Recent research has revealed a more complex pattern. For example, observations along the Fall River, Colorado (Fig. 3.19), confirm the helical core flow, but identify two key additions: first an inward flow (X) across the shallow inner bank, and second an upwelling (Y) against the outer bank. This latter core helps to explain the ledge (Z) found along the outer banks of some meanders. Figure 3.20 summarises our understanding of meander hydraulics.

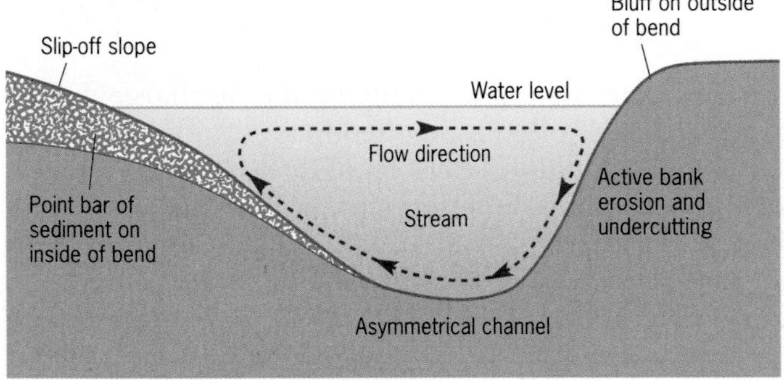

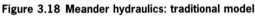

Figure 3.18 Meander hydraulics: traditional model

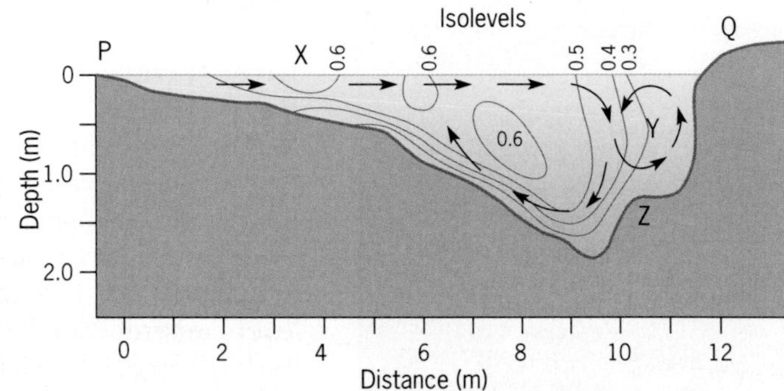

Figure 3.19 Flow patterns in the Fall River, Colorado, USA

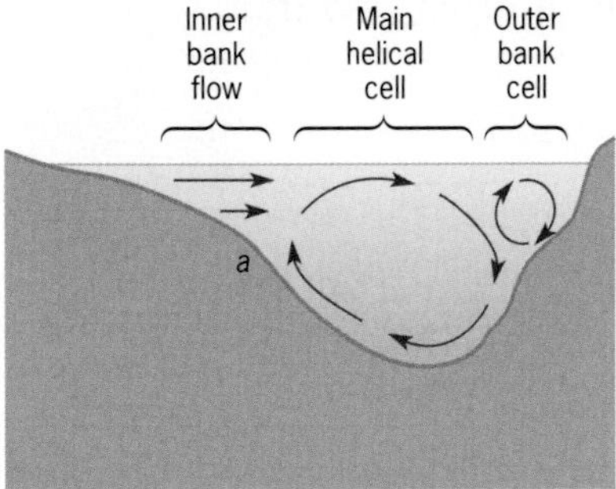

Figure 3.20 Modern meander hydraulics. The main and outer bank flow cells attack the outer bank. This causes bank erosion and a deep thalweg channel. The main flow cell then sweeps bed material towards the inner bank. The sediment is heaped up as a convex bar (*a*) where the outer and inner bed currents converge (*After:* Thornes, 1992)

Meander migration

The important outcome in terms of floodplain formation is that meanders 'migrate', i.e. move location over time. They move not only by lateral extension to broaden the floodplain, but also by downstream motion (Fig. 3.21). The dynamics of meander formation therefore involve the processes of erosion and deposition. Furthermore, the current cycle of erosion and sedimentation may be adding a new layer over an older surface.

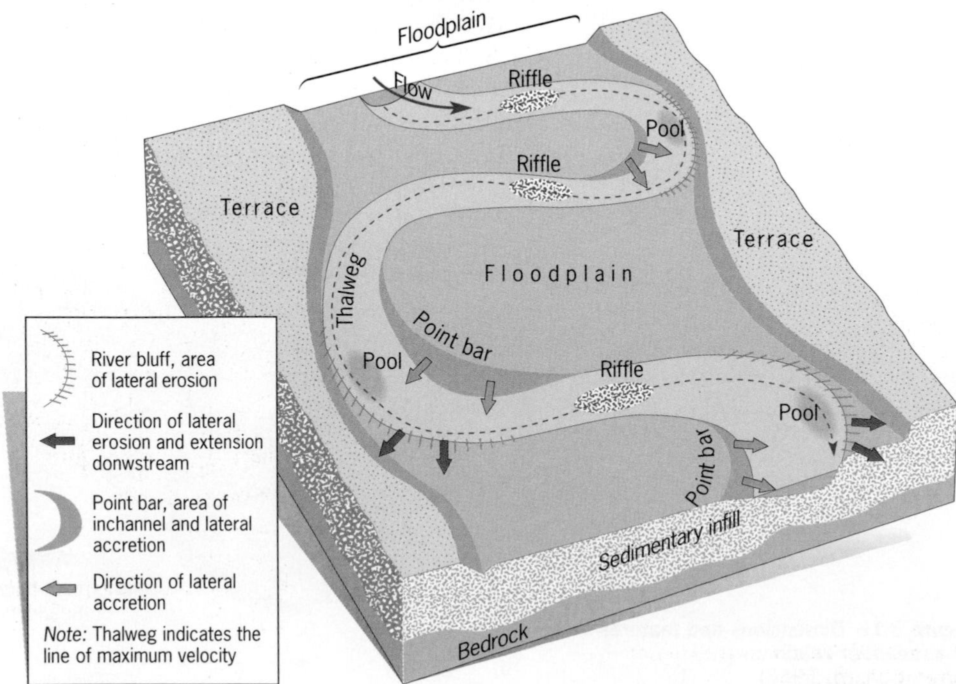

Figure 3.21 The mechanics of meander migration

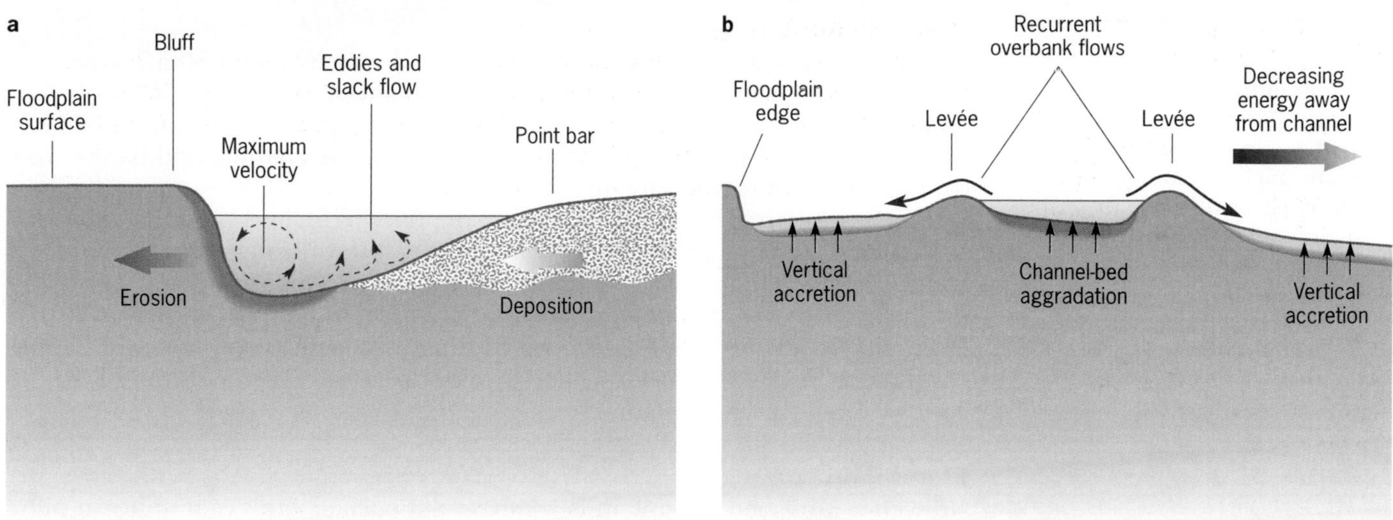

Figure 3.22 Floodplain formation processes

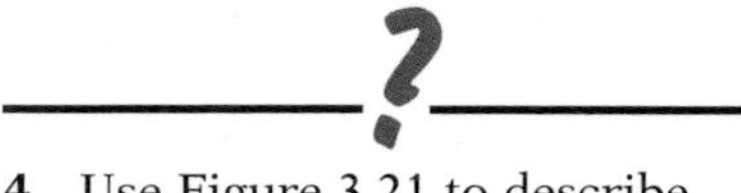

4 Use Figure 3.21 to describe and explain how meanders migrate and so help to construct a floodplain.

5 From Figures 3.22 and 3.23, explain how a floodplain is built by the combination of lateral and vertical accretion.

6 Explain what is meant by the statement that river terraces are remnants of former floodplains.

Where meander migration dominates, the floodplain is formed primarily by lateral accretion, as shown on Figure 3.22. The migration varies widely in speed. For instance, the River Kosi in Bihar, India, has recorded maximum rates of 750 m a year; the lower Colorado near Needles, California, 244 m; the lower Mississippi at Rosedale, Mississippi, 48 m. Where channel bed aggradation produces levées and recurring overbank flows (floods), then vertical accretion is dominant (Fig. 3.23). Most rivers exhibit a combination of both.

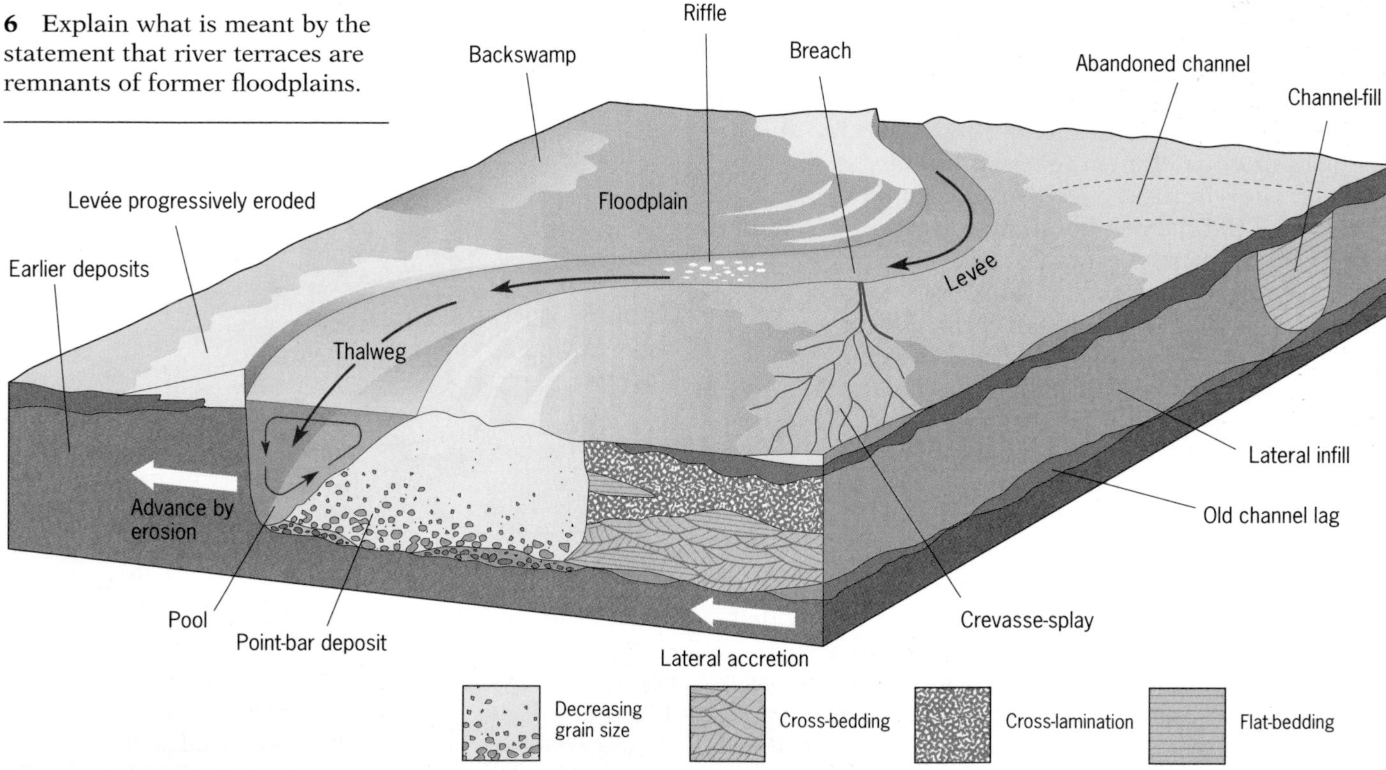

Figure 3.23 Floodplain growth and meander migration

7 Because most deposition during an individual flood event takes place near to the main channel, we might expect that sediment thickness and grain size will decrease as distance from the main channel increases. Yet many floodplains show a uniform spread of sediment across their width. Suggest how the process of meander migration helps to explain this distribution of sediment. (Labelled diagrams will help your answer.)

How a floodplain grows

We have learned that a floodplain is built up slowly by a combination of lateral channel shifts and sedimentation from successive floods. What happens when a river 'bursts its banks', i.e. discharge exceeds bankfull channel capacity, depends upon (a) the nature of the channel and banks, and (b) the nature of the floodplain.

If a river bed has been built up above the floodplain and there are natural or artificial levées, then the water will initially burst through a breach and spill on to the floodplain in a sudden violent surge. This may result in a localised landform known as a *crevasse splay* (Fig. 3.28). When there is no levée, the floodwaters spill out more gradually. In both cases, however, there is a sudden decrease in velocity and available energy. This leads to the deposition of the river's load.

Floodplain complexity

No well-established floodplain is a simple flat surface built of a uniform pile of sediment layers. Close observation reveals a surface dimpled with shallow troughs and hollows which are the remains of former channels, meander curves and backswamps. Beneath the surface is a complex sequence of different materials, including the infill sediment of abandoned channel sections (Fig. 3.23). When floodwaters spill on to the floodplain their movements may be directed by the tracery of former channels. Where there is sufficient energy to incise a channel, the river may keep to this newly exhumed channel after the flood recedes, unless artificial levées, etc., are constructed (see River Tay Case Study below).

Floodplain morphology: the River Tay, Scotland

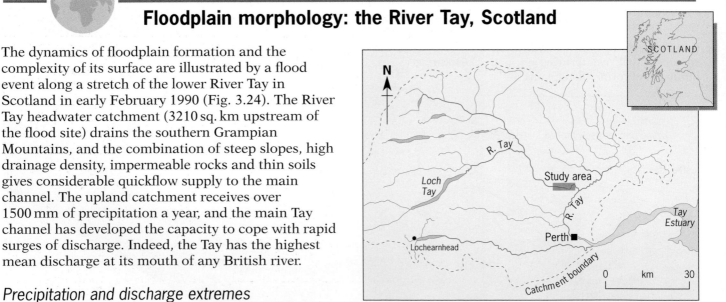

The dynamics of floodplain formation and the complexity of its surface are illustrated by a flood event along a stretch of the lower River Tay in Scotland in early February 1990 (Fig. 3.24). The River Tay headwater catchment (3210 sq. km upstream of the flood site) drains the southern Grampian Mountains, and the combination of steep slopes, high drainage density, impermeable rocks and thin soils gives considerable quickflow supply to the main channel. The upland catchment receives over 1500 mm of precipitation a year, and the main Tay channel has developed the capacity to cope with rapid surges of discharge. Indeed, the Tay has the highest mean discharge at its mouth of any British river.

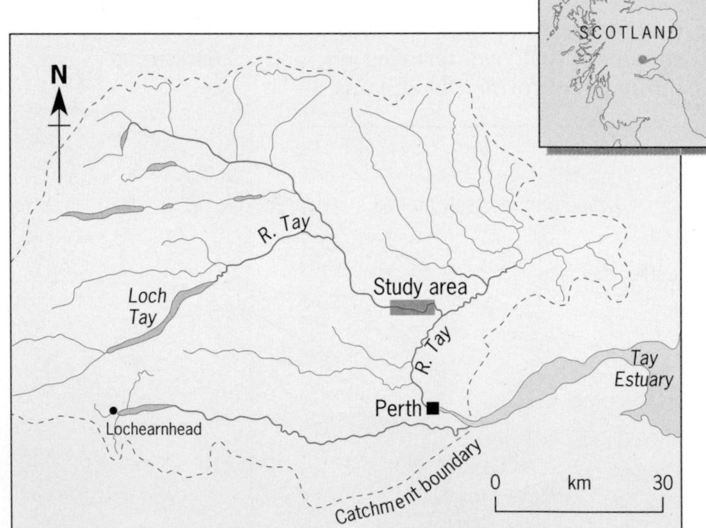

Figure 3.24 The Tay basin

Precipitation and discharge extremes

Despite its capacity, the channel system was unable to cope with the continued heavy rains of January and early February 1990 (Fig. 3.25). The graphs show that the catchment and the channel were able to absorb the January precipitation, part of which fell and was stored as snow. But with the catchment saturated and little storage capacity remaining, further sustained rainfall in early February, making over 500 mm in less than four weeks, plus some rapid snowmelt, made headwater flooding inevitable. The massive runoff and input to the main channel caused severe flooding across the floodplain of the lower course. The peak discharge at the Caputh gauging station (Fig. 3.26c) was 1747 cumecs compared with a mean flow of 132 cumecs and a previous recorded peak of 1196 cumecs.

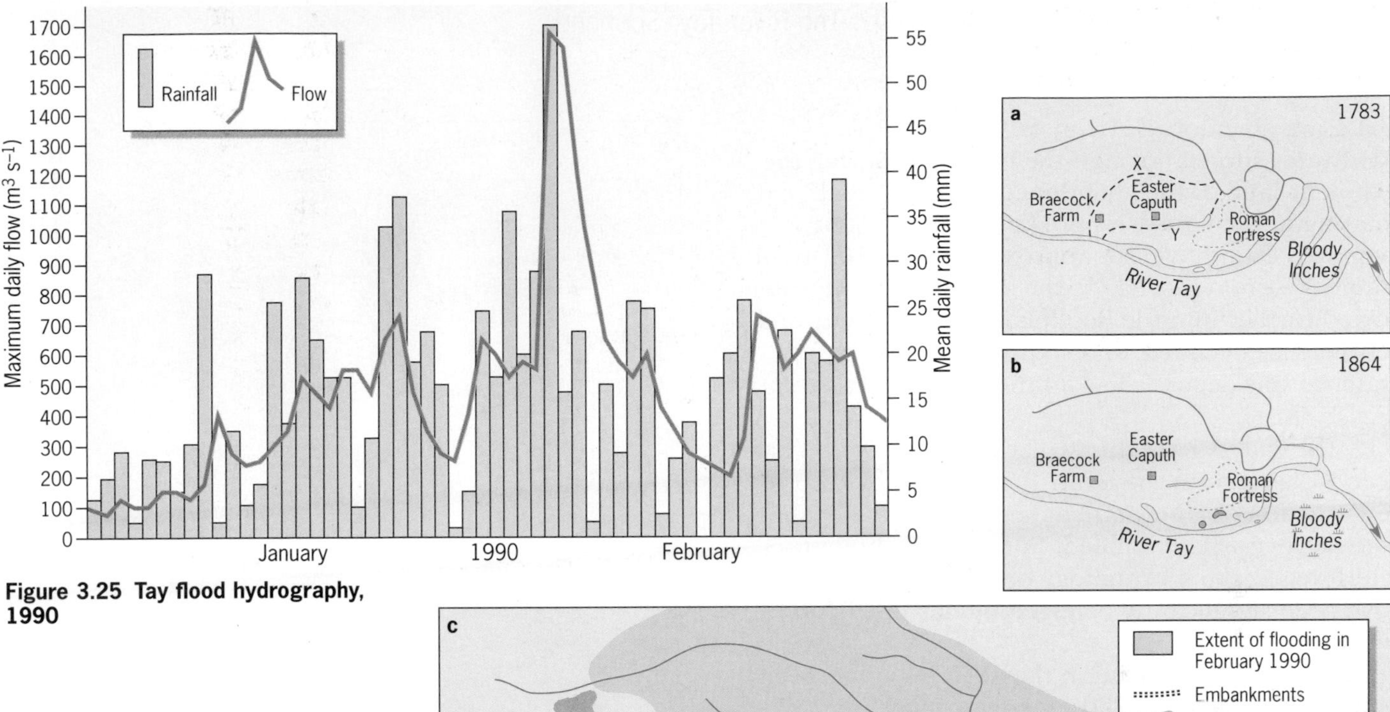

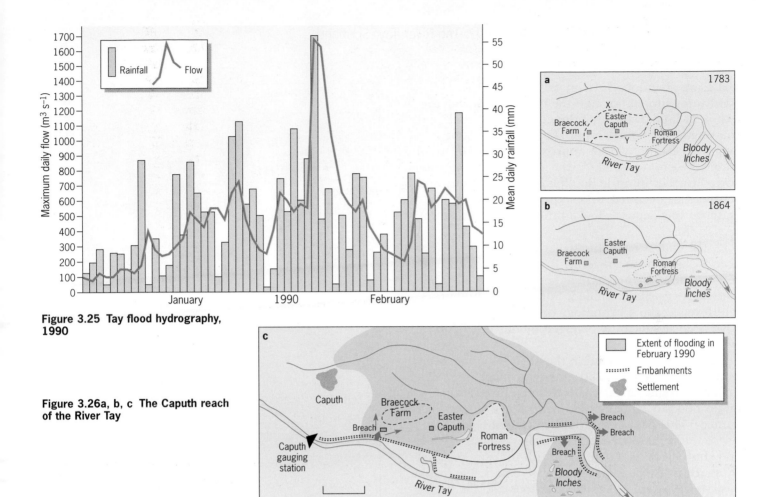

Figure 3.25 Tay flood hydrography, 1990

Figure 3.26a, b, c The Caputh reach of the River Tay

Floodplain impacts

One section of the floodplain which was inundated lies immediately below Caputh (Fig. 3.26c). This stretch of the river has shown several changes of course during the past 200 years (Fig. 3.26a, b), e.g. the meander cutoff at Bloody Inches. Notice, too, the two more ancient channel courses marked on the 1783 map (X, Y), looping round Braecock Farm. As we examine where the breaches took place in the 1990 flood (Fig. 3.26c), it is clear that artificial embankments failed. The breach location showed that the two most vulnerable points on the Bloody Inches meander are the 'leading edge' on the north bank where the velocity and erosive power are greatest, and at the entrance site to the old meander channel on the south bank.

Around Braecock Farm, along a 300 m stretch of the north bank on the outside of the river bend, the river bank was eroded approximately 5 m laterally. There was also severe bed scour. The bank retreat and bed deepening caused the embankment to collapse and so be breached.

Once the floodwaters surged through the breach they followed two shallow depressions around the terrace on which Braecock Farm stands (Fig. 3.27).

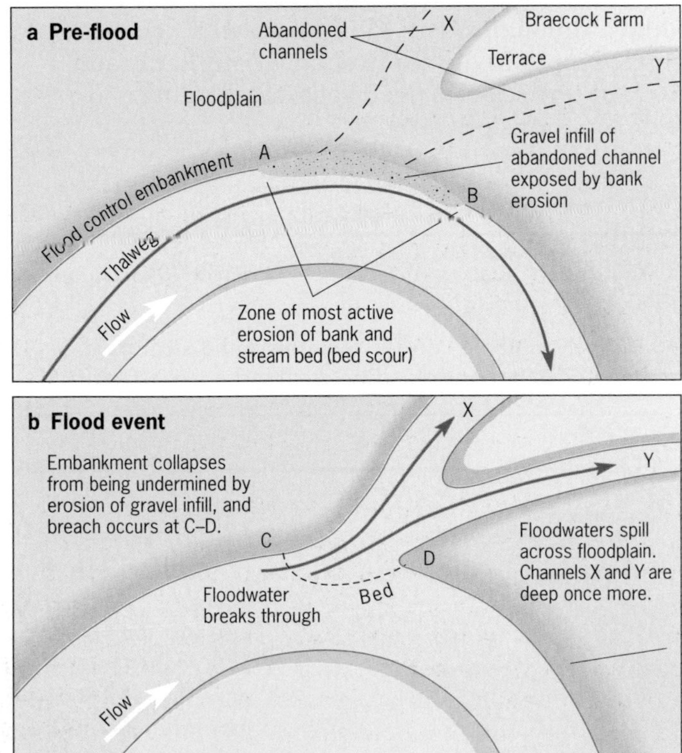

Figure 3.27 The Braecock Farm breach and channel erosion

Floodplain morphology: The River Tay, Scotland

These are abandoned channels X and Y. The floodwater surged through the breach, scouring the sediment infill of the abandoned channels, and creating new channels up to 2 m deep and 500 m long. It is estimated that approximately 31 500 m^3 of material were removed. As the sediment-laden waters flowed further into the floodplain their capacity and competence declined, and so progressively fine material was deposited as a fan-shaped crevasse splay (Fig. 3.28).

By the end of 1991 the two erosional channels had been isolated from the main Tay channel by the construction of a new embankment, and the channels themselves partially infilled. Without this human interference, the morphology of the floodplain would have been significantly changed by this one flood event.

The discharge attained in the 1990 flood has been estimated as having a 200-year recurrence interval. This record discharge was eclipsed, however, during a major flood in January 1993, with flows of over 2000 cumecs. This has caused the Institute of Hydrology to revise the recurrence intervals of flows along the Tay, and shows that we must be careful in using relatively short records. The 1993 flood was remarkable for the basin-wide extent of spate conditions and the fact that the peak flows in all the tributaries occurred almost simultaneously. This produced a huge and sudden input to the trunk channel of the Tay and explains the scale of the discharge, flooding and sediment transfer.

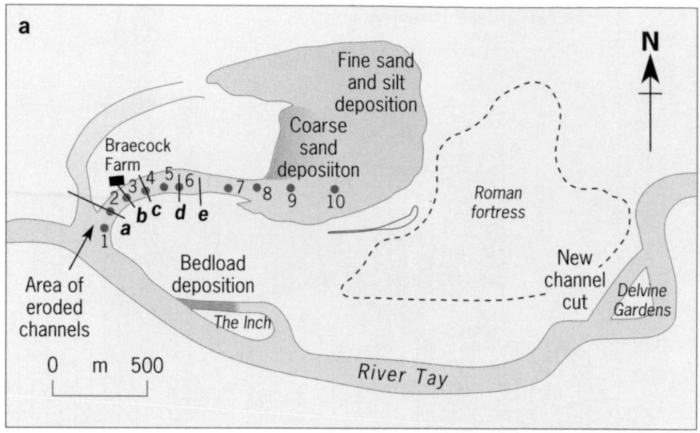

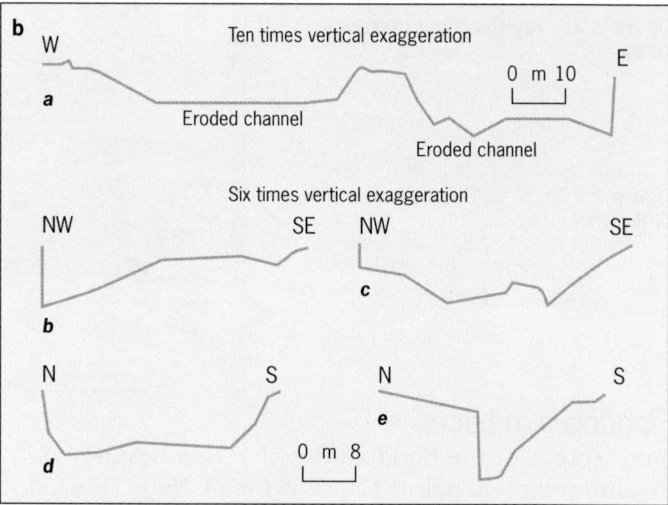

Figure 3.28 The crevasse splay at Braecock Farm

?

8 Outline the factors that caused the River Tay to flood in 1990.

9 Describe and explain the location and distribution of channel breaches and flooding during the 1990 flood.

10 In the account of the 1990 flood, what evidence is given that a floodplain is not a simple flat landform?

11 What unusual situation caused the 1993 flood to be so severe?

Floodplain–channel relationships in seasonally wet environments

In permanently humid environments there is generally a clear distinction between the form and function of the river channel and of its floodplain. The key influence upon channel form is that discharge with a return interval of 3–5 years (see section 2.6). The floodplain is occupied and aggraded by occasional floods with a longer recurrence interval. In regions where marked rainfall seasonality generates highly seasonal flow regimes, the channel–floodplain relationships may be different, as the case study of the River Auranga illustrates.

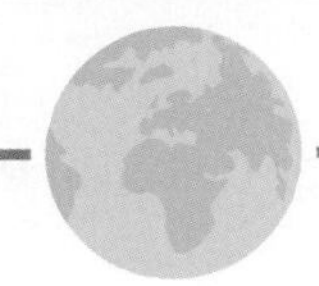

High seasonal flow regimes: River Auranga, India

The Auranga environment

The River Auranga is a southern tributary of the Ganges (Ganga), set into the north-east corner of the Deccan (Fig. 3.29). The river is 85 km long, drains a basin of 1664 sq. km, and falls approximately 900 m as it flows west-north-west to join the River Koel. It is a seventh-order stream, with a dense basin stream network, as indicated by the high **bifurcation ratio** (Table 3.1). The river has a strongly seasonal regime, controlled by the tropical monsoon climate. Rainfall totals range from 1200 mm to 1500 mm across the basin, but everywhere more than 80 per cent arrives between June and September, often in intense storms (Fig. 3.30). After a long period of deforestation, extensive reafforestation is taking place with the aim of slowing down runoff.

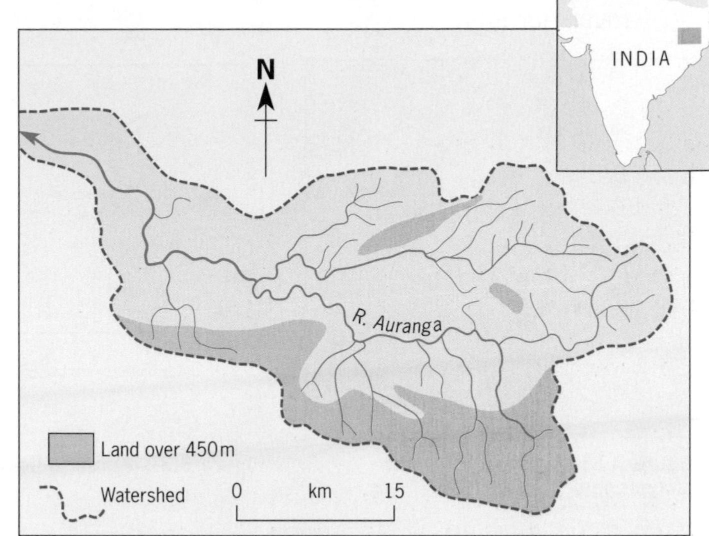

Figure 3.29 The Auranga basin

Table 3.1 Stream orders of the Auranga Basin, using Strahler's classification (*Source:* Gupta and Dutt, 1989)

Order	Number	Bifurcation ratio
1	7747	4.08
2	1898	5.02
3	378	7.41
4	51	3.92
5	13	6.50
6	2	2.00
7	1	
	Average	4.82

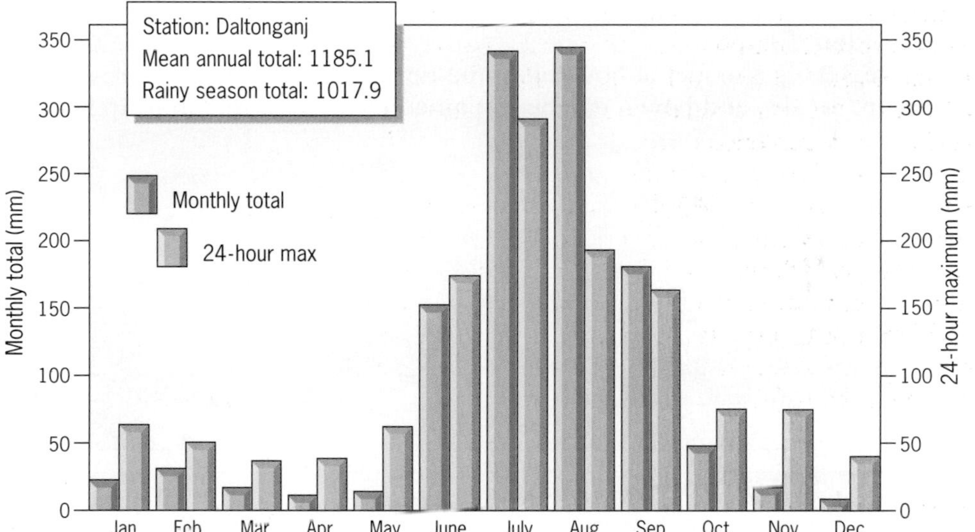

Figure 3.30 Auranga basin rainfall: Daltonganj station (*After:* Gupta and Dutt, 1989)

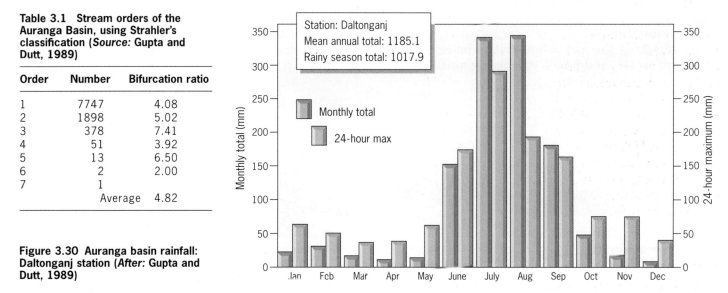

The combination of drainage density, the steep gradients of the headwater catchment, the thin soil cover and the heavy summer storms ensure rapid runoff and a heavy sediment supply. Thus, once the river drops from the edge of the Deccan plateau, the present valley is being formed on an alluvial infill up to 9 m thick. The ample debris input is almost entirely sand and pebbles, and the Auranga appears to be struggling to cope with this supply. In its main valley section, the river course consists of a series of swinging meanders (sinuosity ratio 1.60) separated by straighter reaches (sinuosity ratio 1.14).

The floodplain

The floodplain edge is marked mainly by a bluff cut into an old terrace which is a remnant of an earlier floodplain. In some places the floodplain edge is cut into solid rock. The floodplain has three components (Figs 3.31, 3.32):

1 the braided channel of base flow during the dry season;

2 the point bar which is covered by the bankfull wet season flow;

3 the flood bar, which is inundated periodically each wet season to accommodate flow surges following heavy storms.

The 'grassy flat' shown on the plan of Figure 3.31 is a sedimentary bench formed along the entrance reach of a meander bend. This is evidence of an outer bank cell of upwelling flow during high-discharge periods

River Auranga

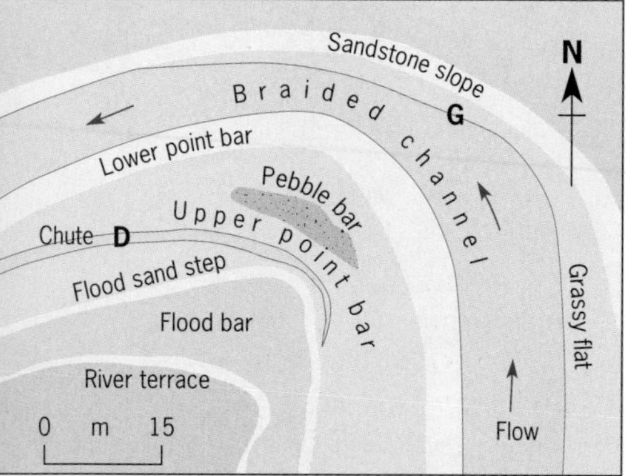

Figure 3.31 Auranga meander components

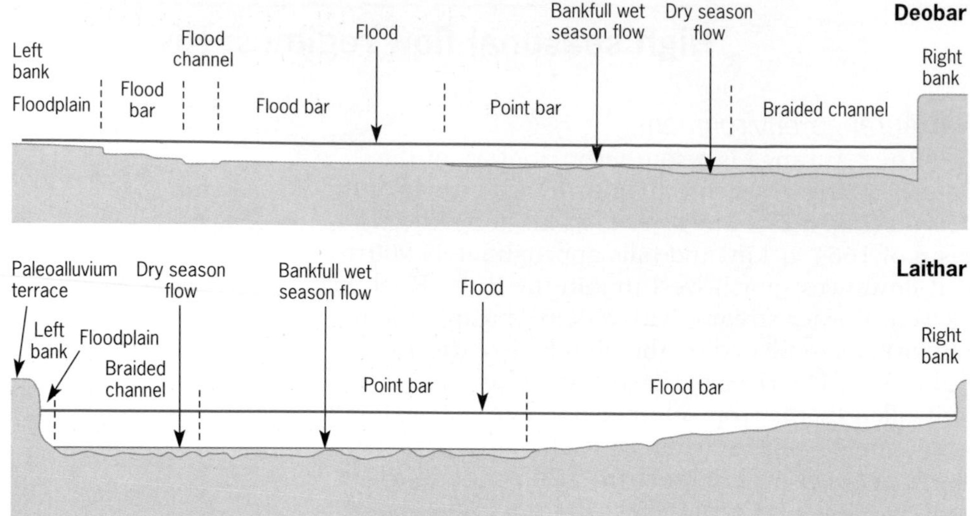

Figure 3.32 Auranga floodplain components

(see Figs 3.19, 3.20). This three-component system is best developed across the meander bends, the straighter crossover reaches being dominated by the braided channels.

Figure 3.33 is a model of how this three-tier landscape works, and how it can be explained in terms of the river regime. It is clear that the 'flood bar' is the equivalent of the 'floodplain' of temperate rivers, except that it is used as a peak flow store *every* year. Thus, in Indian rivers, controlled by the monsoon rhythms, the floodplain–channel relationship is distinctive.

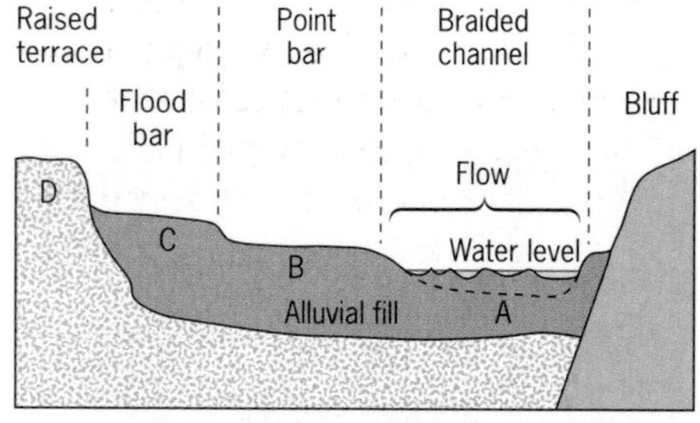

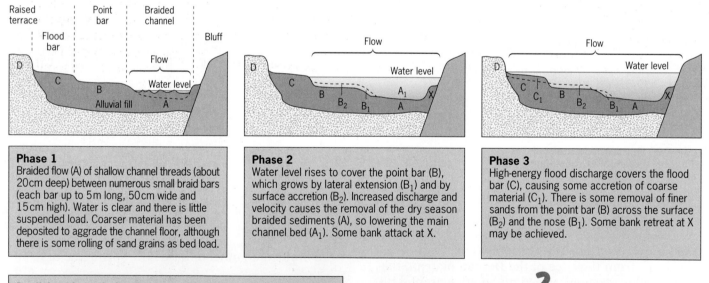

Phase 1
Braided flow (A) of shallow channel threads (about 20cm deep) between numerous small braid bars (each bar up to 5m long, 50cm wide and 15cm high). Water is clear and there is little suspended load. Coarser material has been deposited to aggrade the channel floor, although there is some rolling of sand grains as bed load.

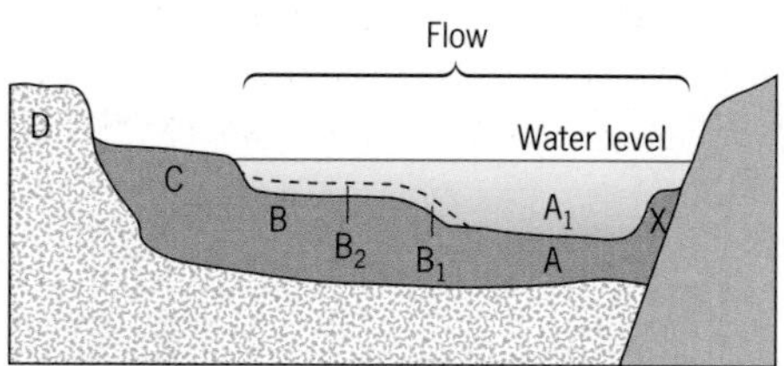

Phase 2
Water level rises to cover the point bar (B), which grows by lateral extension (B_1) and by surface accretion (B_2). Increased discharge and velocity causes the removal of the dry season braided sediments (A), so lowering the main channel bed (A_1). Some bank attack at X.

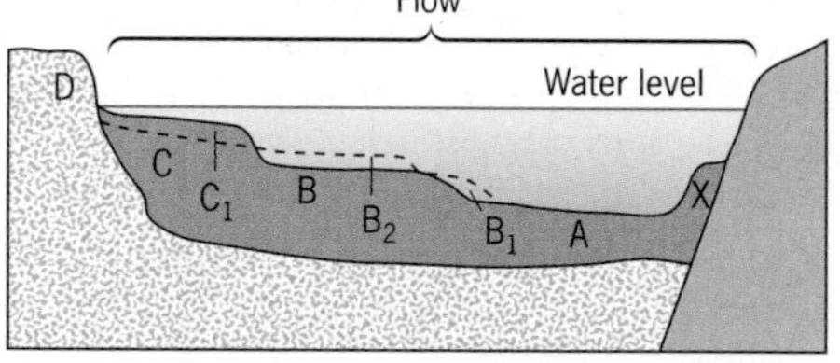

Phase 3
High-energy flood discharge covers the flood bar (C), causing some accretion of coarse material (C_1). There is some removal of finer sands from the point bar (B) across the surface (B_2) and the nose (B_1). Some bank retreat at X may be achieved.

Declining phase during the dry season
From October to February, water level and discharge shrink back to the braided condition of base flow (A). As part of the flood bar (C) and then the point bar (B) become partially exposed, chutes (gravel-bedded channels) may be cut across their surfaces (see D on Fig. 3.31). As available energy declines so the main channel floor (A) is infilled with sand once more.

Figure 3.33 River Auranga flow regime and landforms

?

12 Describe the form and location of each of the three main landform components of the Auranga valley.

13 Explain how each component may change in form and function from season to season.

14 In what ways is the 'flood bar' similar to and yet different from a 'floodplain' of British rivers?

3.4 Braiding: the multiple-channel approach to valley formation

What is braiding?

We say that a stream is braided when the flow moves through a series of interlocking channels or threads rather than down a single thread. Some rivers exhibit a braided form only at periods of base flow, but others retain this multiple-thread form at all stages except during extreme flood discharges.

What causes braiding?

The answer appears to lie in the relationship between gradient, discharge and load. While meandering streams create floodplains through lateral migration of the main channel thread (Fig. 3.21), a braided channel is far less stable, and carves a broad valley floor by frequent shifts. Braided systems are, in fact, among the least stable of all channel types. The planform of the broad valley floor has low sinuosity, although the rapidly shifting channel threads are much more sinuous, as Figure 3.35 (5) shows.

Research on rivers crossing the High Plains of North America has revealed that there is a critical threshold gradient at which a river tends to change from a meandering to a braided form of much lower sinuosity (Figs 3.36, 3.37). Perhaps even more important is the load–discharge relationship. Braided channels always appear to have a very heavy supply of sediment which the streams have difficulty in removing. Many glacier-fed streams are excellent examples, with copious sediment supplies from morainic debris and past valley floor infill (Fig. 3.34).

Figure 3.34 A braided channel, Exit Glacier, Alaska, USA

Braided channels are common, too, in semi-arid regions. This would suggest that streams with well-defined seasonal regimes are susceptible to braiding, in combination with appropriate gradient and sediment supplies. The significance of abundant debris supply means that we must take into account slope weathering as well as channel processes. The Howgill Fells case study, from the uplands of northern England, allows us to test some of these relationships in an area with a dense drainage network of small streams.

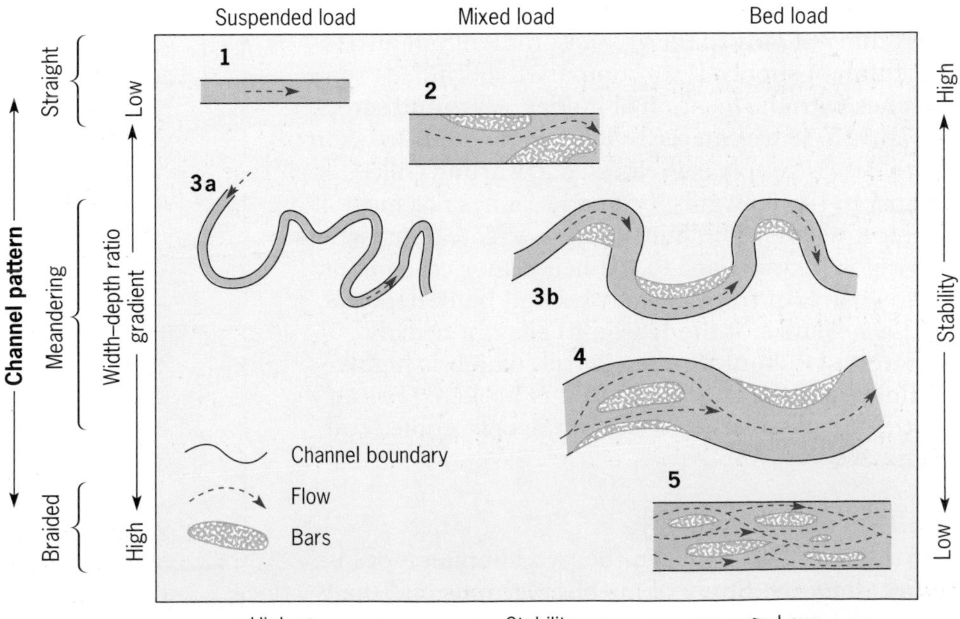

Figure 3.35 Relationships between channel planform, load and stability (*Source:* Thornes, 1992)

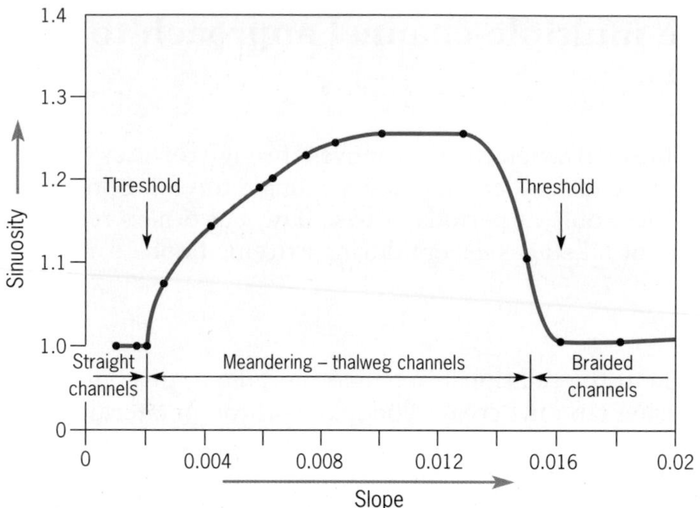

Figure 3.36 Relationships between sinuosity and channel form (*After:* Embleton and Thornes, 1979)

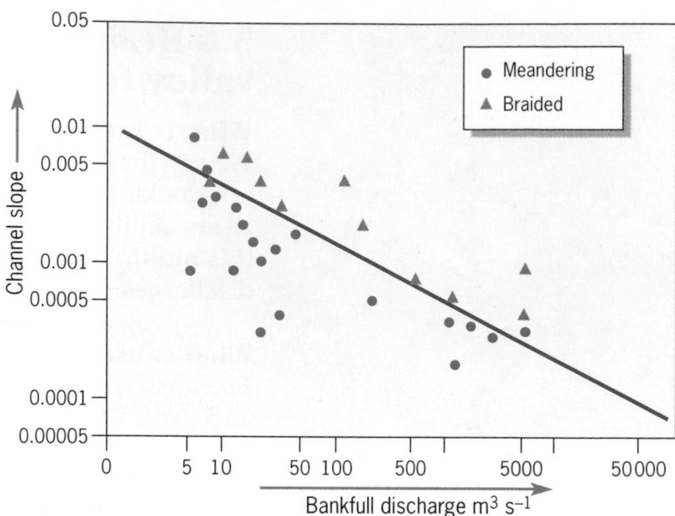

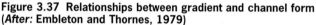

Figure 3.37 Relationships between gradient and channel form (*After:* Embleton and Thornes, 1979)

Channel braiding: the Howgill Fells, Cumbria, England

In many upland areas of Britain, despite apparently similar topography, two distinct types of stream channel are often found: single-thread and multiple-thread (braided) channels. The braided channels are less sinuous, wider, and have steeper gradients than the single channels. Figure 3.39 (a–c) illustrates the influence of bankfull discharge on width and slope and the slope–width relationships for streams in the Howgill Fells, Cumbria (Fig. 3.38).

The graphs show the distinctiveness of the two channel types and that the braided streams have coarser bed materials than the single channels. But this is due not only to the greater gradient but also to the sediment supply. If we compare the braided stretches with the location of gullies, scars and screes on Figure 3.38 we can see that braiding tends to occur where debris supply is greatest, e.g. Carling Gill.

Thus, in the Howgills, braided reaches are most common where sediment supply is greatest: (a) where gullying is taking place in the headwater catchment; (b) downstream from actively eroded bank deposits. The lower flanks of the Howgill Fells are heavily plastered with bouldery glacial till, which is highly erodible once the vegetation mat is broken. (For an illustration of this smoothed, but deeply gullied, till veneer, see Fig. 7.36.)

Relative channel stability

A further contrast between the two channel types is in their *stability*. Study of air photographs and maps shows that between 1948 and 1982 single-thread

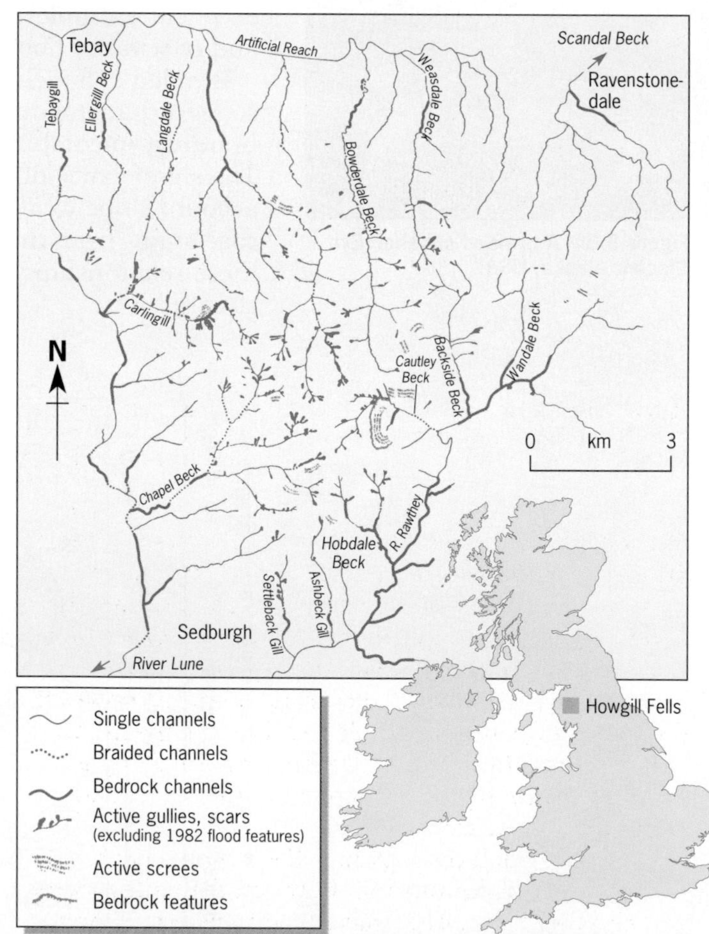

Figure 3.38 Howgill Fells: stream network and channel forms (*Source:* Harvey, 1991)

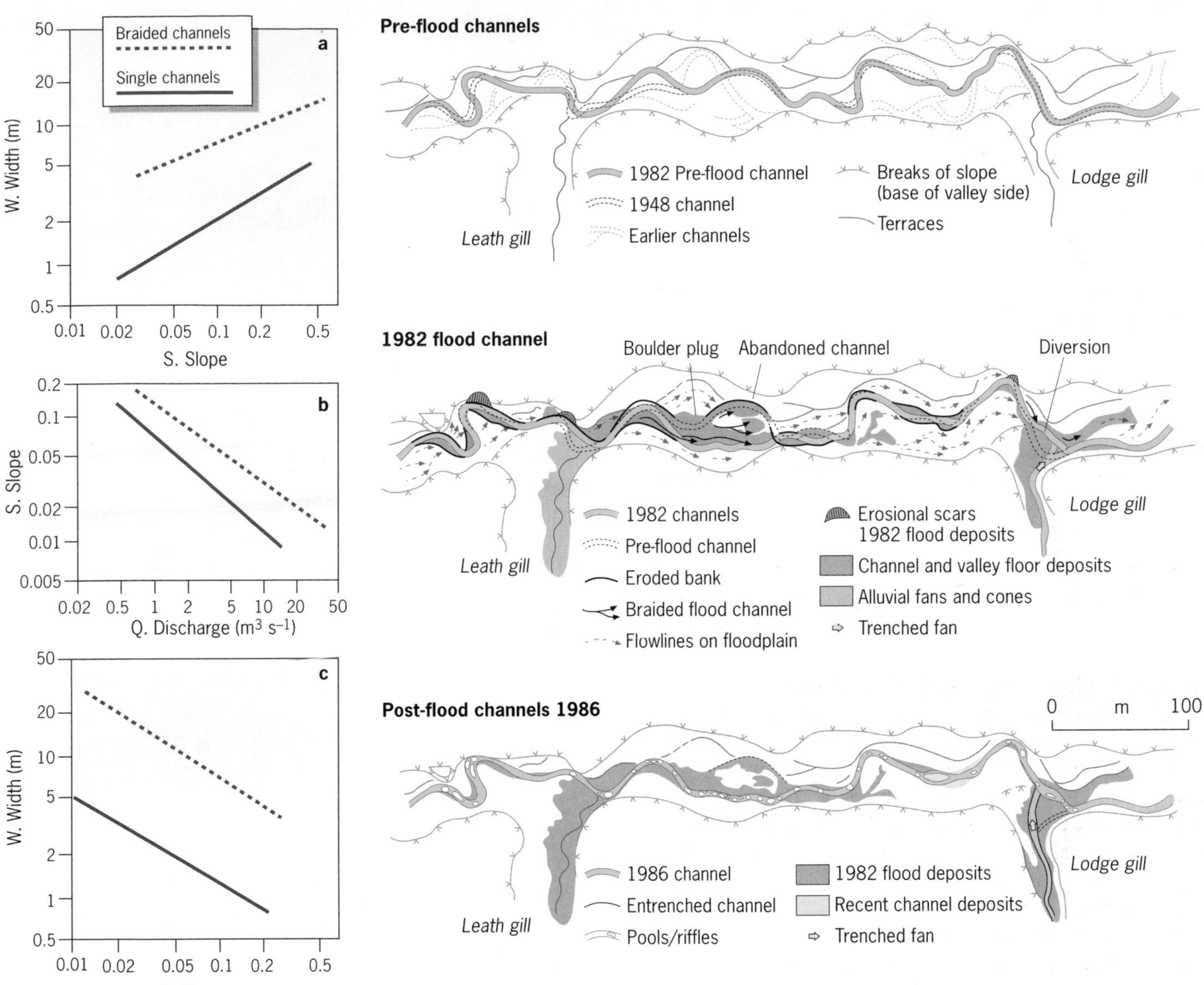

Figure 3.39 Braided and single channel relationships (*Source:* Harvey, 1991)

Figure 3.40 Bowderdale. The equilibrium–impact–response–recovery cycle at work for two streams (see also Fig. 3.41) after a major storm in June 1982. Over 70 mm of rain fell across the catchment in about two hours (a 100-year return period), causing rapid runoff, widespread slope failure and a dumping of debris into the stream channels.

channels experienced little change of course, whereas braided channels were far less stable. This is typical of braided streams as they attempt to cope with the abundant debris supply. Some Howgill streams have stretches where the regular supply maintains a braided channel form. Along other streams, however, the normal form is single thread, but is capable of being 'kicked' into braided form temporarily by a surge of debris input, e.g. slope failure associated with an unusual storm.

Following such disturbance, the stream sets to work to restore equilibrium between the valley slope and channel form. The relaxation (recovery) time may be several years, as Figures 3.40 and 3.41 illustrate.

?

15 Use the regression (best-fit) lines of Figure 3.39 to support the idea that braided channels in upland streams have wider and steeper channels than single-thread channels at given bankfull discharges.

16 For *either* Bowderdale (Fig. 3.40) or Langdale (Fig. 3.41), describe the equilibrium–impact–response–recovery model at work.

17 Produce evidence to support the following statement: With high slopes, high sediment concentrations and/or high discharges, the channel conditions shift from the stable single-thread forms to braided forms in which individual channel threads change position over quite short periods of time.

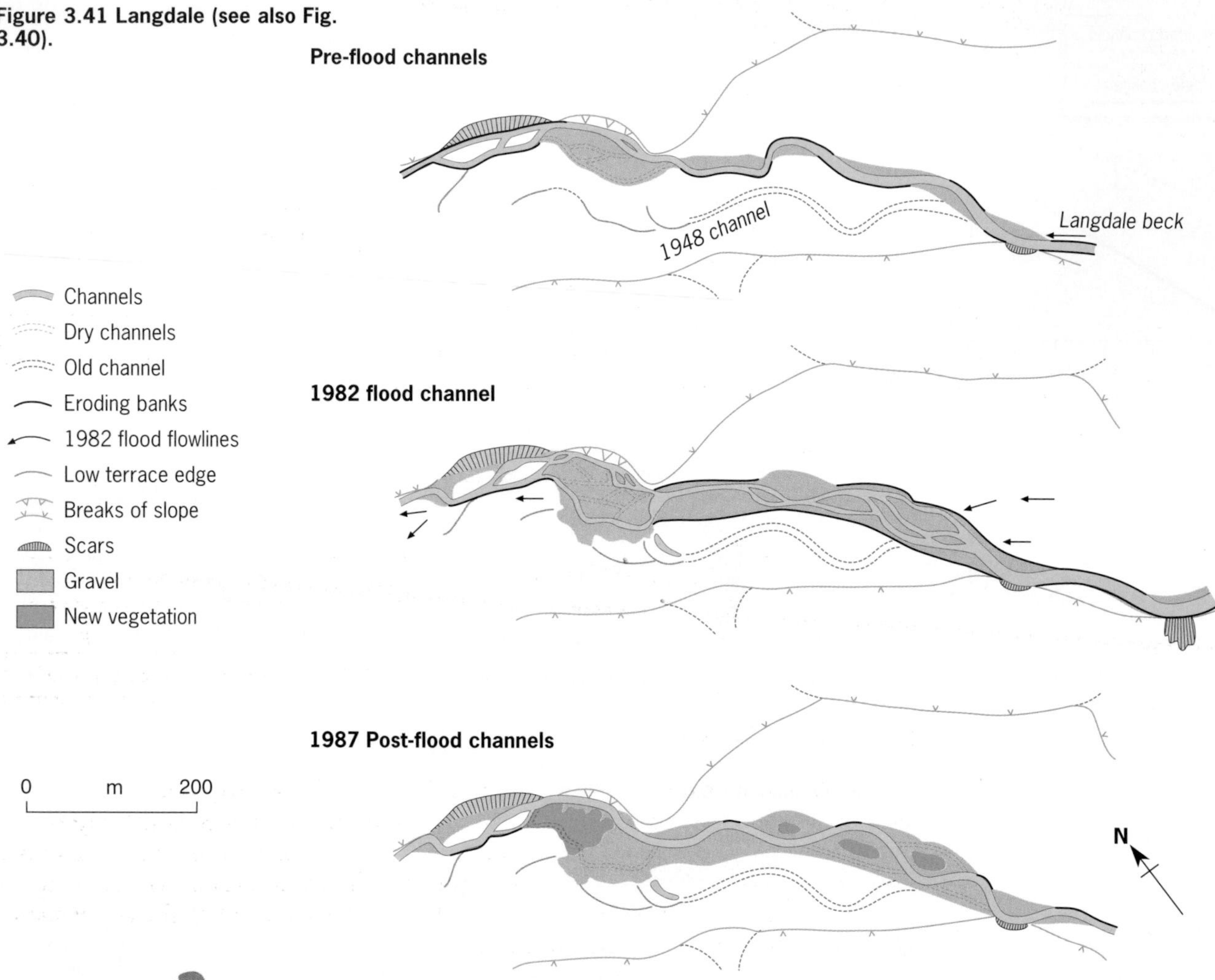

Figure 3.41 Langdale (see also Fig. 3.40).

18 Essay: Use the evidence presented in this chapter to discuss the assertion that fluvial landforms are the result of work done during spasmodic extreme events.

Summary

- Fluvial landforms are the result of the work of running water, and involve the interaction of erosional and depositional processes over time.
- Vertical and lateral forces are applied to create a valley long profile related to base level, and a valley cross-profile related to stream and slope processes.
- Fluvial landforms are controlled by the changing balance between erosional and depositional processes over space and time, and both erosional and depositional landforms can occur at any location within a drainage basin.
- Where landform formation is dominated by processes of vertical incision and slope retreat, valley forms are generally V-shaped.
- Many fluvial landforms are the result of both erosional and depositional processes, e.g. meanders, river terraces.
- Floodplains are the main landforms of fluvial deposition and are created by a combination of meander migration and overbank flooding.
- A meander is an extreme form of a sinuous river course.
- A stream is most likely to become braided, i.e. to flow via a series of interlocking channels, when inputs of debris load exceed the capacity of the stream to remove it. There seems, too, to be a critical gradient threshold at which a stream may change from a meandering to a braided form.

4 Coastal processes

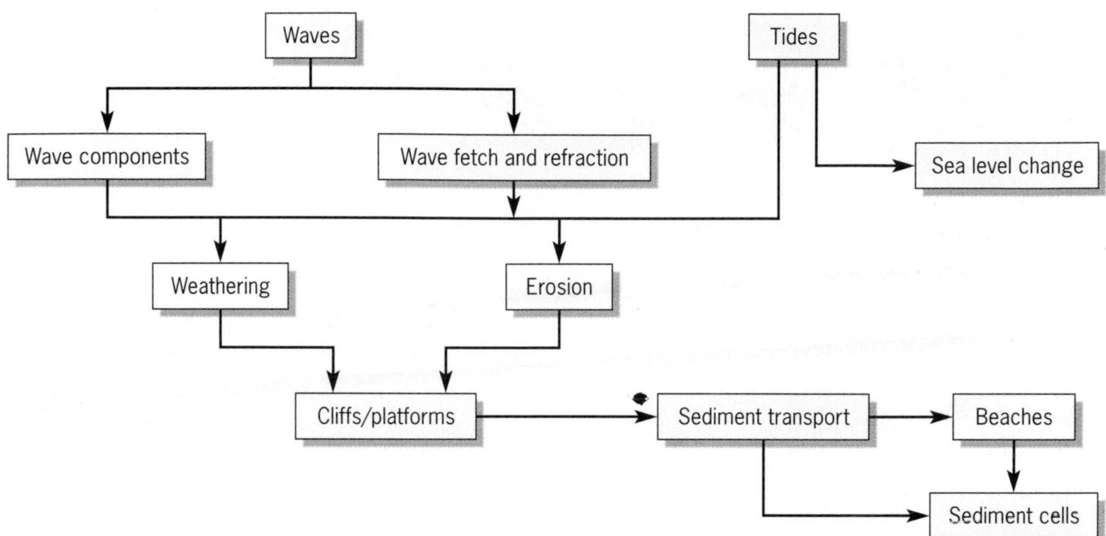

4.1 Introduction

Coastal landforms occur where wave and tidal activity carry out geomorphological work on the fringes of a landmass. This coastal strip is highly attractive for settlement and economic activity. As a result, the narrow land/sea interface is the setting for strong and direct interactions between natural and human processes. Figure 4.1 illustrates several of the key issues and questions raised by this interaction:

- What landforms are involved?
- Why is the Holderness coastline eroding so rapidly?
- Why has Spurn Head been washed away and reformed?
- How will a reef of tyres help to solve the problems of rapid erosion?
- Why have the sea walls caused increased rates of erosion and so much controversy along this coastline?
- Should we be building engineering structures to reduce coastal erosion?

The coastline is a complex and dynamic environment, with change occurring over a range of timescales from twice-daily tidal rhythms, through seasonal storms, to long-term sea-level fluctuations. The complexity and dynamism are illustrated, too, in the zonation across this strip (Fig. 4.2). Each zone has its distinctive processes, landforms, and interactions with people.

The attractiveness of coastal environments for living, working and playing, combined with an incomplete understanding of the natural processes at work, has led to decisions which have often proved increasingly risky, dangerous and costly. The growing number of coastal management schemes throughout the world illustrate this reality. They can be effective only if they are based upon a good understanding of how natural processes work, which processes are active in any locality, and what happens if we disturb them. All face the fundamental issue: people want to impose static conditions upon a naturally dynamic environment.

Erosion-hit resorts pin hopes on reef of tyres

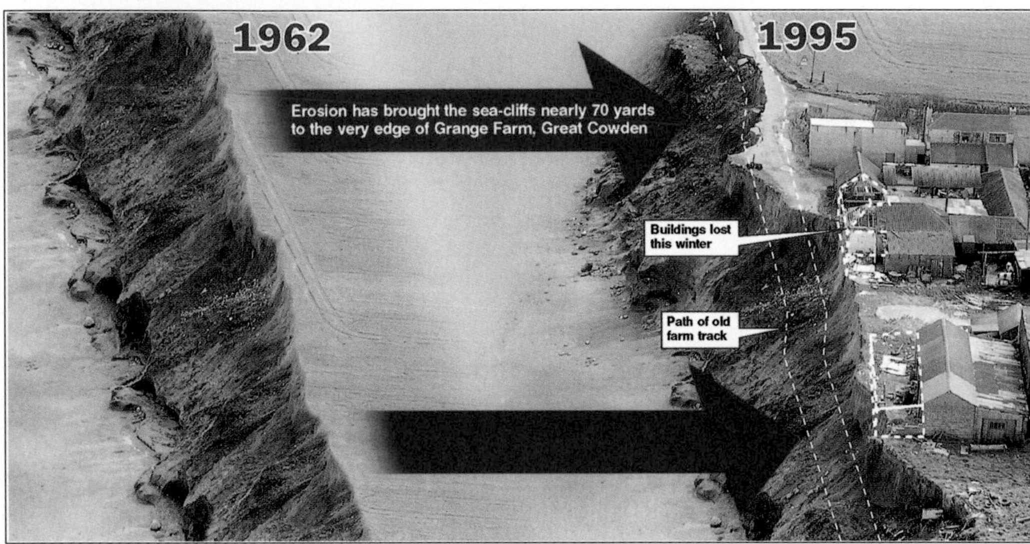

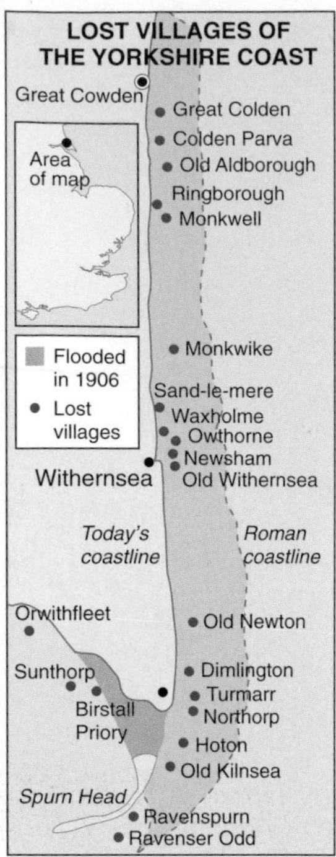

Mr David Earle and his niece, Susan, have become the flag-bearers of the campaign for Government action on the erosion that will almost certainly end their way of life later this year.

Until the £2 million sea wall was built in 1991 at Mappleton, a mile or so north of where they live, Mr Earle and Miss Earle lost their land at the rate of about a metre a year.

Now they say 10 metres a year are falling into the sea. In February, Holderness Borough Council gave them seven days to demolish the dairy, which was teetering on the clifftop.

Sheds housing 47000 broiler chickens stand some way from the edge but they fear erosion could claim them within a few years. Miss Earle is chairman of the Holderness Coast Protection Committee which is campaigning for changes to the Coast Protection Act to allow more defences to be built.

There are three existing schemes – at Withernsea and Hornsea as well as Mappleton – along Holderness's 30 miles of coast. The committee wants financial help for those forced to move.

Grange Farm's 50 acres have been reduced to about 40 and Miss Earle says 20 of those will be lost to them soon, when stretches of clifftop road – the Earle's only access to this parcel of land – disappear.

She would like assistance in replacing £15000 worth of dairy equipment and their home. She says: 'Flood victims get help, so why not erosion victims?'

RESIDENTS along the fastest eroding coastline in Europe are hoping a plan to dump millions of tyres in the sea as a protective reef will get the go-ahead by the Government.

Villages and the resorts of Withernsea and Hornsea on the Holderness coast in Humberside are in danger of slowly falling into the sea.

If the Ministry of Agriculture grants a licence for the trial tyre-reef scheme, it could lead to one of the most ambitious coastal engineering projects in Europe since the Dutch reclaimed its polders from the other side of the North Sea.

The area from Hull to the low, muddy cliffs of the Humberside coast has always suffered from erosion. Spurn Head, the spit of land which juts out into the Humber estuary, has been washed away and re-formed six times in recorded history, while many villages already lie underwater.

But, in the past five years, the pace of change has rapidly increased. Some homes have been abandoned and farmers are seeking compensation for loss of land and buildings.

The Humberside trial would submerge a bank of 1.5 million compressed tyres bound with nylon and concrete into a tangle of ropes six or seven metres high, 110 metres long and 60 metres wide.

Placed up to 1000 metres offshore, it would be tested for its stability, effects on local currents and pollution. If it worked, the full scheme could place more than a billion tyres in seven, two-kilometre strips all the way up the coast.

Humberside County Council accepts that such an ambitious project is unlikely to go ahead quickly — possibly not even this decade.

In the meantime, the coast depends on smaller schemes under the supervision of Holderness Borough Council.

The most recent, at the village of Mappleton, was opened with fanfares four years ago but, while it has saved the village, it has also caused resentment.

Other villages say that it has accelerated the rate of erosion elsewhere by preventing the protective sand that drifts down the coast from reaching the beaches.

It raised expectations that other schemes could be put in place, hopes the Government dashed in 1993 with a review of policy imposing new environmental and financial demands.

The Department of Environment is expected shortly to approve a controversial £4.5 million, 1000-metre sea wall around the North Sea gas terminal run by BP and British Gas near Easington. A full plan, which would also have protected the village, was turned down by the department.

Mr Robin Taylor, Holderness's director of development, said this appeared to be because under the new guidelines schemes had to prove not just 'cost-beneficial' but to be in the national interest. Saving gas supplies probably was, saving villages not.

Mr Ambrose Larkham, who owns the Easington Beach Caravan and Leisure Park, is demanding a public inquiry. 'The ludicrous thing is it is almost as cheap to build 1600 metres while the equipment's there as it is 1000,' he said.

Mr Taylor said: 'The question of why we are protecting the terminals and not the people of the village is likely to become very controversial. The issue is whether we should be protecting multinational companies and not our own residents.'

But Mr Geoffrey Twizell, terminal manager for British Gas and himself a resident, said: 'We are happy to contribute to any scheme that meets everyone's aspirations. Nobody would be talking about any protection at all for Easington if it weren't for the gas terminals here.'

Figure 4.1 Newspaper report of severe erosion on the Holderness coast in Humberside (*Source: Daily Telegraph*, 1 April 1995)

Onshore zone

Terrestrial environment

Backshore zone

Only affected by waves under storm conditions. Cliffs, beaches and sand-dunes may or may not be present.

Sand dunes

Beach ridge

Berm

Mean sea level

Intertidal and nearshore zone

Waves are modified and sediment movement occurs

Swash zone

Varies with tides. Water moves up as swash and down as backwash.

Surf zone

Broken waves travel towards the shore. Sediment movement may be onshore/offshore or along the shore.

Breaker zone

This varies in width depending upon the type of wave approaching the shore.

High tide level

Low tide level

Intertidal zone

Bar

Coastal sediments

Offshore and subtidal zone

Water depths are greater than half the wave length of the incoming waves. Limited sediment movement by waves. Tidal currents most effective.

Tidal range

Terminology

Type of coast and human uses

Onshore zone	Backshore zone	Intertidal and nearshore zone			Offshore and subtidal zone
Coastal towns and reclaimed land Ports and harbours Marinas Housing Industry Agriculture Nature reserves Tourism	*Dunes* Recreation Golf courses Nature reserves Building Water extraction Army manoeuvres *Cliffs* Shore protection works Building on cliff-top Mining Conservation coasts	*Sand and gravel beaches* Recreation Sand and gravel mining Back beach buildings Coastal protection schemes.	*Wetlands* Aquaculture Oyster beds Reclamation Grazing Reed and timber extraction Canals Pipelines Nature conservation	*Shore platforms* Seafood hunting/collecting Quarrying	*Continental shelf* Fishing, Oil, Mining, Sand dredging Dumping of waste Sewage outfalls *Coral reefs* Tourism, fishing Quarrying *Estuaries* Tidal barrages Coastal protection

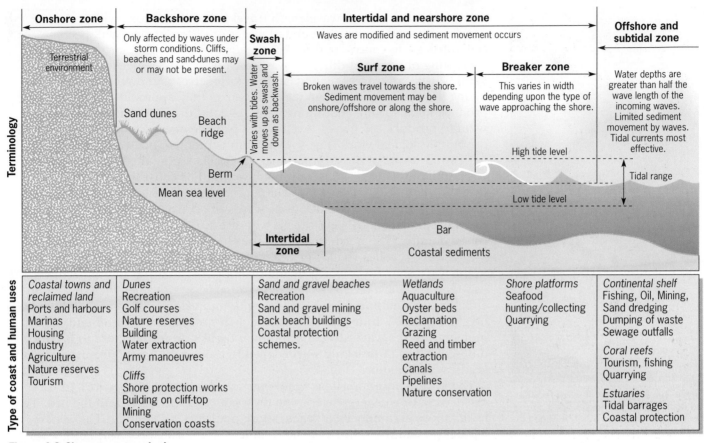

Figure 4.2 Shore zone terminology and human use

4.2 Waves and coastal processes

Landforms are the result of inputs of energy to surface and near-surface materials. At the coast, the main source of energy is waves generated by the wind. The friction effect of wind at the seawater surface produces motion in the upper surface of the water (Fig. 4.3). This motion is only a wave shape on the water surface and not an actual flow or forward movement of water, much like a wave flicked along a piece of hand-held rope. Individual water particles move in a circular path as the wave shape moves across the water surface. *Sea waves* are caused by the wind directly, and the stronger and more persistent the wind, the higher the waves and the more energy there is available to do geomorphological work.

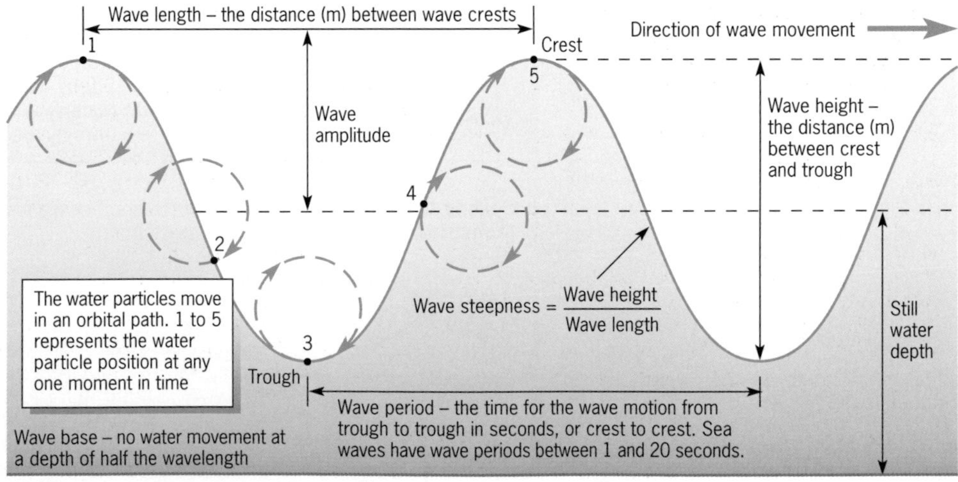

Figure 4.3 Wave terminology

Figure 4.4 A large swell wave reaches the coastline at Barton on Sea, Hampshire. This was a hot July day in 1995 when there was no wind blowing. The wave had been generated offshore and is breaking with a huge release of energy on to the beach at Barton.

Waves with a greater wavelength travel faster than those with a shorter wavelength, and so wind-generated waves of different sizes will become separated as they move at different speeds over the ocean surface. This separating of waves of different wave periods produces *swell waves*. These swell waves can travel for thousands of kilometres and affect coastlines facing the open ocean. When they steepen as they approach the shore they produce ideal surfing conditions. Thus, even at times when there is little wind to produce sea waves, swell waves provide energy at the coastline to do geomorphological work (Fig. 4.4). Britain is mainly affected by sea waves, with swell waves of most importance on coasts facing the open ocean, e.g. Cornwall and its well-known surfing beaches. For coastlines which do not face the open ocean, the waves will not have had time to separate, and the result is choppy wave conditions as waves of several wavelengths arrive at the shore together. This is the case with the North Sea, although ocean swell waves do affect the northern part of the North Sea.

As a wave reaches shallower water at the coast, the circular water motion in the wave shape is affected by the seafloor. As the water depth becomes less than the wave base (Fig. 4.3), the water path movements change from a circular to an elliptical shape, wavelength and velocity decrease, and wave height increases. Individual particles achieve a forward motion and momentum, the wave steepens and then breaks on to the shore to produce different types of breaking wave, depending upon the wave steepness and the gradient of the shore (Fig. 4.5).

The lower shore intertidal zone (Fig. 4.2) does most of the work in dissipating wave energy in friction drag between the water and the bed. This lower shore area may be a beach, shore platform, or mudflat. The upper backshore and the saltmarsh, sand-dune, shingle ridges or cliff landforms form a second line of defence against wave energy during extreme wave energy events. Spilling breakers occur more frequently on gently sloping shorelines and plunging breakers on steeper shorelines. In deep water at the coastline, especially with steep, hard rock cliffs, the waves do not break and their energy is reflected. This effect is called a *standing wave clapotis* (see Fig. 5.4).

?

1 Use Figure 4.3 to define wavelength, wave height, wave amplitude, wave period and wave base.

2 What are the implications for beach morphology of the different types of breaking waves (Fig. 4.5)?

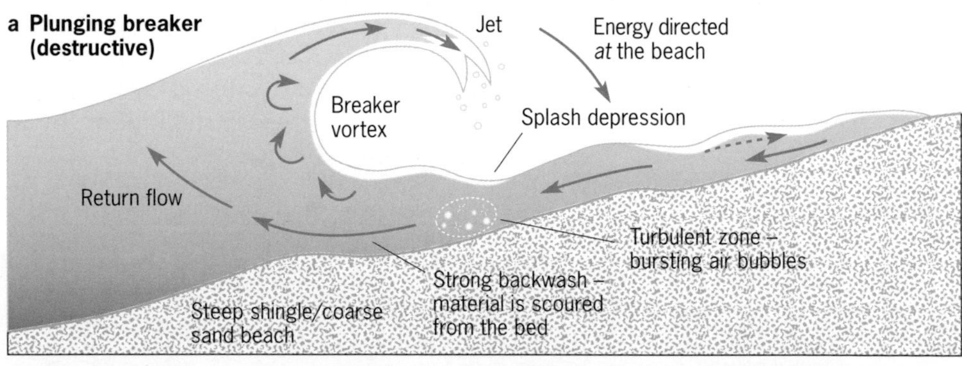

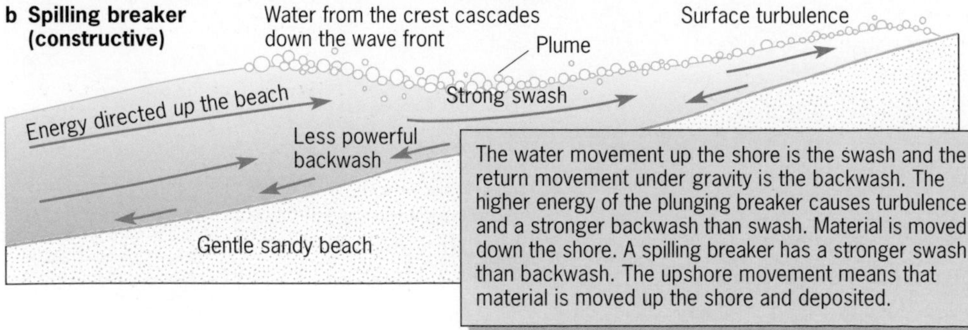

Figure 4.5 Main types of breaking wave

Figure 4.6 Wave refraction diagram

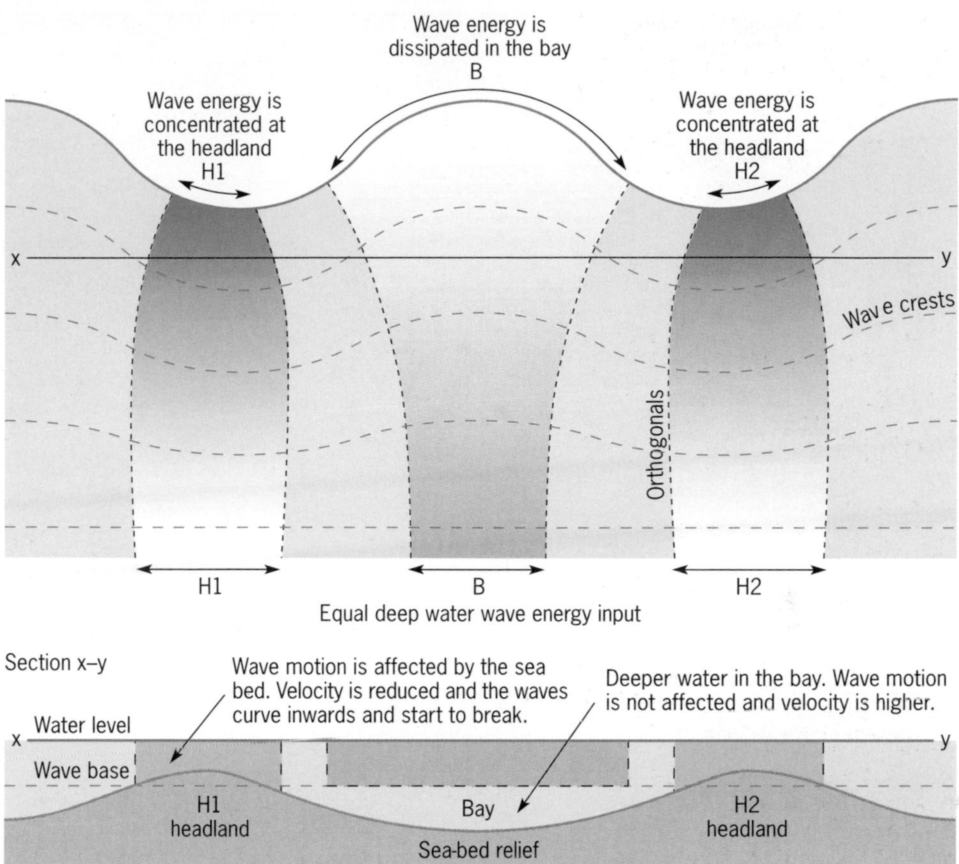

The shape of the sea bed reorients approaching waves, because the waves reach shallower water at different times. As water depth decreases there is a reduction in wave velocity and the waves are bent towards that part of the shoreline where they are moving most slowly. The wave crests roughly mirror the contours of the sea bed. The higher sea-bed relief, and therefore shallower water off the headlands, slows down the approaching waves as the wave base is affected by the sea bed. This slows down the wave approach and increases the wave height, and they start to break. The wave crests converge on to the headlands, increasing the energy released by the breaking wave. In the bay (B) the wave-base depth is reached later in the deep water of the low-relief areas, and the wave crests diverge in the bay releasing less energy as they break. Lines drawn at right angles to the wave crests called orthogonals show this effect most clearly. Where orthogonals diverge, energy per unit wave crest decreases, and where orthogonals converge, energy increases.

3 From Figure 4.6, suggest how the form of the coastline will change in the future due to wave refraction effects. A sketch map may help your answer.

4a Use the wind rose diagrams (Fig. 4.7) to accurately describe the wind frequency and direction for the Holderness coastline and Christchurch Bay, Hampshire.
b Which wind directions will be responsible for most sea wave erosion at each of the two locations? Think carefully about wind frequency and the orientation of the coastlines.
c The offshore topography refracts wave energy to produce the orthogonals shown. What are the geomorphological implications for Aldbrough, Dimlington and Barton?

Wave fetch and refraction

Since wave generation involves the transfer of energy from the wind to the ocean, atmospheric conditions are a critical variable, e.g. the stronger and more persistent the winds from a certain direction, the higher the waves, and the more energy is applied to the coastal system. A further important variable is the distance of uninterrupted ocean over which the wind can blow. This is called the *fetch*, and influences wave height and period, and hence energy. The configuration of the coastline determines which winds will be most effective, e.g. the great swell waves approaching the surfing beaches of Cornwall from the south-west have a long fetch across the Atlantic Ocean.

As waves approach a coastline, the undersea relief becomes important. The direction of wave approach is modified by the bottom topography of the sea-bed (Fig. 4.6). This *wave refraction* means that wave energy from breaking waves varies along the coastline. As Figure 4.6 shows, the wave energy becomes concentrated at headlands and disperses around bays. The complex interaction of coastal configuration, offshore topography, exposure, wave fetch and wind characteristics help to explain local variations in processes and landforms (Fig. 4.7).

Currents resulting from wave activity

Wave action results in water currents (Fig. 4.8) which have important geomorphological implications. Where wave crest approach is parallel to the shore a cell circulation with rip currents develops as the backwash becomes concentrated by variations in the beach surface under the force of gravity. Rip currents used to be called 'undertow', and they can scour a channel up to 3 m deep (Fig. 4.9). This results in the development of minor beach

Figure 4.7 The influence of wave fetch and coastline exposure

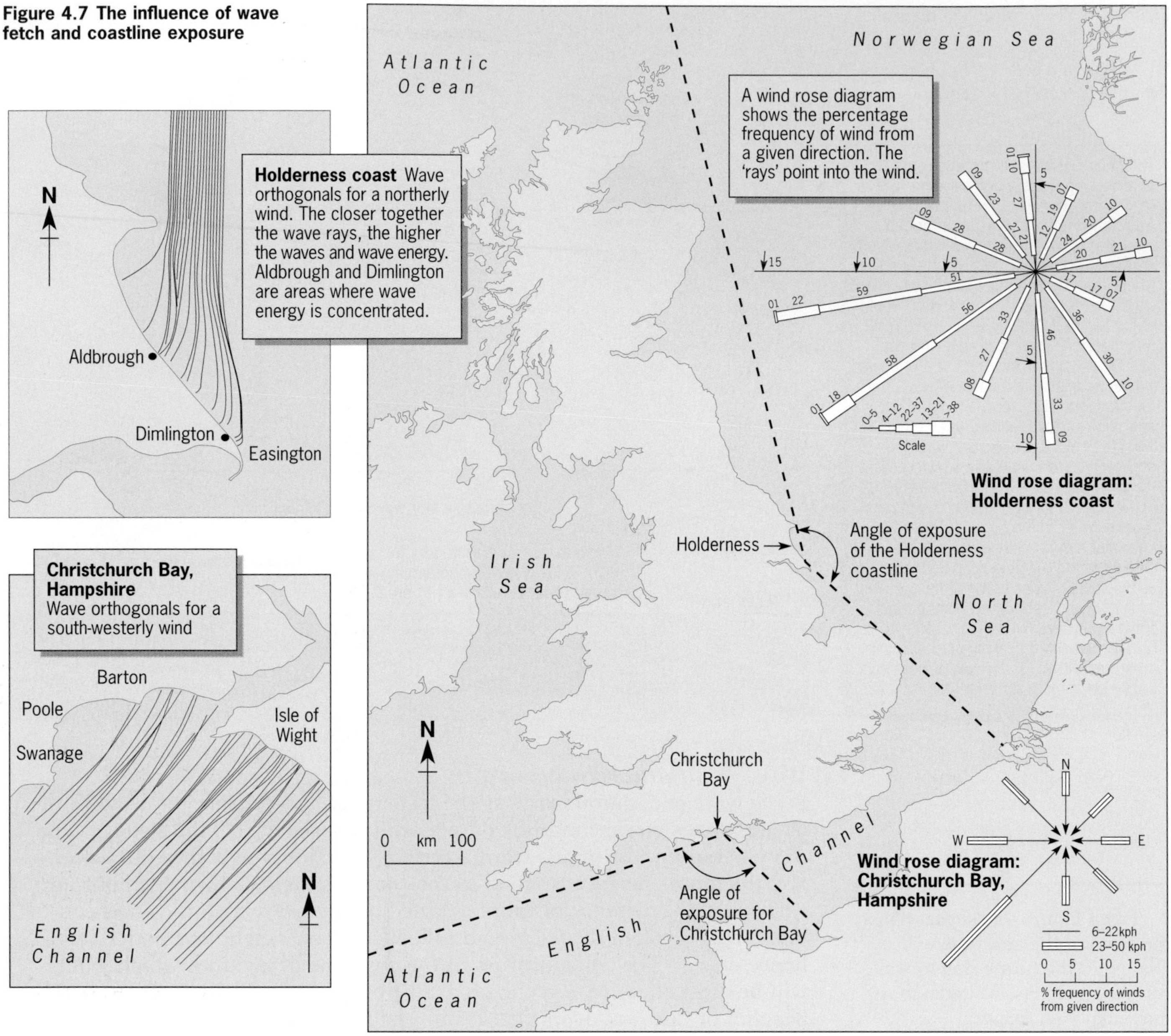

Holderness coast Wave orthogonals for a northerly wind. The closer together the wave rays, the higher the waves and wave energy. Aldbrough and Dimlington are areas where wave energy is concentrated.
N
Aldbrough
Dimlington
Easington
Christchurch Bay, Hampshire
Wave orthogonals for a south-westerly wind
Barton
Poole
Swanage
Isle of Wight
English Channel
N
Atlantic Ocean
Norwegian Sea
A wind rose diagram shows the percentage frequency of wind from a given direction. The 'rays' point into the wind.
Scale
Wind rose diagram: Holderness coast
Holderness
Angle of exposure of the Holderness coastline
Irish Sea
North Sea
N
0 km 100
Christchurch Bay
Angle of exposure for Christchurch Bay
English Channel
Atlantic Ocean
Wind rose diagram: Christchurch Bay, Hampshire
N
E
S
W
6–22 kph
23–50 kph
0 5 10 15
% frequency of winds from given direction

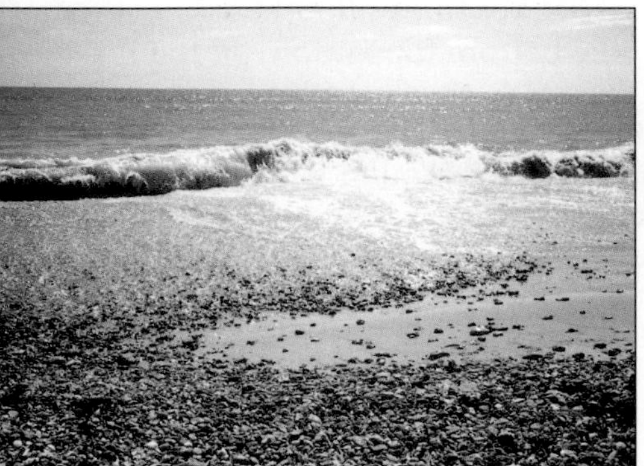

Figure 4.8 A rip current on the beach at Barton on Sea, Hampshire. The flow of water is concentrated in the rip current area and interrupts the incoming breaking waves.

landforms (see Fig. 5.27). However, most of the time, waves will approach the shore obliquely as determined by the wind direction. This oblique wave approach means that water is forced along the shore (longshore) as well as perpendicular to it (i.e. shore normal). There is a resulting longshore current and sediment movement called *longshore drift* (Fig. 4.9b). The angle of wave approach determines the amount of wave energy available, with an angle of about 30° being the most effective in longshore movement. This longshore movement of sediment is crucial in an understanding of landforms and in coastal management.

Tides

Tides are changes in water levels due to the gravitational attraction exerted on the oceans mainly by the moon, and to a lesser extent by the sun. Tides therefore represent extraterrestrial energy transfer into the coastal system. British coastlines experience semi-diurnal tides where there are two low and two high tides in each day. The different positioning of the sun, moon and

ner in the
n Figure
Fig. 4.9.)

ement of
s approach
with an
ach (Fig. 4.9).
tuations is
st common
why?

are upwelling tides and have nothing to do with the season. They are produced when sun, moon and earth are in alignment. The range or height between low water and high water is greatest at the time of a spring tide. Neap tides are produced when the sun and moon are in opposition to one another. Neap tides have a small tidal range. The sequence from spring through neap tides and back to spring again takes one lunar month, i.e. 29 days. The interval between two high tides is 12 hours, 24 minutes, 30 seconds.

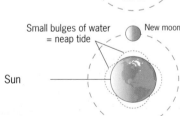

a Currents and sediment movement when waves approach parallel to the beach. The water motion sets up a cell circulation in the nearshore zone.

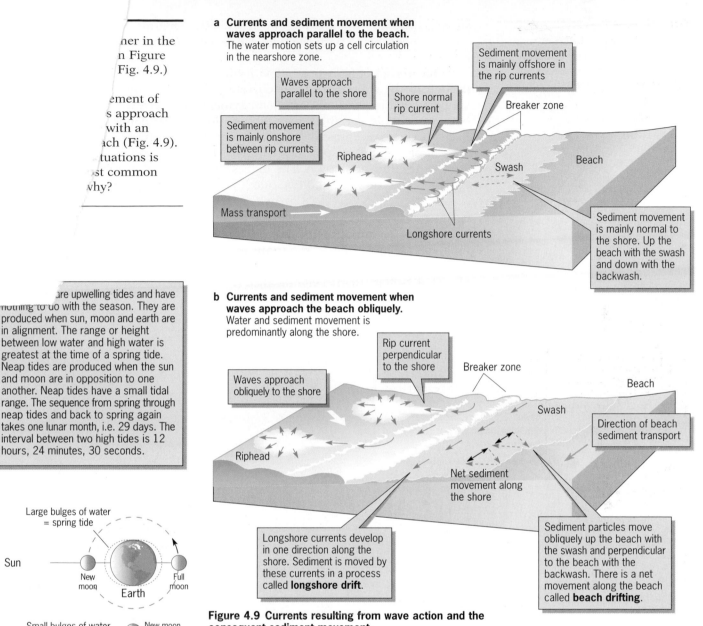

Waves approach parallel to the shore
Shore normal rip current
Sediment movement is mainly offshore in the rip currents
Breaker zone
Sediment movement is mainly onshore between rip currents
Riphead
Swash
Beach
Mass transport
Longshore currents
Sediment movement is mainly normal to the shore. Up the beach with the swash and down with the backwash.

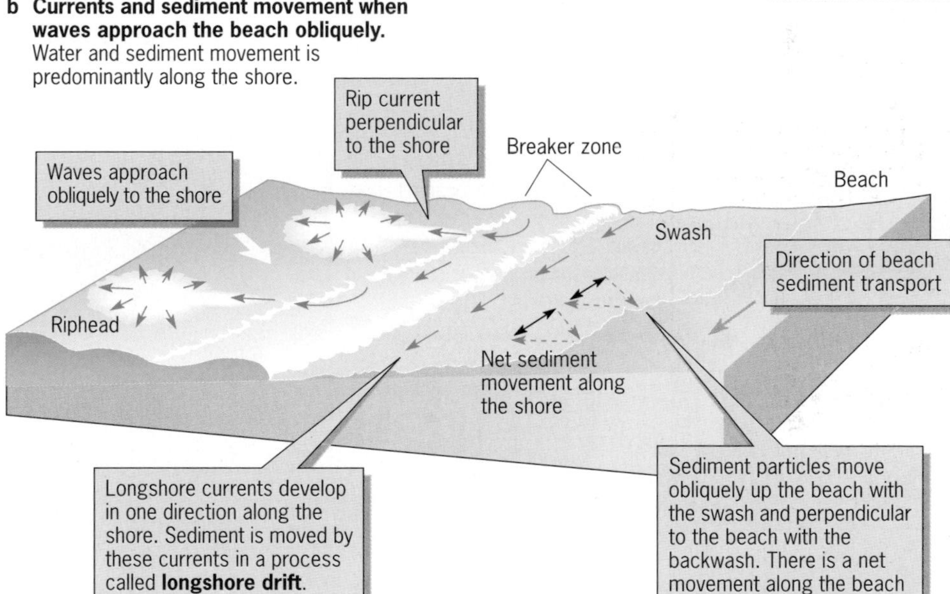

Figure 4.9 Currents resulting from wave action and the consequent sediment movement

Large bulges of water = spring tide
Sun
New moon
Earth
Full moon

Small bulges of water = neap tide
New moon
Sun
Full moon

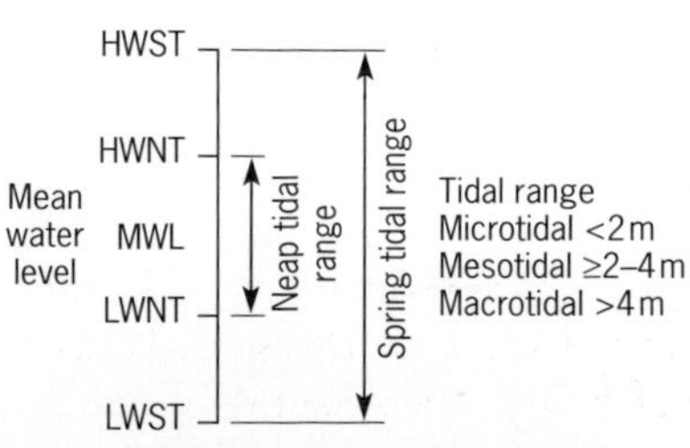

Figure 4.10 The generation of tides (*Source:* Hanson)

earth produce upwelling or high-range *spring tides* and downwelling, low-range *neap tides* (Fig. 4.10).

The water movements associated with tides have important geomorphological implications. *Tidal range* is the vertical distance between low and high tide and controls the vertical distance over which waves and currents can carry out geomorphic work. A small tidal range will result in wave energy being concentrated across a relatively narrow zone with increased erosion potential as wave energy is focused. Depositional outcomes include the production of a relatively narrow beach – for example, the Mediterranean Sea has a small tidal range and many popular holiday resorts there have narrow beaches. The tidal range around the British coastline is variable, reaching a maximum in large estuaries, e.g. over 6 m in the Severn Estuary and Morecambe Bay. In Christchurch Bay, Hampshire, the tidal range is 1.8 m, and along the Holderness coastline of Yorkshire the range is between 3.6 m and 4.2 m. In the intertidal zone between high and low tides the rocks and sediment are exposed for a few hours each day, when

7 Draw annotated diagrams to illustrate how the relationship between shoreline gradient and tidal range influences the nature of the intertidal zone.

subaerial weathering and biological processes operate. The gradient of the shoreline combined with the tidal range will influence the actual area exposed.

Water movements associated with tides produce tidal currents. Tidal range influences the strength of these currents, which become particularly concentrated in narrow coastal embayments, estuaries, and narrow channels between islands, e.g. the Rance estuary tidal power generation site in France. This concentration of tidal flow produces strong tidal currents which are important in transporting sediment. They are rarely strong enough to achieve significant erosion. The highest water velocities usually occur midway between high and low tide.

Long-term sea-level change

Tidal and wave movements cause short-term changes in sea level, but it is long-term change that is important for coastal landforms and management. There are two causes of long-term sea-level changes, both of which have global as well as British significance (Fig. 4.11).

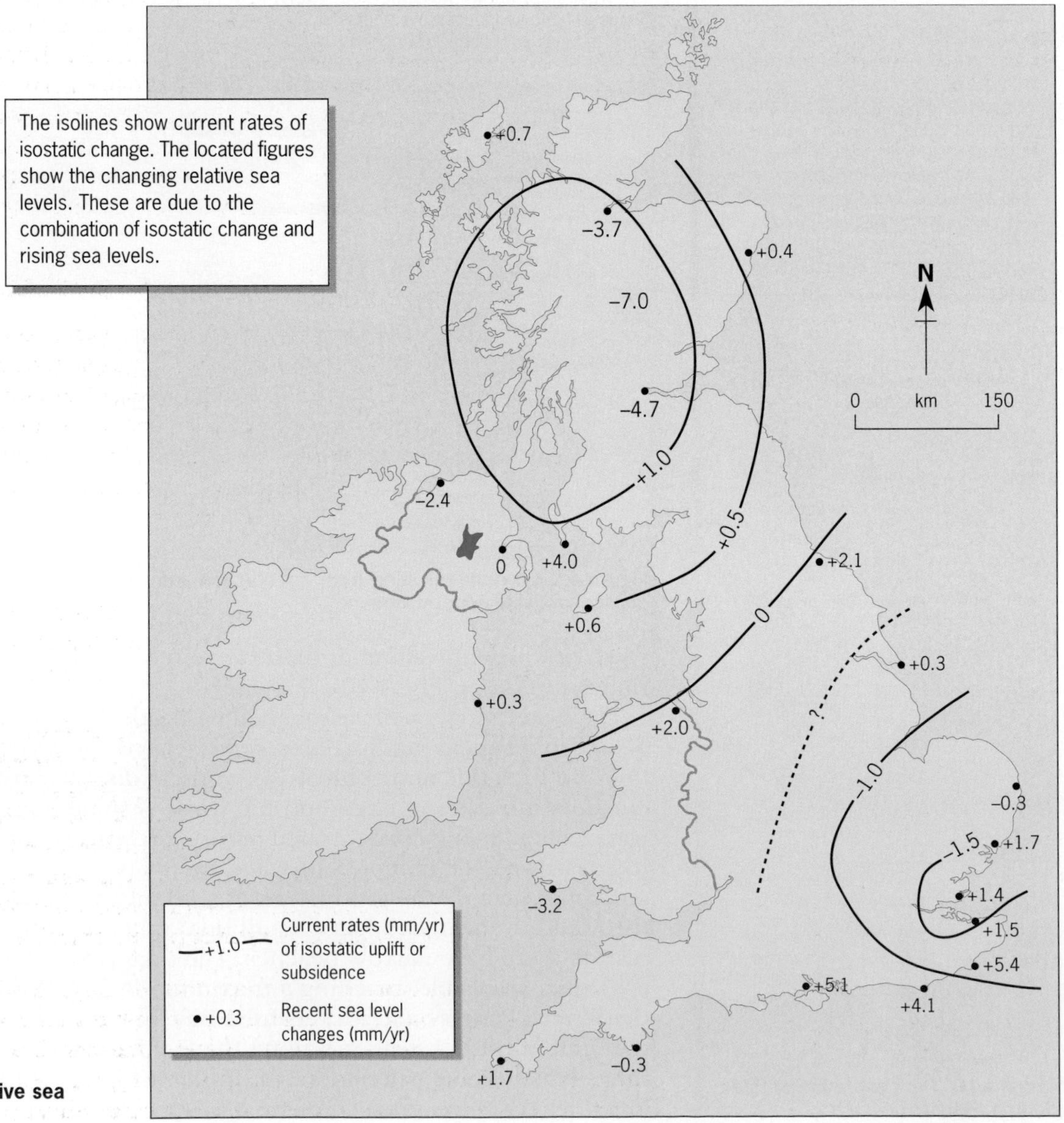

The isolines show current rates of isostatic change. The located figures show the changing relative sea levels. These are due to the combination of isostatic change and rising sea levels.

Figure 4.11 Changing relative sea levels in the UK

Eustatic changes

These are changes in the size of the global ocean store (a worldwide change in sea level), caused by climatic change resulting in the growth or decay of the global ice caps (glacio-eustatic). During the last glaciation (see Chapter 7), global sea level fell to as much as 150 m and 175 m below present levels.

Of increasing concern is the eustatic change which may result from global warming, caused by increasing atmospheric pollution from carbon dioxide and other greenhouse gases which trap outgoing solar radiation and increase the earth's surface temperatures. This is likely to result in some melting of the polar ice caps and raise global sea levels. Current estimates put the sea-level rise by 2100 to be 50 cm. Already since 1900, sea levels have risen 10–15 cm. This has serious implications for low-lying coastlines and for flooding and erosion rates along all coastlines. In regions where climatic change results in increased storm magnitude and frequency, the higher wave energy will result in adjustment throughout the coastal system.

Isostatic change

This is the change in the level of the land relative to the sea, with a resulting apparent change in sea level. This can be the result of local earth movements, but the most important of these changes in Britain and other areas has resulted from the Pleistocene glaciation. During glaciation, the ice exerts a great pressure on the earth's crust which causes the land to be lowered. This is the case for Antarctica and Greenland today, and for Britain during the Pleistocene. When the ice load is removed, *isostatic readjustment* causes the landmass to rise by up to 20 mm per year. Thus today, Scotland is still rising, with a compensating depression in southern England.

4.3 Weathering and erosion in the coastal zone

It is clear that in the **littoral zone** energy is highly concentrated. This energy, when applied to geomorphic processes, can achieve considerable 'work', i.e.can create landforms. Erosional landforms produced in the coastal zone are the result of the interaction of weathering and erosional processes which will vary in their relative importance from place to place and from time to time. Erosional landforms are produced by a combination of marine and subaerial processes (see Chapter 5).

Marine erosion

Direct erosion by the sea occurs in a restricted zone between wave splash and wave base (see Fig. 4.2). The shore area affected will vary according to fluctuations in tidal range. There are three main erosional processes at work.

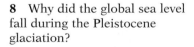

8 Why did the global sea level fall during the Pleistocene glaciation?

9 Use Figure 4.11 to describe and explain the changing relative sea levels in northern and southern Britain.

10 What implications do these changes have for human activity in the coastal zone? Refer to Figure 4.2.

Quarrying (hydrostatic pressure)

The force of a breaking wave can exert considerable hydrostatic pressure within joint systems by air compression as the wave breaks on to the rocks, followed by rapid air expansion as the wave subsides. This can pull away, or *quarry*, blocks of rock from a cliff face, or remove smaller weathered fragments. The force of the breaking wave can also hammer a rock surface. At high velocities, there is a cavitation effect where bubbles form in the water, and as they collapse they erode by hammer-like pressure effects.

Abrasion or corrasion

Breaking waves armed with sand and shingle can cause the rock-hitting-rock (ballistic impact) process of abrasion or corrasion. The size and amount of sediment available to be used as abrasives, along with the type of breaking wave, will determine the relative importance of this process.

Both quarrying and abrasion are more effective in storm wave conditions, and thus the frequency of storm conditions and plunging breakers along a coastline will be important in their effectiveness. These mechanical erosive processes will pick out any differences in rock strength, or weaknesses within a rock such as joints, cracks, bedding planes and fault-lines. This selective erosion is called *differential erosion*.

Attrition
This is the gradual wearing down of rock particles by impacts and abrasion during transport. Attrition reduces the size and causes the rounding of sediment particles as they are moved by waves, tides and currents. If you stand on the swash zone of a sandy or even pebble beach, and watch how the waves and backwash move the materials back and forth, you may be able to hear the noise made by the impacts.

Subaerial processes

The denudational processes normally operating on any land surface, i.e. subaerial weathering and mass movement, also operate in the coastal zone. Subaerial mass movement processes will be considered in detail with respect to cliffs (see Chapter 5, pp. 72–80). Weathering processes are determined by the climatic conditions and lithology of the coastline. For example, if climatic and lithological factors favour frost action weathering then this will also occur in the coastal zone, e.g. freeze–thaw weathering is a significant factor in coastal cliff retreat in Alaska, USA. However, the coastal zone provides conditions for other weathering processes to operate.

Corrosion
This is chemical weathering by the dissolving of rock and sediment in saline water. The area of coastline affected will be determined by the water level and the height reached by splashes and spray from breaking waves.

Water-layer weathering
Tidal movements allow the alternate wetting and drying of rock exposed at the coast. The collective effect of the chemical weathering processes of hydration, oxidation, salt crystallisation, and alternate wetting and drying (slaking) of rock minerals combine to weather the rocks. As this weathered layer is removed by erosional processes, a fresh surface is exposed and so the rock face retreats progressively.

Clay minerals are particularly susceptible to wetting and drying effects, but the combined action of the water-layer weathering processes is effective on most rock types. Research during January 1985 on the chalk rock shore platform near Brighton suggests that saline seawater and frost action are active in combination under severe weather conditions. The salts enhance the frost action in the chalk rock pores. As the salts crystallise in the pore spaces, they expand, just as water expands as it changes to ice. Most weathering occurs where the tidal changes expose the rock to freezing air temperatures for six or more hours.

Biological weathering and bioerosion
Plants and animals weather rocks by boring, abrading, or by chemical action. Boring algae, lichens, sponges and molluscs may leave fine tubes and pits in the rock surface. Molluscs may bore into soft rocks leaving holes up to 10 mm wide and 100 mm deep. If you have struggled barefoot across a rocky shore, you are painfully aware that seaweeds, barnacles, mussels, lichen and algae may encrust the rock surface. They form some protection to the rock from wave action. However, if they are removed by storm waves, there may be rock material attached. Rocky shorelines provide a wide range

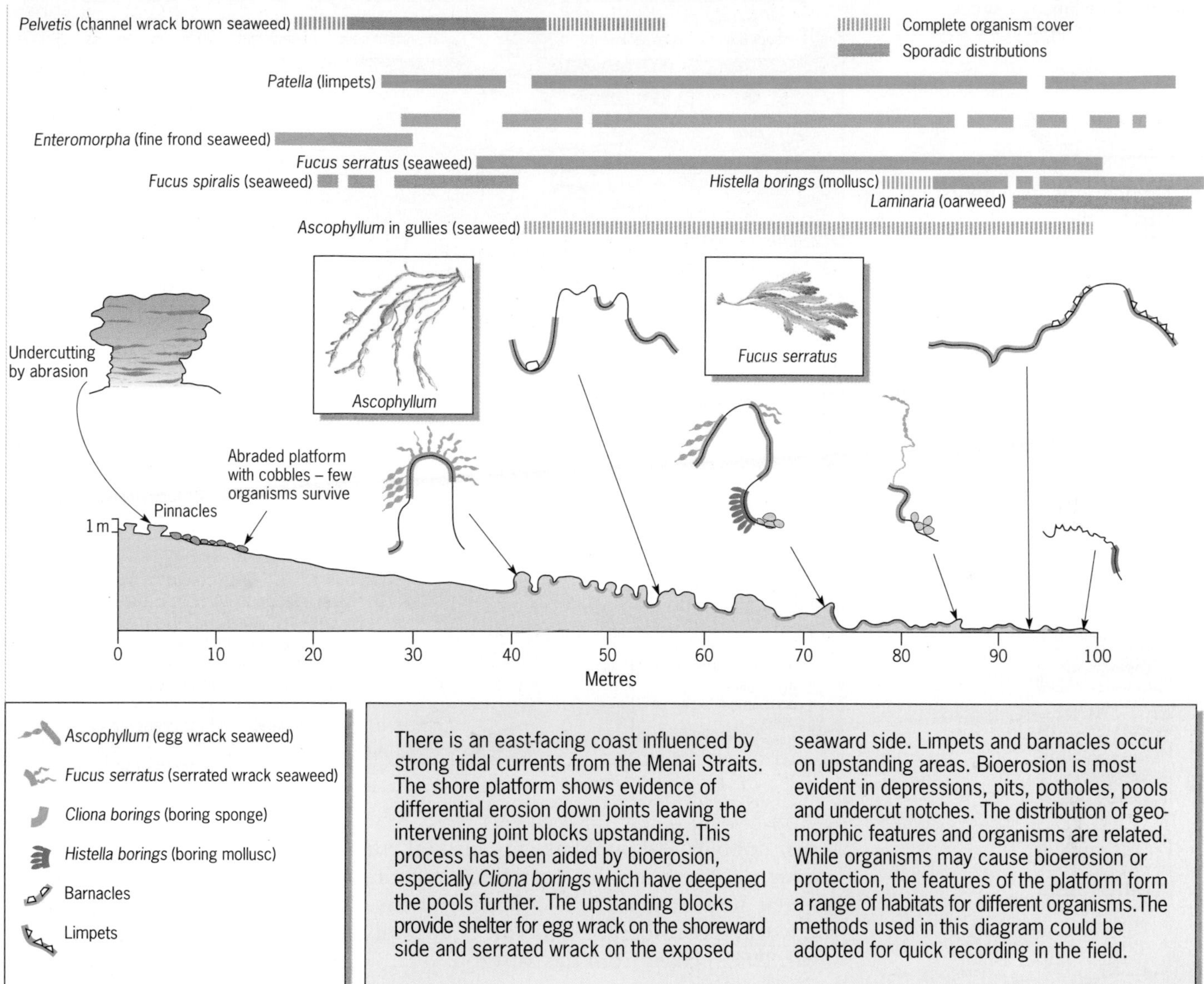

Pelvetis (channel wrack brown seaweed)
Complete organism cover
Sporadic distributions
Patella (limpets)
Enteromorpha (fine frond seaweed)
Fucus serratus (seaweed)
Fucus spiralis (seaweed)
Histella borings (mollusc)
Laminaria (oarweed)
Ascophyllum in gullies (seaweed)
Undercutting by abrasion
Ascophyllum
Fucus serratus
Abraded platform with cobbles – few organisms survive
Pinnacles
1 m
0
10
20
30
40
50
60
70
80
90
100
Metres
Ascophyllum (egg wrack seaweed)
Fucus serratus (serrated wrack seaweed)
Cliona borings (boring sponge)
Histella borings (boring mollusc)
Barnacles
Limpets
There is an east-facing coast influenced by strong tidal currents from the Menai Straits. The shore platform shows evidence of differential erosion down joints leaving the intervening joint blocks upstanding. This process has been aided by bioerosion, especially Cliona borings which have deepened the pools further. The upstanding blocks provide shelter for egg wrack on the shoreward side and serrated wrack on the exposed seaward side. Limpets and barnacles occur on upstanding areas. Bioerosion is most evident in depressions, pits, potholes, pools and undercut notches. The distribution of geomorphic features and organisms are related. While organisms may cause bioerosion or protection, the features of the platform form a range of habitats for different organisms. The methods used in this diagram could be adopted for quick recording in the field.

Figure 4.12 Coastal section at Penmon, Anglesey, North Wales, showing the distribution of marine organisms and platform features

of habitats for marine organisms which play a significant role in weathering and protection (Fig. 4.12). Softer rock types such as sandstones, shales, mudstones, clays and limestones are particularly susceptible to organic activity.

4.4 Sediment in the coastal system

Sediment sources

We have seen that erosion processes within the coastal system itself can provide direct sediment inputs, but there are other important sources of sediment, i.e. the coastal system is an open system (Fig. 4.13). These sources will vary in importance at the local scale, but in all areas the sediment inputs are stored as depositional landforms, or as nearshore features. Alternatively, they are transported (throughputs) and become outputs from the coastal system as they are moved into the deeper waters of the ocean and away from the immediate coastal zone. The sediments can be divided into two main groups. *Clastic sediments* are from rock weathering and erosion. These vary in size from minute clay and mud particles to sand and

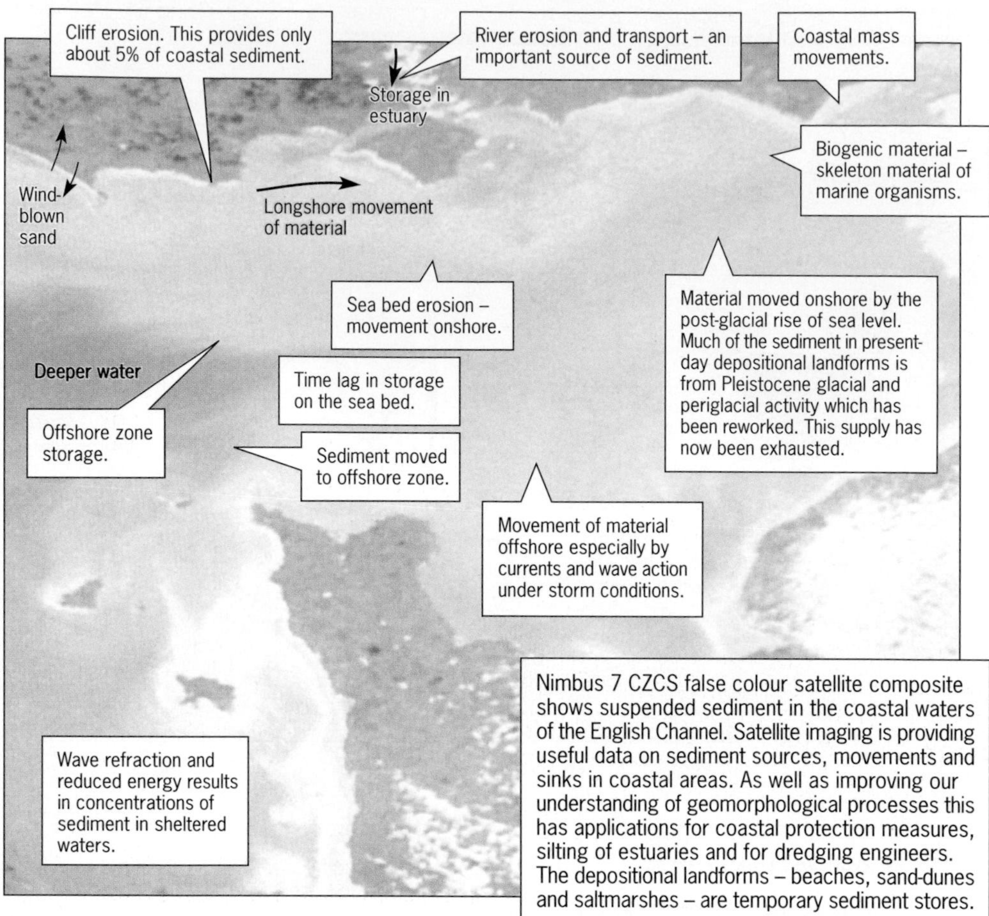

Cliff erosion. This provides only about 5% of coastal sediment.
River erosion and transport – an important source of sediment.
Coastal mass movements.
Storage in estuary
Biogenic material – skeleton material of marine organisms.
Wind-blown sand
Longshore movement of material
Sea bed erosion – movement onshore.
Material moved onshore by the post-glacial rise of sea level. Much of the sediment in present-day depositional landforms is from Pleistocene glacial and periglacial activity which has been reworked. This supply has now been exhausted.
Deeper water
Time lag in storage on the sea bed.
Offshore zone storage.
Sediment moved to offshore zone.
Movement of material offshore especially by currents and wave action under storm conditions.
Wave refraction and reduced energy results in concentrations of sediment in sheltered waters.
Nimbus 7 CZCS false colour satellite composite shows suspended sediment in the coastal waters of the English Channel. Satellite imaging is providing useful data on sediment sources, movements and sinks in coastal areas. As well as improving our understanding of geomorphological processes this has applications for coastal protection measures, silting of estuaries and for dredging engineers. The depositional landforms – beaches, sand-dunes and saltmarshes – are temporary sediment stores.

Figure 4.13 Sediment sources, movement and stores in the coastal system

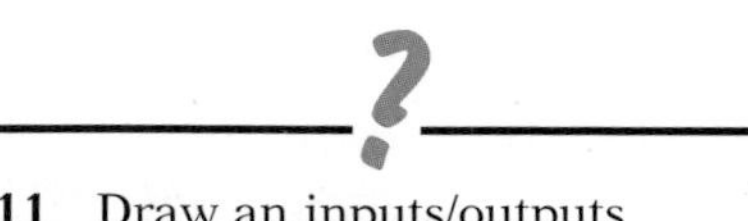

11 Draw an inputs/outputs diagram for the coastal system using Figure 4.13.

12 Where is sediment stored in the coastal system?

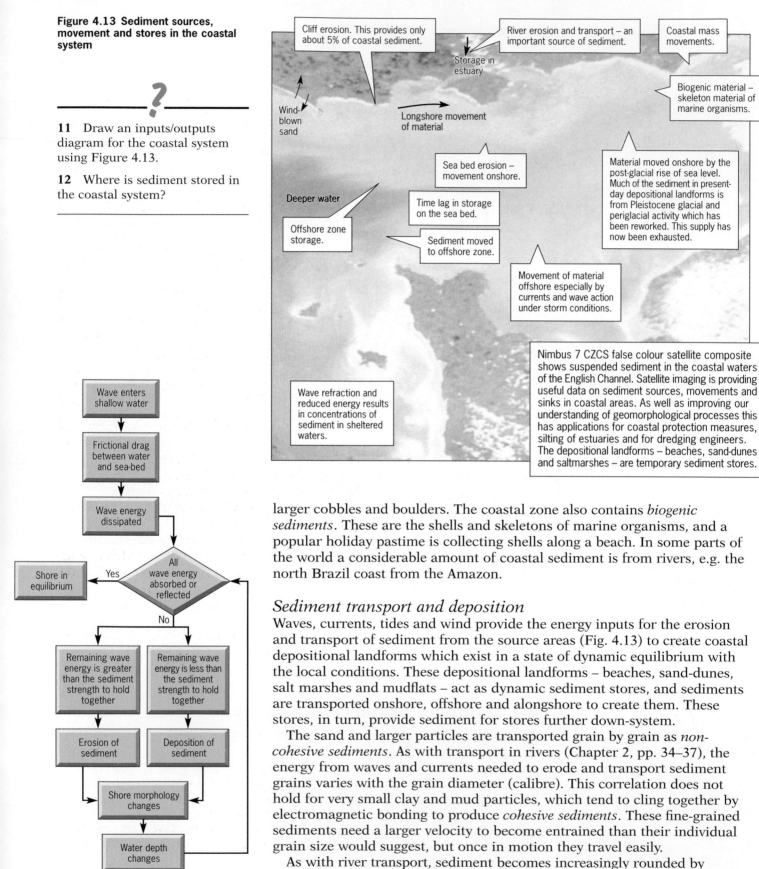

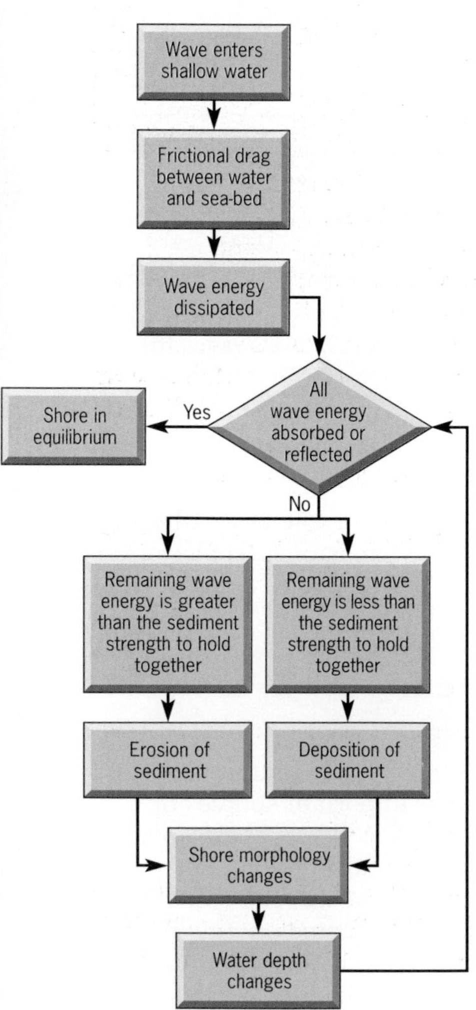

Figure 4.14 Sediment transport and shore morphology

larger cobbles and boulders. The coastal zone also contains *biogenic sediments*. These are the shells and skeletons of marine organisms, and a popular holiday pastime is collecting shells along a beach. In some parts of the world a considerable amount of coastal sediment is from rivers, e.g. the north Brazil coast from the Amazon.

Sediment transport and deposition

Waves, currents, tides and wind provide the energy inputs for the erosion and transport of sediment from the source areas (Fig. 4.13) to create coastal depositional landforms which exist in a state of dynamic equilibrium with the local conditions. These depositional landforms – beaches, sand-dunes, salt marshes and mudflats – act as dynamic sediment stores, and sediments are transported onshore, offshore and alongshore to create them. These stores, in turn, provide sediment for stores further down-system.

The sand and larger particles are transported grain by grain as *non-cohesive sediments*. As with transport in rivers (Chapter 2, pp. 34–37), the energy from waves and currents needed to erode and transport sediment grains varies with the grain diameter (calibre). This correlation does not hold for very small clay and mud particles, which tend to cling together by electromagnetic bonding to produce *cohesive sediments*. These fine-grained sediments need a larger velocity to become entrained than their individual grain size would suggest, but once in motion they travel easily.

As with river transport, sediment becomes increasingly rounded by attrition as it is transported by waves, tides and currents. Larger sediments – sand, shingle and pebbles – are deposited in relatively high-energy

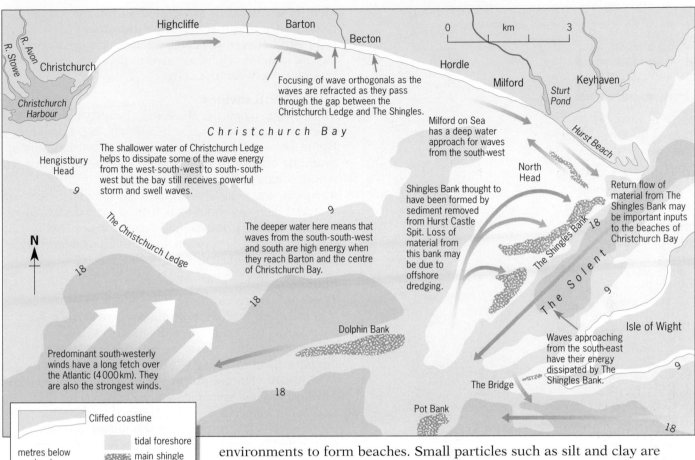

Isle of Wight
Dolphin Bank
Waves approaching from the south-east have their energy dissipated by The Shingles Bank.
Predominant south-westerly winds have a long fetch over the Atlantic (4000 km). They are also the strongest winds.
The Bridge
Pot Bank
Cliffed coastline
tidal foreshore
main shingle banks
movement of materials
Rivers and streams
metres below sea level
0–9
9–18
> 18

Figure 4.15 Sediment transport and sea-bed morphology in Christchurch Bay, Hampshire

environments to form beaches. Small particles such as silt and clay are carried in suspension, and settle as depositional landforms in areas of reduced wave energy but high tidal energy to form saltmarshes and mudflats. It can be seen, therefore, that the shoreline morphology becomes adjusted to local wave and tidal energy conditions (Fig. 4.14).

The landforms which reflect these local conditions are in a state of dynamic equilibrium. In areas of mobile clastic sediments these adjustments take place continuously in response to short-term wave and tidal conditions as well as reaching a long-term equilibrium. The erosion, transport and deposition processes sustain the equilibrium both in the onshore and offshore zone (Fig. 4.15).

At low tide and on sandy beaches, the wind can act as a transport agent by entraining and transporting small sand and silt-sized particles. The sand particles are quickly deposited again and may form areas of sand-dunes (see p. 99).

13 What are the main sources of sediment for the beaches of Christchurch Bay (Fig. 4.15)?

14 Describe the sediment movement pattern within Christchurch Bay and explain why research is under way to investigate the impacts of dredging the Shingles bank.

15 How does offshore relief and the coastline orientation affect wave energy inputs at the shoreline in Christchurch Bay?

Sediment cells

The movement of sand and shingle-sized sediment in the nearshore zone by littoral (longshore) drift has been found to occur in discrete, functionally separate sediment cells. Around the coastline of England and Wales, eleven main sediment cells, with smaller subcells, have been identified (Fig. 4.16). These major cells are defined as 'a length of coastline and its associated nearshore area within which the movement of coarse sediment (sand and shingle) is largely self-contained. Interruptions to the movement of sand and shingle within one cell should not affect beaches in an adjacent sediment cell' (MAFF, 1995).

Highcliffe
Barton
Becton
0 km 3
R. Avon
R. Stowe

Christchurch Harbour
Hordle
Milford
Sturt Pond
Keyhaven
Focusing of wave orthogonals as the waves are refracted as they pass through the gap between the Christchurch Ledge and The Shingles.

Milford on Sea has a deep water approach for waves from the south-west
Hurst Beach
Hengistbury Head
The shallower water of Christchurch Ledge helps to dissipate some of the wave energy from the west-south-west to south-south-west but the bay still receives powerful storm and swell waves.
North Head
Shingles Bank thought to have been formed by sediment removed from Hurst Castle Spit. Loss of material from this bank may be due to offshore dredging.
Return flow of material from The Shingles Bank may be important inputs to the beaches of Christchurch Bay
The Christchurch Ledge
The deeper water here means that waves from the south-south-west and south are high energy when they reach Barton and the centre of Christchurch Bay.

The Solent

These sediment cells, as functional systems, are increasingly forming the basis for coastal management schemes (see Chapter 6). Subcell boundaries

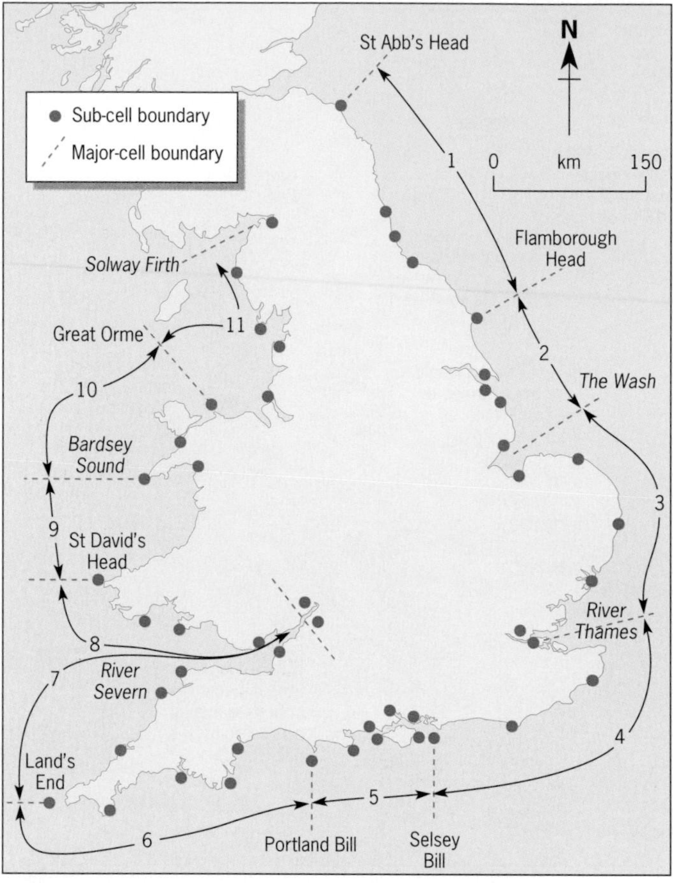

Figure 4.16 Major sediment cells in England and Wales

identify smaller cells associated within the major cells. Notice that there is some movement of sediment between the cells, i.e. they are open, not closed, systems. There are two main types of boundary between the cells.

Littoral drift divides

These occur when the coastline abruptly changes direction such as at major headlands, e.g. Portland Bill in Dorset. Drift divides also occur without any dramatic change in the shape of the coastline, but where wave conditions cause a change in drift direction, e.g. near Sheringham in Norfolk. Since material is moved outwards from a drift divide, there is a net output of sediment from the area. This results in a dominance of erosional processes and landforms, e.g. eroding beaches and cliffs.

Sediment sinks

These are where sediment transport paths meet so that sediment builds up in major depositional environments. Sediment sinks occur in deeply indented bays and estuaries, although **spits** and **cuspate forelands** may form subcell sinks.

Sediment cells are an important concept in coastal geomorphology, but there are problems with this approach. The cells are based upon the movement of coarser particles, rather than the movement of suspended fine material which can be moved long distances. Cell boundaries in such a dynamic system cannot be fixed and static.

16 Using Figure 4.16, how many cell boundaries are sinks and how many are drift divides?

17 Use an atlas to identify the sediment sinks from Figure 4.16.

18 Describe the movement of sediment along the whole south coast of England. Suggest which areas are dominated by sediment erosion and which areas are dominated by sediment deposition.

Summary

- The coastline is a complex and dynamic environment, with change occurring over a range of timescales.
- Waves are the main source of energy for geomorphological processes. Wave characteristics and type are variable in time and space, and influence the processes acting along the coastline.
- Sea-level change occurs over a range of timescales. Short-term change occurs as a result of wave and tidal action. Long-term sea-level change can be divided into eustatic and isostatic change.
- Weathering and erosion in the coastal zone are the result of a range of subaerial and marine processes.
- Sediment in the coastal zone results from a range of sources and includes both clastic and biogenic sediments. The sediment is deposited according to size in high-energy wave environments such as beaches, or low-energy wave environments such as salt marshes and mudflats.
- The English and Welsh coastline can be divided into eleven main littoral cells which are largely self-contained. These cells are a useful framework for understanding and managing coastlines at the regional scale.

5 Coastal landforms

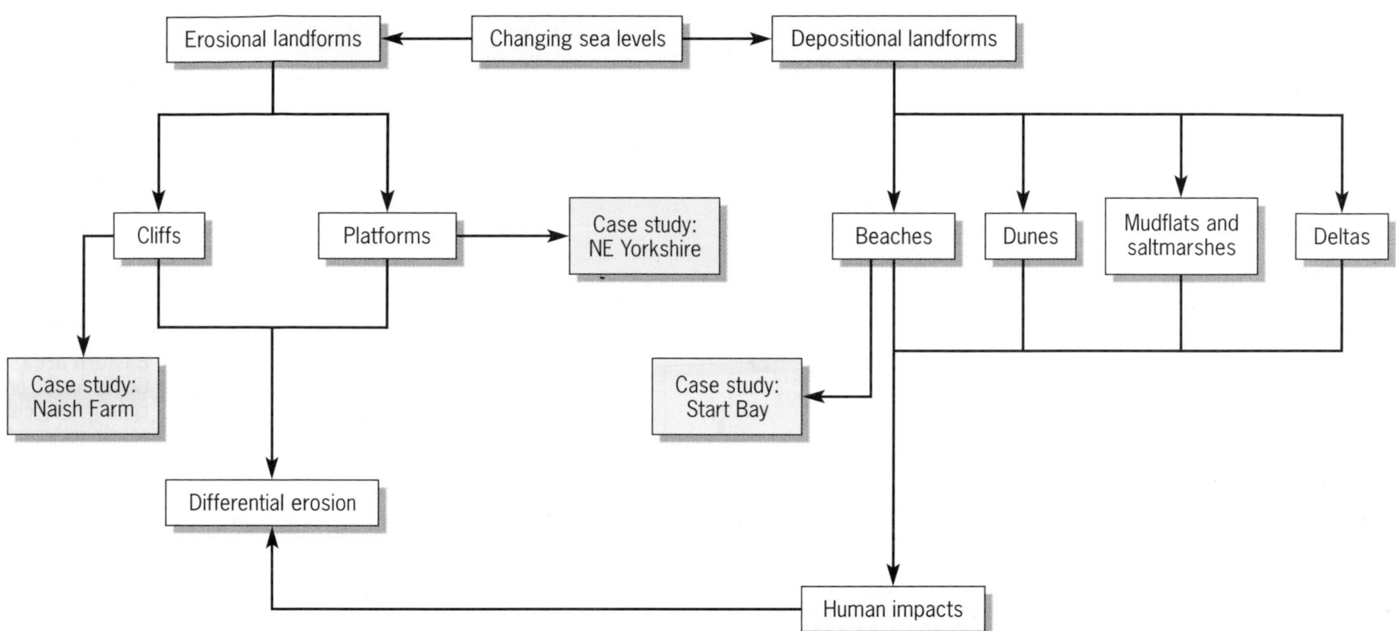

5.1 Introduction

In this chapter we shall study the main coastal landforms and the processes which produce and change them. Although erosional and depositional landforms are considered separately here, it is important to remember throughout this chapter that in reality erosion and deposition are not completely unconnected. Areas of mainly erosional processes are affected by deposition of sediment, and depositional landforms themselves are eroded during periods of high energy input.

5.2 Landforms of coastal erosion

Erosional landforms produced in the coastal zone are the result of the interaction of the subaerial processes of weathering and mass movement, and marine erosional processes (see Chapter 4) which will vary in their relative importance from place to place and time to time. The key erosional landforms we shall study in this chapter are cliffs, shore platforms, and headlands and bays; minor landforms include caves, arches and stacks.

Cliffs

Cliffs are the most important coastal erosional landform in terms of their spatial extent and because human activity is affected by cliff development processes (Fig. 4.1). For instance, buildings and roads along cliff tops are threatened by cliff retreat. A cliff can be defined as 'a marked break of slope between the hinterland and the shore' (Pethick, 1984). Cliffs are a very diverse class of landform in terms of their height and profile and the processes operating on them. Hard, resistant rocks tend to form the highest and steepest cliffs, especially in near-horizontal rock structures. For example,

Wave action

Wave action is concentrated at the cliff base and produces the break of slope which defines the cliff platform. This action has two important roles in cliff development: direct undercutting to produce a **wave-cut notch**, and the removal of cliff-foot debris. These activities are controlled by local variation in the availability and application of wave energy (Fig. 5.4), which vary constantly over time and space. For example, along the Holderness coast of east Yorkshire, rip-current channels called **ords** run seawards from the cliff foot and cut into the shore platform (Fig. 5.5). The head of each ord exposes the cliff foot to wave attack, and as the ord channels shift slowly southwards, they allow accelerated cliff recession.

The wave-cut notch deepens progressively in the zone of maximum erosion between the tidal limits. Eventually, sections of the overhanging cliff will collapse, creating a debris pile at the cliff foot. Waves quickly remove the finer material, but coarser blocks may remain to protect the cliff from wave impact. This **toe armouring** persists until either a high-energy storm or gradual attrition of the block causes its removal and exposes the cliff foot once more. This illustrates a strong negative feedback between cliff erosion and the storage of material at the cliff foot.

The type of wave approaching the cliff foot is important in determining the amount of wave energy available for marine erosion. Remember that water depth will vary at the same location with incoming and outgoing tides. Wave characteristics will also vary with wind strength and direction.

a Standing (reflective) wave

Reflected wave form
MSL
Advancing wave form
Water depth

The depth of the water is greater than the depth of the incoming wave. The cliff reflects the wave energy to give a standing wave effect called a clapotis. Little cliff erosion occurs.

b Breaking (plunging) wave

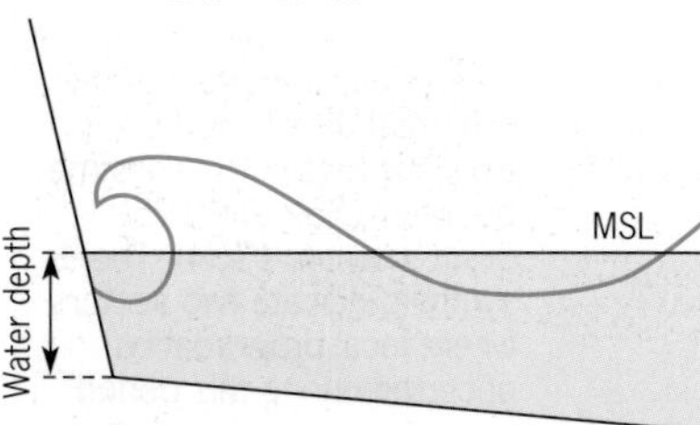

If the water depth and breaking wave depth are equal then the wave energy is directed at the cliff foot and erosion occurs. Very high pressures will occur if the crest and trough hit the cliff at the same time.

c Broken (spilling) wave

MSL = mean sea level
Surf zone
MSL
Water depth

If the water depth is less than the breaking wave depth, the wave energy is rapidly reduced by turbulence and friction with the bottom before it reaches the cliff foot. Erosion is reduced by the energy loss.

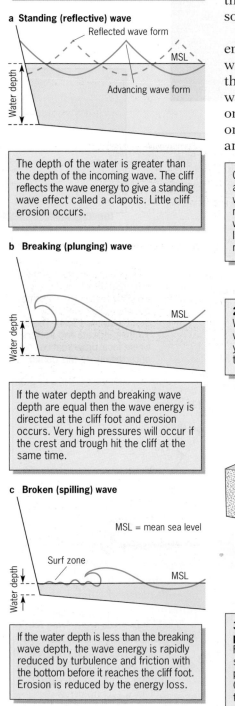

Figure 5.4 Wave types in front of a cliff (*After:* Sunamura, 1978)

Ords move along the coast mainly in a southerly direction at an average rate of 0.5km per year. Northerly onshore winds over 10ms^{-1} and their waves developed in the maximum fetch for the Holderness coast have high energy which deepens the ord, i.e. the ord length is long and a large area of the till shore platform is exposed. Ords also move most rapidly with these northerly winds.

1 Steep, rapidly eroding till cliff
A lack of beach material means that the cliff is undermined by wave erosion. The cliff profile rapidly steepens by rotational slumping, mud flows and mass falls. Between 1974 and 1982 over four times the volume of sediment was eroded from 'ord cliffs' compared with 'inter-ord' cliffs. Beach lowered by 2–4m enables high neap tides to reach the cliff foot. Mean volume of till eroded 72m^3 per metre per year.

2 Lower angled, more stable cliff
With some vegetation cover. Mean volume of till eroded 9m^3 per metre per year. Length of inter-ord cliff is three times the cliff affected by ords.

NW

SE

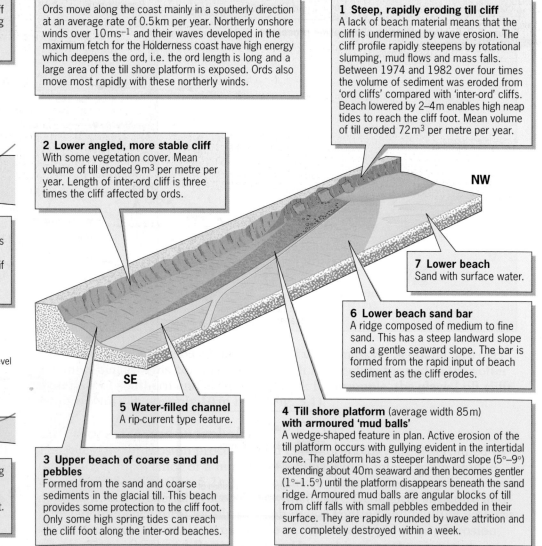

7 Lower beach
Sand with surface water.

6 Lower beach sand bar
A ridge composed of medium to fine sand. This has a steep landward slope and a gentle seaward slope. The bar is formed from the rapid input of beach sediment as the cliff erodes.

5 Water-filled channel
A rip-current type feature.

4 Till shore platform (average width 85m) **with armoured 'mud balls'**
A wedge-shaped feature in plan. Active erosion of the till platform occurs with gullying evident in the intertidal zone. The platform has a steeper landward slope (5°–9°) extending about 40m seaward and then becomes gentler (1°–1.5°) until the platform disappears beneath the sand ridge. Armoured mud balls are angular blocks of till from cliff falls with small pebbles embedded in their surface. They are rapidly rounded by wave attrition and are completely destroyed within a week.

3 Upper beach of coarse sand and pebbles
Formed from the sand and coarse sediments in the glacial till. This beach provides some protection to the cliff foot. Only some high spring tides can reach the cliff foot along the inter-ord beaches.

Figure 5.5 The characteristic features of a Holderness ord (*Source:* Pringle, 1985)

2 Using Figure 1.18 as a guide, draw a systems diagram for a cliff showing how toe armouring produces a negative feedback effect on the cliff.

3 Use Figure 5.5 to describe the main features of a Holderness ord.

4 Explain why cliff erosion is increased at point (1) and reduced at point (2) on Figure 5.5.

5 Use Figure 5.6 to describe and explain how geological structures can affect the form of a cliff.

6 Place the seven cliff profiles in Figure 5.6 in rank order of susceptibility to erosion. Give reasons for your ordering.

Cliff lithology and structure

Rock type and geological structure are important controls on the form (morphology) of a cliff. Hard-rock cliffs, e.g. basalt and granite, will erode slowly and tend to produce steep, high cliffs, while softer rocks and glacial material will be weathered more quickly and mass movement processes will dominate, resulting in less steep slopes. Joints and bedding planes determine cliff form and the amount of movement of material to the cliff foot (Fig. 5.6). Rock type and structures influence the nature of the subaerial processes which deliver material to the cliff foot (Fig. 5.7). The rate of marine erosion at the cliff foot, the effectiveness of subaerial processes working on the cliff face and the strength of the cliff materials all interrelate to control the cliff profile and how the cliff evolves (Fig. 5.8). For both hard-rock and soft-rock cliffs, their evolution is *episodic* and *cyclical*: brief spells of rapid activity where material accumulates at the cliff foot, separated by longer periods of slower erosion, attrition and sediment transport as the accumulated material is made smaller by attrition, then removed by marine action which is itself episodic. As the toe armour is removed the cliff base is exposed to wave erosion and the erosion cycle is repeated.

There is a direct relationship, therefore, between rock type, erosion rate and cliff morphology. For example, the high rates of cliff recession along the Holderness coast, noted in the report of Figure 4.1, are due mainly to the nature of the glacial materials forming these cliffs. The till has a high clay and mud content (75 per cent). This fine material is removed rapidly by

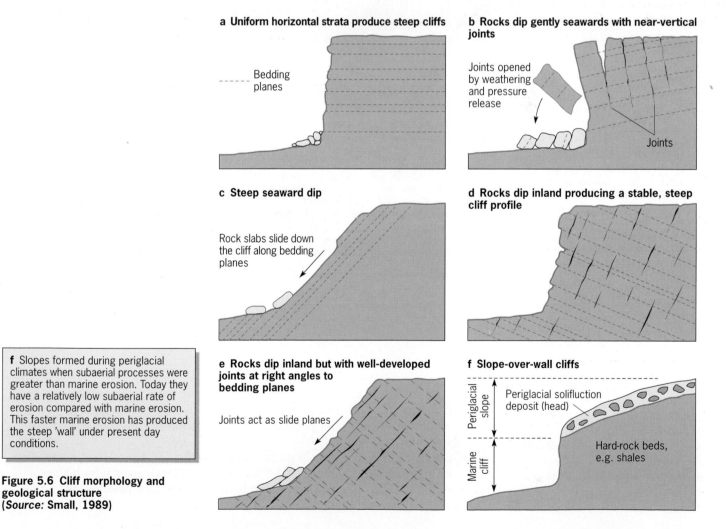

a Uniform horizontal strata produce steep cliffs

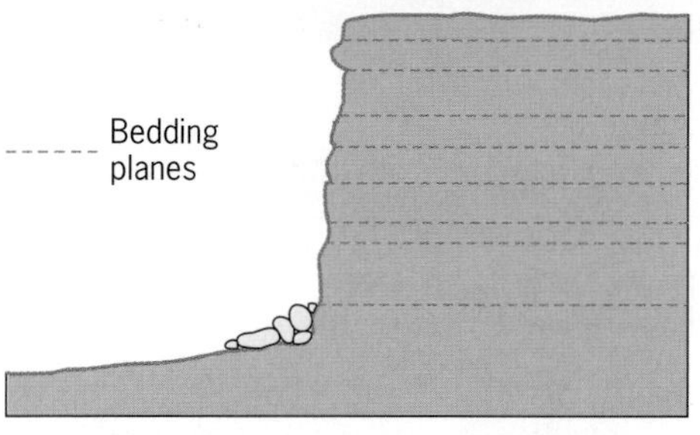

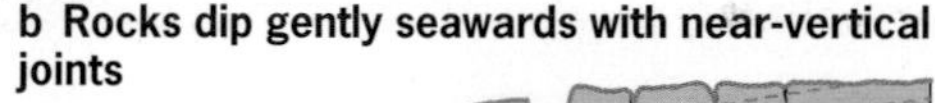

b Rocks dip gently seawards with near-vertical joints

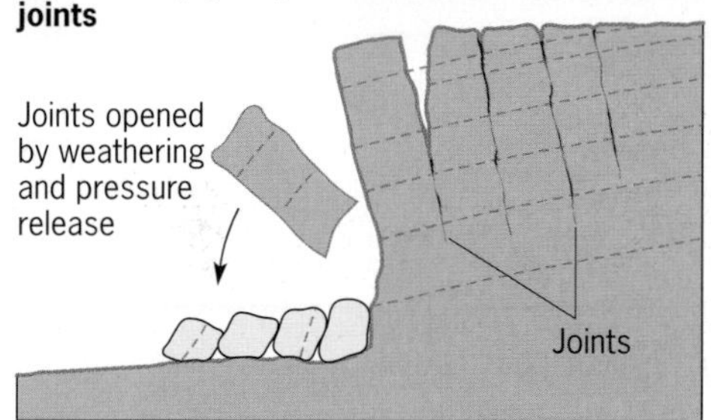

c Steep seaward dip

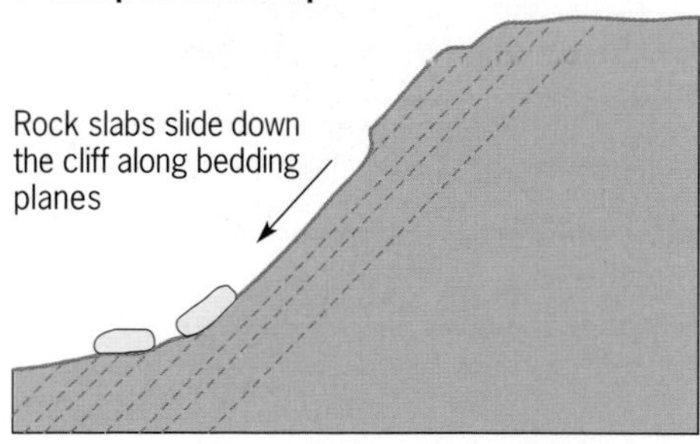

d Rocks dip inland producing a stable, steep cliff profile

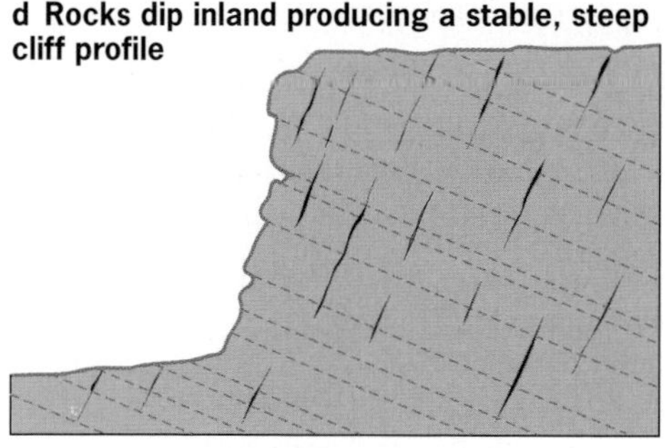

e Rocks dip inland but with well-developed joints at right angles to bedding planes

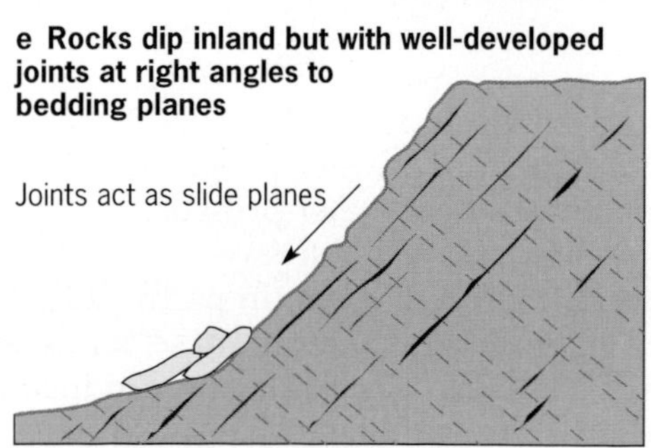

f Slope-over-wall cliffs

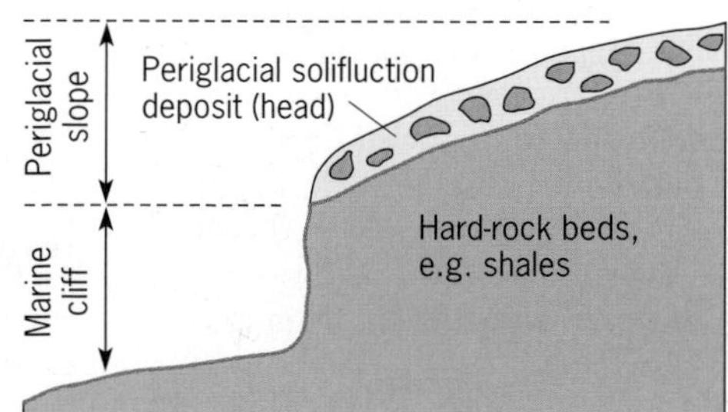

f Slopes formed during periglacial climates when subaerial processes were greater than marine erosion. Today they have a relatively low subaerial rate of erosion compared with marine erosion. This faster marine erosion has produced the steep 'wall' under present day conditions.

Figure 5.6 Cliff morphology and geological structure (*Source:* Small, 1989)

Figure 5.7 Types of mass movement affecting cliffs

Rockfalls

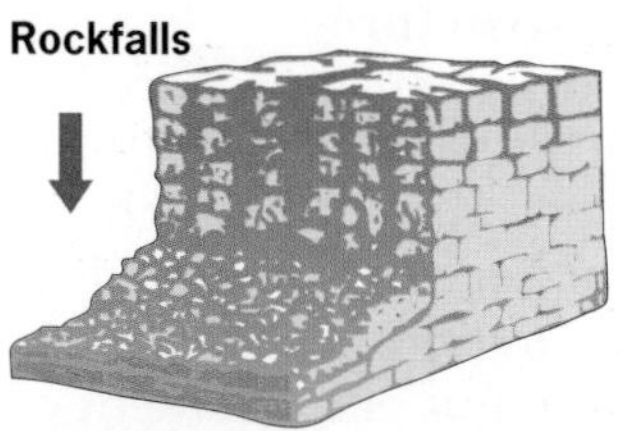

Rockfalls are most important on hard-rock cliffs. Rock blocks are dislodged by weathering and fall to the cliff foot, e.g. the limestone cliffs of South Wales

Toppling

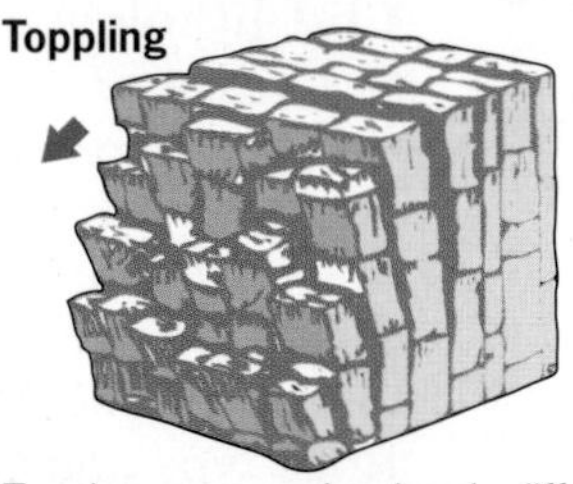

Topples occur on hard-rock cliffs with a joint structure producing columns in the rock, e.g. Carboniferous sandstones and shales of North Devon

Wedge failure

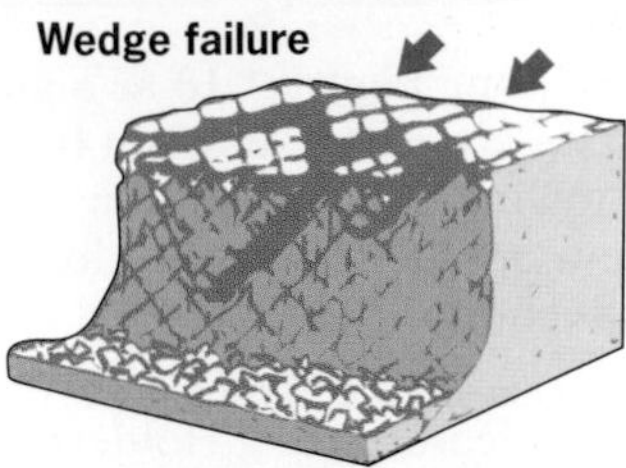

Hard-rock cliffs with a diagonal jointing pattern

Mudslide

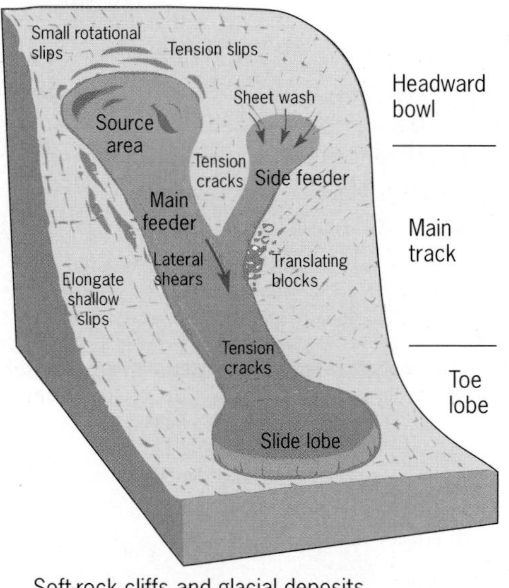

Soft-rock cliffs and glacial deposits, e.g. North Norfolk and Holderness

Rotational slides

Mudflow

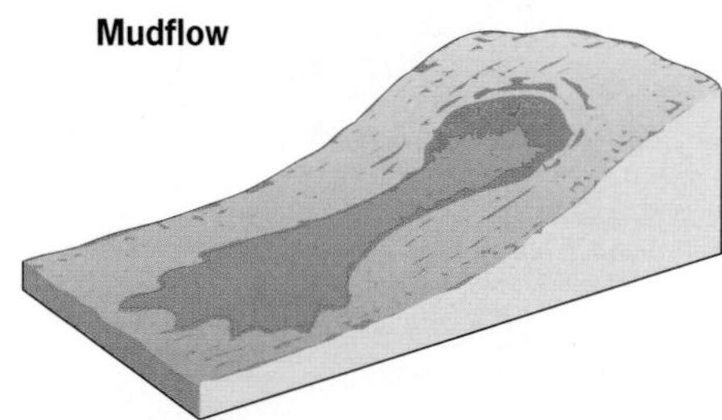

Soft-rock cliffs and glacial deposits, e.g. North Norfolk and Christchurch Bay

Cliffs of hard-rock types, e.g. granite, limestone and shale, tend to have very low rates of cliff erosion. Weathering proccesses reduce the strength of the rocks and mass movement by rockfall and toppling occurs. In softer-rock cliffs and glacial materials, the role of groundwater in mass movement processes becomes very important. Notice that the softer-rock cliffs have a lower angled profile compared with hard-rock cliffs. It is important to understand that several of theses processes may operate along the same cliff, and rockfalls, toppling and wedge failures may also occur on steeper sections of soft-rock cliffs (see Naish Farm case study).

?

7 How do rock type and structure influence mass movement processes on cliffs (Fig. 5.7)?

8 Use Figure 5.8 to draw input/output diagrams for the three relationships between subaerial and marine processes.

9 Use Figures 5.6, 5.7 and 5.8 to evaluate the following generalisations:

a On hard-rock cliffs, e.g. basalts, marine processes work faster than subaerial processes, and cliff recession occurs mainly by rock fall and toppling collapse.

b On soft-rock cliffs, e.g. mudstones and glacial deposits, subaerial processes work faster than marine processes, and cliff recession is dominated by slumping, rotational sliding and debris flows.

wave and tidal action, leaving only 25 per cent to form a protective sandy beach. Furthermore, water penetrating the glacial till fabric during wet spells increases pore water pressure and saturation levels, triggering mass movement and slope failure.

When cliff materials vary in permeability, saturated zones may occur along one or more bedding planes. These act as shear planes over which the overlying materials may slide and produce complex cliff profiles, as the Naish Farm case study illustrates.

ie
rial and

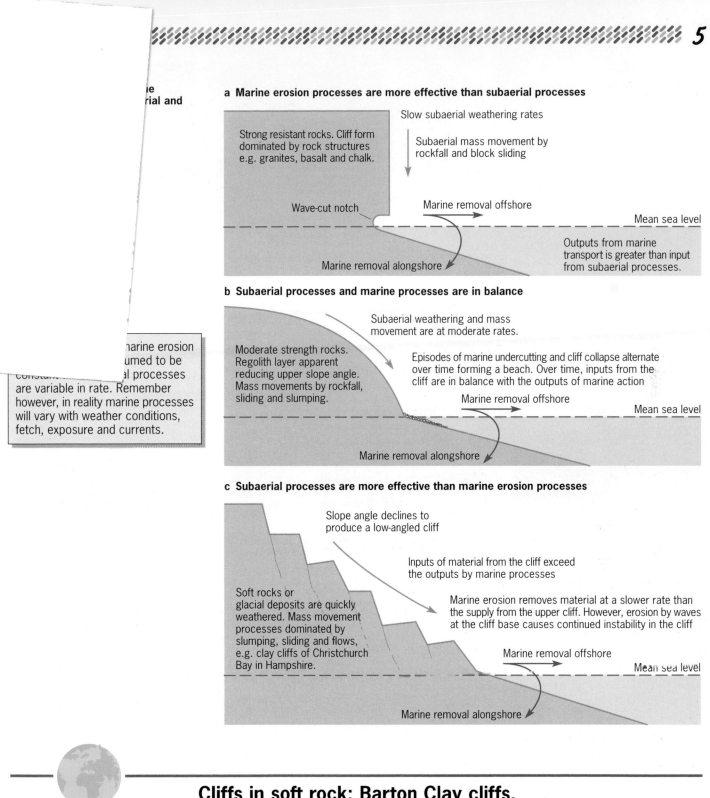

a Marine erosion processes are more effective than subaerial processes

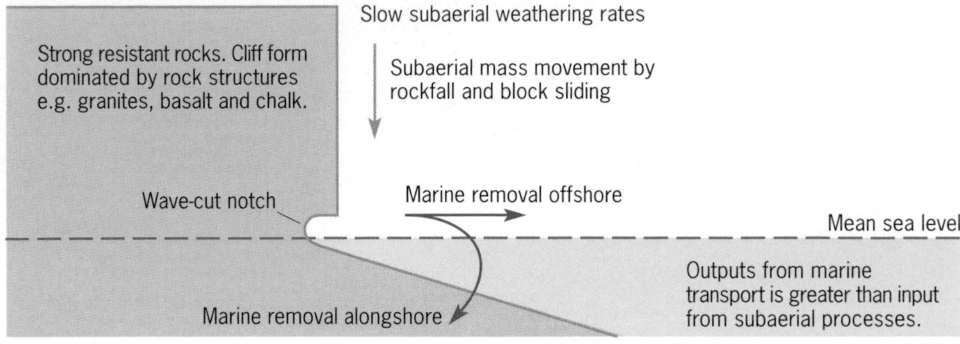

b Subaerial processes and marine processes are in balance

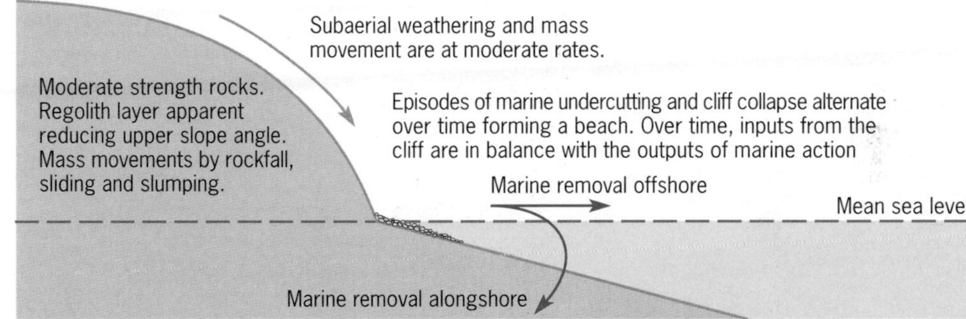

c Subaerial processes are more effective than marine erosion processes

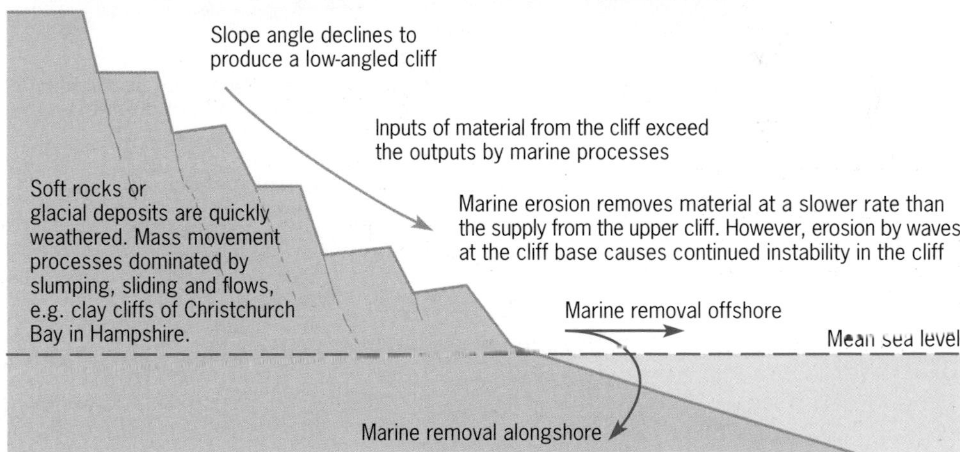

Slope angle declines to produce a low-angled cliff
Inputs of material from the cliff exceed the outputs by marine processes
Soft rocks or glacial deposits are quickly weathered. Mass movement processes dominated by slumping, sliding and flows, e.g. clay cliffs of Christchurch Bay in Hampshire.
Marine erosion removes material at a slower rate than the supply from the upper cliff. However, erosion by waves at the cliff base causes continued instability in the cliff
Strong resistant rocks. Cliff form dominated by rock structures e.g. granites, basalt and chalk.
Moderate strength rocks. Regolith layer apparent reducing upper slope angle. Mass movements by rockfall, sliding and slumping.
Episodes of marine undercutting and cliff collapse alternate over time forming a beach. Over time, inputs from the cliff are in balance with the outputs of marine action
Marine removal offshore
Mean sea level
Marine removal alongshore

…arine erosion …umed to be …al processes are variable in rate. Remember however, in reality marine processes will vary with weather conditions, fetch, exposure and currents.

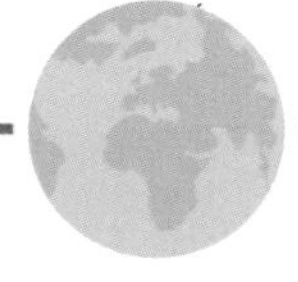

Cliffs in soft rock: Barton Clay cliffs, Naish Farm, Hampshire, England

The Barton Clay cliffs in Hampshire show clearly the complex mass movements which occur on soft-rock cliffs (Fig 5.9). Between Barton on Sea and Highcliffe there is a section of cliff which has no human management structures to control cliff erosion (Fig. 5.10), so the natural processes operating can be seen clearly. A length of 200 m of cliff at Naish Farm was studied between July 1981 and July 1982.

Cliff form and mass movement processes

The cliff has a multi-benched profile controlled by three main bedding planes within the Barton Clay (F, D, A3 on Fig. 5.12). The concentration of groundwater movement along these planes encourages shearing and slope failure. Mass movement of the saturated clays by bench sliding along the shear planes has produced three distinct benches in the cliff profile.

Barton Clay cliffs, Naish Farm, Hampshire

Figure 5.9 General view of Naish Farm cliffs, Hampshire, looking westwards towards the Chewton Bunny outfall and strongpoint in the extreme top left

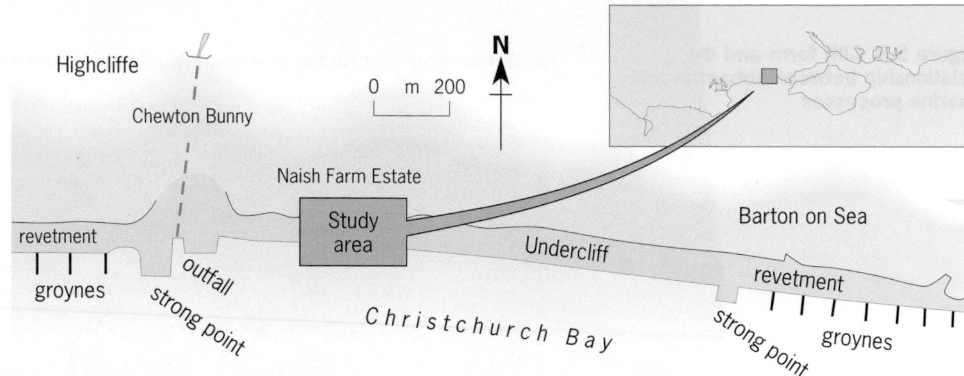

Figure 5.10 The location of the study area, a 200-m stretch of unprotected cliff line in Christchurch Bay fronting the Naish Farm Estate at Highcliffe (*Source:* Barton et al., 1983)

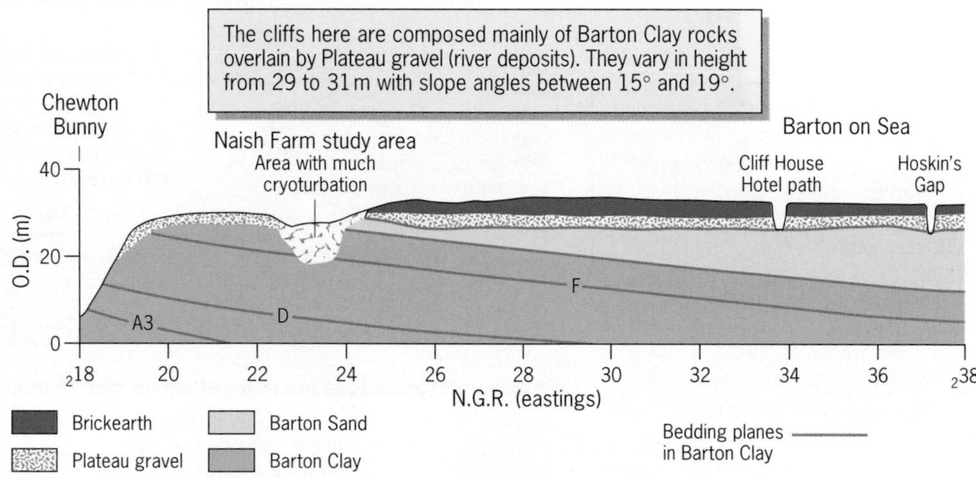

Figure 5.11 Longitudinal section (i.e. parallel to cliff line). Lines marked A3, D and F are bedding plane shear surfaces in the Barton Clay. (*Source:* Barton et al., 1983)

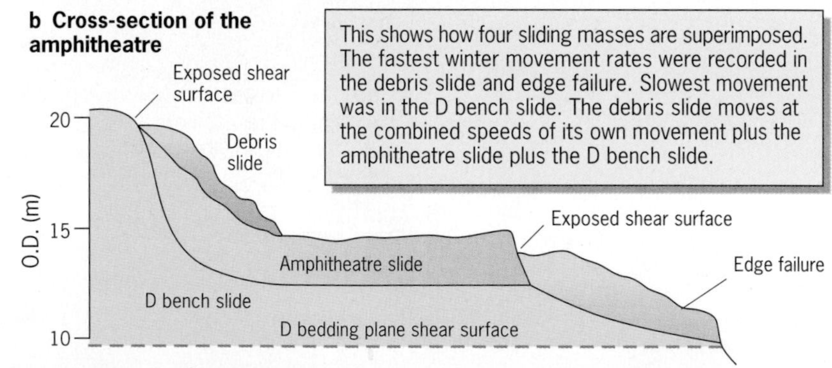

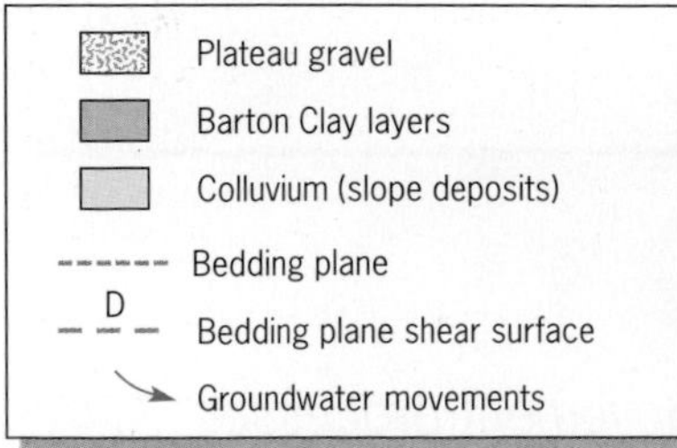

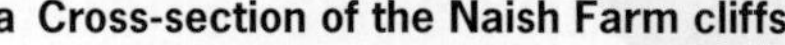

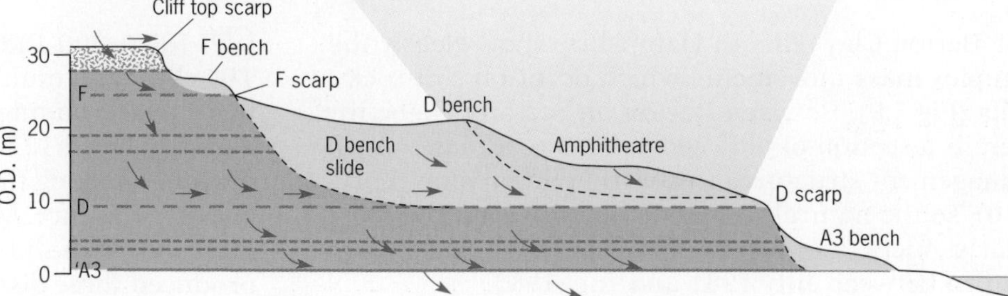

Figure 5.12 Cross-section of the Naish Farm cliffs to show the sliding masses. The main slides and benches are shown in the lower diagram. Smaller-scale movements in the large D bench slide are shown in the upper diagram.

Each bench is named after the plane along which the materials have moved. The benches are separated by steeper scarps, with a further scarp below the cliff top. Only along these scarp faces is the clay bedrock exposed. The rest of the slope surface is composed of loose, weathered material known as *colluvium*.

The D bench is the dominant profile component, and is itself a multiple feature, e.g. the 'amphitheatre slide' on Figure 5.12b. All the benches have smaller slides, slumps and flows superimposed on them, e.g. mudslides, which continue to produce the overall complex structure of the cliff (Fig. 5.13). Water is the

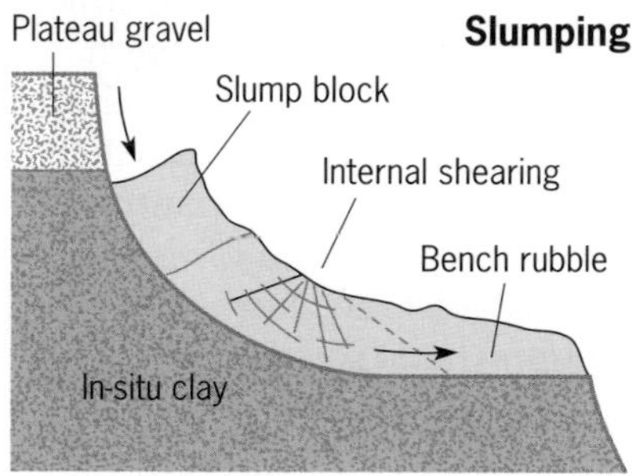

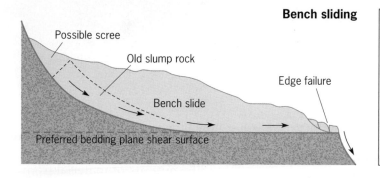

Bench slides are large-scale, relatively slow movements of dry material. Large amounts of slope deposits (colluvium) may be moved including former slump blocks. Movement follows a bedding plane in the horizontal part of the shear surface. This movement means that bench slides can extend far along the rock outcrop. Most of the material moved is colluvium. Here, 93% of the cliff colluvium moves by bench sliding. Separate rotational failure may occur at the edge. The average winter moisture content of bench rubble was 37% and the seaward movement of material varied from 0.9m to 10.1m in one year.

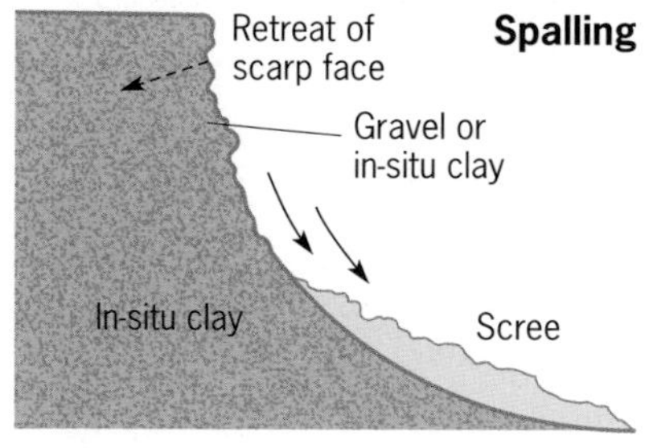

Slumping is a discontinuous or catastrophic process. The material moved is in-situ rock on the slope. Slumping affects all the steeper slopes or scarp faces on the cliff, but is most easily seen on the cliff-top scarp. This is rotational failure, i.e. the material is rotated around a radius of curvature. The shearing in the centre of the block breaks up the rock into finer debris. The upper parts of the rock mass may remain intact until it is broken up by weathering. Slumping is smaller scale than bench sliding and results in the parallel retreat of the cliff face, i.e. the slope recedes at the same angle. Movement is relatively slow and the process operates with relatively dry rock. The winter moisture content of slumps was 28% and movement of 1.1m seawards was recorded in one year.

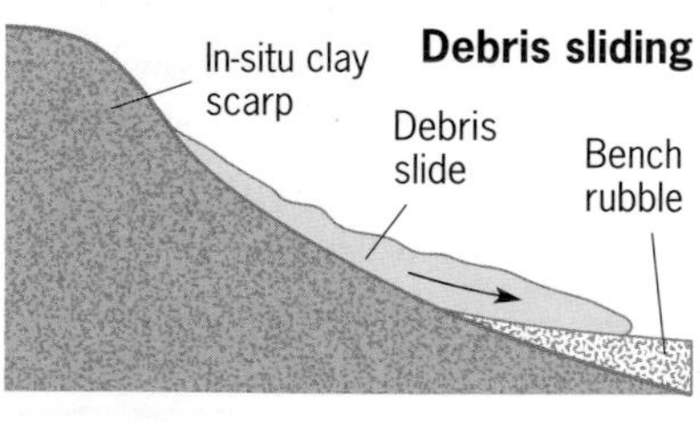

Spalling is the weathering, movement and eventual fall of a small mass of material which is then added to the scree below. The process is continuous on all the steeper scarp faces including the cliff top. Weathering processes, especially frost action, rainwash and clay shrinkage, loosen rock material which then falls off the cliff face. The process is fastest on freshly exposed scarp faces after slumping. The rapid release of stress in the exposed rock as the weight of the slump block rock is removed helps the weathering processes to be effective. During the observation year, spalling resulted in scarp recession of 0.46m on average.

Debris sliding

In-situ clay scarp

Debris slide

Bench rubble

Debris slides are the small-scale movements of loose debris, e.g. a scree on a steep slope. The steep clay scarp slope is between 25° and 40°, and the material slides down this until it runs into a bench where movement is slowed down. The debris mass is loose material and so there is much deforming of this during the movement. A thin tongue of debris usually less than 1m thick is involved in the slide. The winter moisture content was 38% and the average seaward movement was 4.6m in one year.

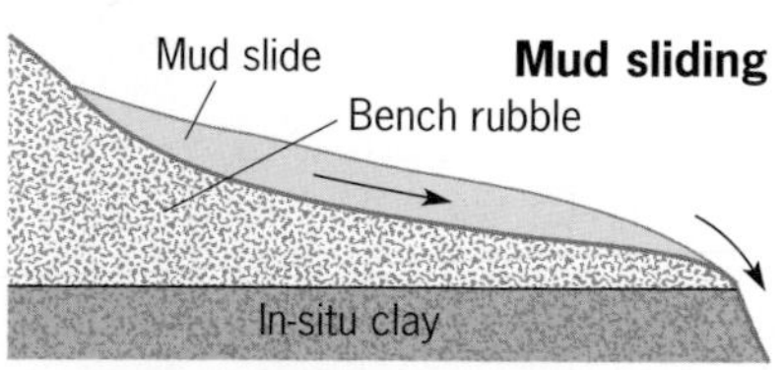

A mudslide does not have a shear surface along a bedding plane, as with the bench slides, the failure surface can cut through bedding planes. Mudslides involve the movement of very wet material in near-fluid state. An increased flow of groundwater after a wet spell, into the colluvium will result in mudsliding. In winter the mudslides are very wet, but in summer they dry out to form a hard crust. Two mudslides were observed: one is small, and movement is of colluvium only; the other is much larger and has eroded a gully through D scarp and forms a wide fan across the A3 bench on its way to the beach. The shear surface cuts through both the D and A3 bedding planes. The winter moisture content was 46% and the average seaward movement was 12.6m in one year.

Figure 5.13 Mass-movement processes operating on the Barton Clay cliffs at Highcliffe (*After:* Barton et al., 1983)

Barton Clay cliffs, Naish Farm, Hampshire

key element in all of these processes. For example, saturated surface debris and mudslides move fastest on top of the D bench slide, which is itself sliding over the *in situ* clay beneath. Such mass movements are more active during the wetter winter months when water loading of the cliff materials is greatest. In winter too, surface runoff (overland flow) is more likely, spilling mudruns of slurry across the lower slopes. In contrast, during summer, downslope surface movement may be triggered by holiday-makers, e.g. mudslides being used as a path.

The Naish Farm cliff, therefore, functions as a *cascading system*, with the colluvium being transferred through a set of stores in each of the scarps or benches on its way to the beach store (Fig. 5.14).

Rates of recession

Cliff recession varies over time and space. Cliff-top recession in recent decades has averaged 0.4 m/yr, but in places reached 5 m/yr. The cliff-foot advances and retreats as slip debris pushes the cliff toe forwards, to be removed by wave and tidal erosion. Between the mid-1950s and mid-1970s mean cliff recession rates increased from 0.4 to 1.9 m/yr, especially in the lower sections. This caused an increase in average slope angle from 13.3° to 18.8°. This steepening triggered a massive slump along the D shear surface in 1977–8 in the centre of the Naish Farm section. Later slides occurred along the F plane. The result was a reduction of the overall slope profile to a more stable gradient.

Such improved stability is proving short-lived. From the late 1960s, coastal management schemes updrift at Highcliffe, e.g. groynes and revetments, have reduced the west–east supply of sediment by longshore drift into the Naish Farm beach store (see the Barton case study in Chapter 6). As a result of the loss of protective beach material, the marine erosion at the cliff foot has increased once more, the lower profile has steepened, and more colluvium is being removed from the cliff than is being supplied. For instance, during 1981–2, the outputs of material from the cascade system to the beach store were five times the inputs from weathering and cliff-top recession. This represents an 8 per cent reduction in the volume of colluvium on the slope, i.e. a debris budget deficit. The cliff is functioning as a **process–response system**. A change in the balance and rate of the processes operating, i.e. increased basal erosion, generates a response through the system, and a renewed cycle of slope steepening–instability–failure–accelerated cliff recession–reduced slope gradient.

Figure 5.14 A landslide flowing on to the beach at Naish Farm

10 Draw a sketch of Figure 5.9. Annotate the sketch with the main slope elements and any evidence you can see of mass movement processes.

11 How important is water in the mass movement processes operating on the cliffs at Naish Farm (Fig. 5.13)?

12 Classify the mass movement processes by speed and size (Fig. 5.13).

13 With the aid of annotated diagrams, explain how human activity has increased cliff recession at Naish Farm.

14 Suggest why human management of the cliffs to reduce rates of cliff recession would be difficult.

Shore platforms

A shore platform is a relatively flat, gently sloping (1°–3°) expanse of rock at the foot of a cliff and extending out to sea (Fig. 5.15). In the UK shore platforms have their flat surface between high and low tide levels, i.e. they are intertidal platforms. They are formed along solid rock coastlines by cliff recession. The platform surface represents the former position of the wave-cut notch. The platform is progressively lowered by weathering and erosion processes as the cliff retreats. Thus, shore platform development can only be understood in relation to the associated cliff development.

The platform–cliff sediment system relates the production, transport and store of sediment. The platform is formed in basalt at more than one level. There is a general lack of sediment. The basalt is quarried, spalled or abraded. Quarrying occurs mainly at the seaward edge of the lower platform and abrasion in the channels, rockpools and on the landward edge of the surface zone of breakers. The platform is covered at high tide and exposed at low tide, i.e. main surface is at intertidal level. Spalling (wetting and drying) is important on the cliff face. Loose sediments move landwards, although some becomes trapped in potholes and pools, others enter sub-marine chutes as water runs off the platform. Most material enters embayment beaches which cover part of the upper platform and cliff foot. The large size of this material provides protection from erosion to the cliff foot and platform.

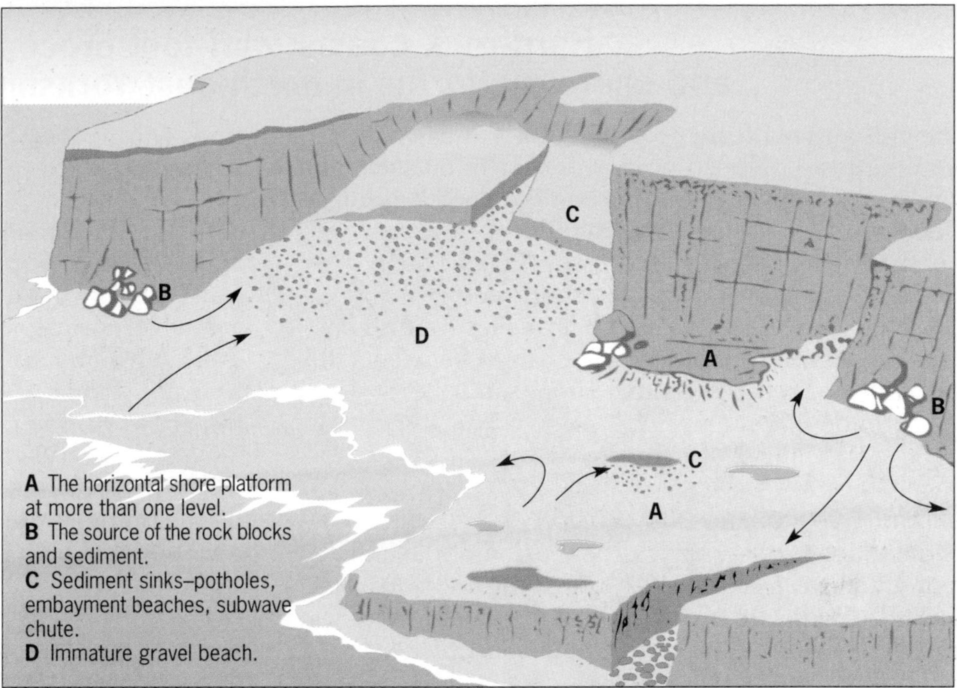
C
B
D
A
B
C
A
A The horizontal shore platform at more than one level.
B The source of the rock blocks and sediment.
C Sediment sinks–potholes, embayment beaches, subwave chute.
D Immature gravel beach.

Figure 5.15 The basalt cliff–platform system on the north-central coast of Ireland (*Source:* Carter, 1988)

All the processes of mechanical wave erosion and weathering operating at the cliff foot operate on the shore platform. For example, salt weathering is likely to be important in the splash and spray zones. Alternate wetting and drying as the tides rise and fall allows salt crystallisation to occur, causing stress to rocks. Hydration and salt solution cause chemical weathering. If there is a source of suitable material, abrasion will be an important erosional process. Bioerosion of the platform by boring and grazing organisms will also be of importance (see Fig. 4.12). The relative importance of these weathering and erosional processes determines the rate of platform formation and changes to its morphology. As with cliff morphology, it is the interaction of these marine processes and the lithology and structure which influence the detailed nature of the platform. The type of wave reaching the cliff foot influences the rate of cliff undercutting and thus the width of the platform (see Fig. 5.4). However, platform width is self-limiting. As the platform widens, and despite the gradual lowering of the platform surfaces, waves will break further out to sea. Increasingly the waves will be the less erosive spilling wave type, their energy dispersed across the platform before reaching the cliff foot.

Figure 5.16 Coastal erosion at Flamborough Head, Yorkshire

Lithology and structure will influence the efficiency of the processes operating on platforms and the type of material available for corrasion at the cliff foot. Generally the widest platforms occur on unresistant rocks, while resistant rocks produce higher and narrower platforms. Horizontal rock strata produce wide platforms (Fig. 5.16) and vertical strata narrow platforms. Where the platform is cut across differing strata, joints and bedding planes will be differentially eroded to produce a micro-relief on the platform surface (see the north-east Yorkshire case study on p.82).

15 Draw an annotated sketch of Figure 5.16. Label the landforms and add details of the coastal processes operating and evidence of rock structural controls on the landforms.

Most British platforms are estimated at between 100 and 1000 years old and are produced by present-day processes and sea levels, although some platforms have a more complex structure associated with former sea levels. Not all shore platforms are intertidal as they are in Britain. In other parts of the world, such as the Mediterranean and the tropics, platforms occur at different tide levels due to the varying tidal environments and the higher importance of rock weathering and bioerosion than is found in the UK.

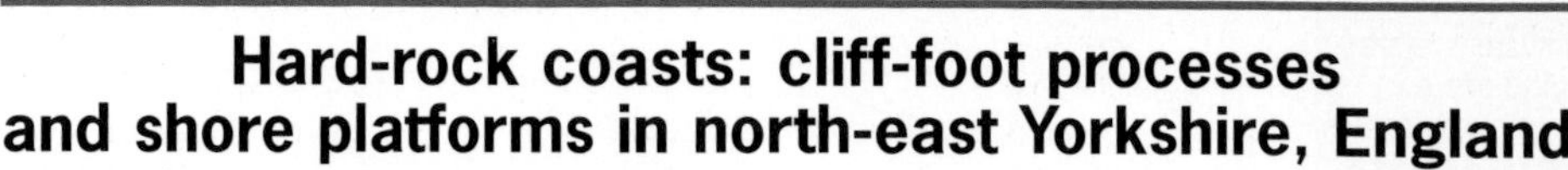

Hard-rock coasts: cliff-foot processes and shore platforms in north-east Yorkshire, England

The cliff and platform coastline of north-east Yorkshire is developed upon sub-horizontal shale and sandstone strata. There are well-developed bedding plane and joint systems which influence the weathering and erosion processes (Fig. 5.17). In places, glacial till overlaps the solid geology. L.A. Robinson studied a number of sites over a two-year period to explore the relationships between cliff-foot erosion, beach sediment supply and shore platform development.

Cliff-foot processes and erosion

The dominant processes in the erosion of the cliff-foot notch were found to be quarrying and corrasion acting episodically. Two-thirds of the geomorphological work was achieved by high-energy waves which occurred in only 12.5 per cent of the study time. During these storm conditions, erosion rates equivalent to 6.0 cm a year were recorded in the shale beds.

Notch development was dominated by quarrying at sites with little or no beach material, e.g. Saltwick Bay (Fig. 5.18). Where a sand beach supply was available, corrasion became increasingly important, especially in a narrow zone up to 8 cm above beach level, e.g. Lingrow (Fig. 5.19). Quarrying by wave impact worked to greater heights and hence produced a wider notch. We must remember, however, that as the beach sediment store gets larger, it absorbs and dissipates more of the wave energy. Thus, beyond a certain threshold, notch development rates begin to slow down.

The influence of rock character was illustrated by the finding that at cliff notches where quarrying was dominant, maximum erosion rates occurred during

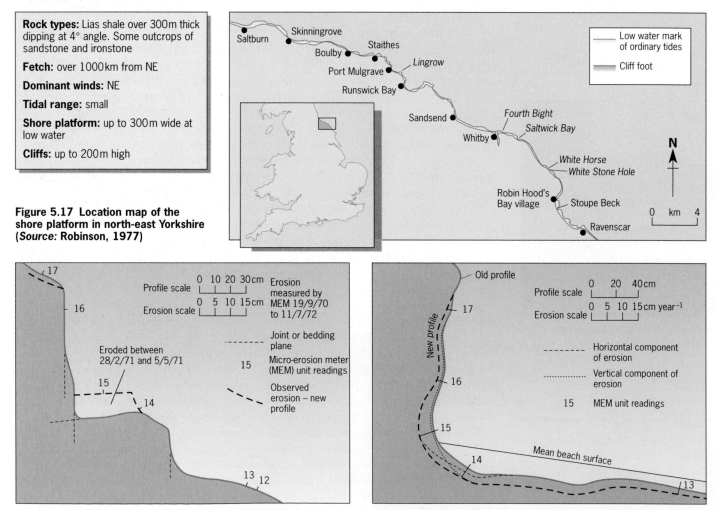

Rock types: Lias shale over 300m thick dipping at 4° angle. Some outcrops of sandstone and ironstone

Fetch: over 1000km from NE

Dominant winds: NE

Tidal range: small

Shore platform: up to 300m wide at low water

Cliffs: up to 200m high

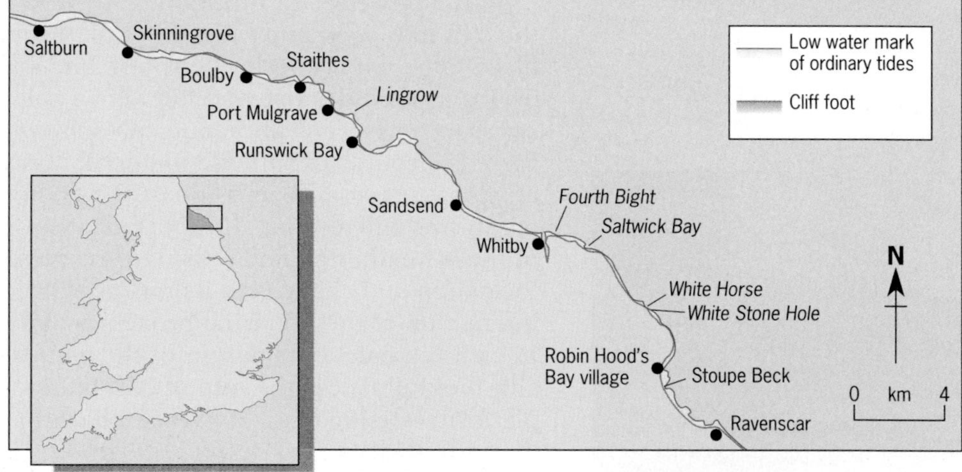

Figure 5.17 Location map of the shore platform in north-east Yorkshire (*Source:* Robinson, 1977)

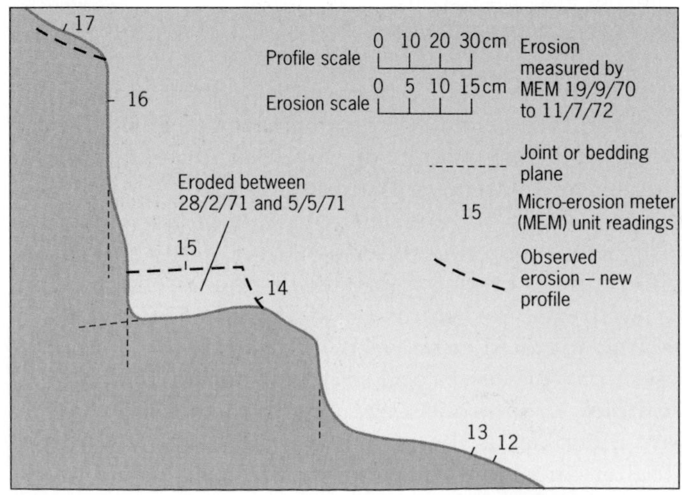

Figure 5.18 Vertical profile of the cliff foot in Saltwick Bay (*Source:* Robinson, 1977)

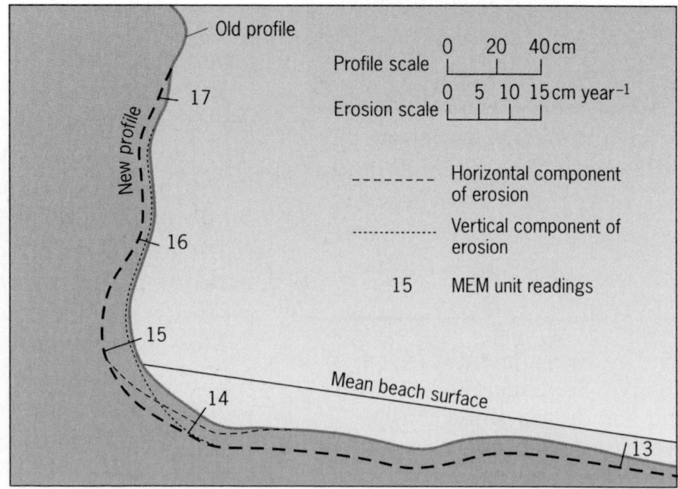

Figure 5.19 Vertical profile of the cliff foot at Lingrow (*Source:* Robinson, 1977)

summer, e.g. at Saltwick Bay; summer rates were up to seven times those during the winter. The well-bedded and jointed shales are permeable, and have a high clay content which swells on wetting. When they are subjected to repeated wetting and drying by the diurnal tidal rhythms, the shale structure begins to disintegrate. The wetting and drying cycle is more vigorous during the warmer summer months, and so produces a greater supply of loosened material for erosion by quarrying action. Conversely, notches where corrasion was significant, e.g. Lingrow, showed more rapid erosion during the stormy winter months.

Overall, the highest erosion rates were recorded during the winter months at locations with a small beach, i.e. sites where corrasion and quarrying were both active during the months of high-energy storms.

?

16a How is the form of the cliff at Saltwick Bay (Fig. 5.18) influenced by the presence of joints and bedding planes?
b Draw an annotated sketch to show how the cliff-foot profile may change in the future.

17 Draw a copy of the cliff foot at Lingrow (Fig. 5.19). Label the wave-cut notch, the area of maximum corrasion. Annotate your sketch to describe the processes operating.

Shore platform gradient and morphology

The shore platform profile gradients show peaks at around 1° and 6° (Fig. 5.20). This bimodal distribution tells us that the platforms along this stretch of coast possess two main components: an almost flat, gentle **plane**, and a steeper rock surface or **ramp**. Except where there are strong geological controls (Fig. 5.21), the platforms fall into three classes: plane only; ramp only; plane and ramp (Fig. 5.22).

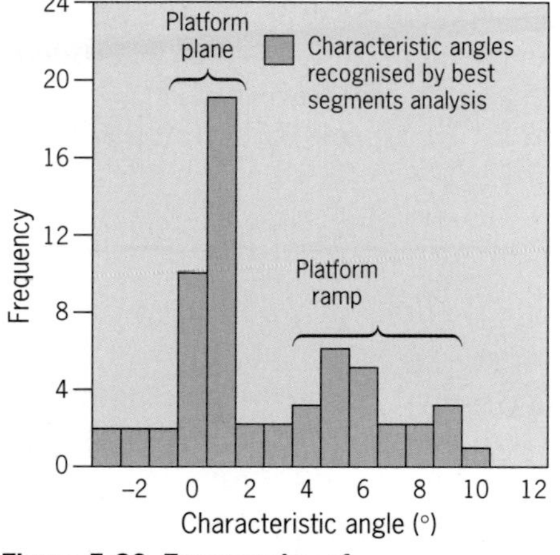

Figure 5.20 Frequencies of characteristic angles on the north-east Yorkshire shore platforms (*Source:* Robinson, 1977)

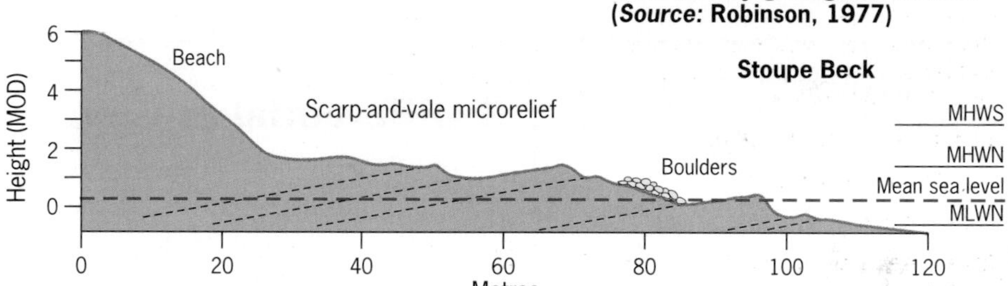

Figure 5.21 Shore platform profile controlled by geological structure (*Source:* Robinson, 1977)

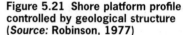

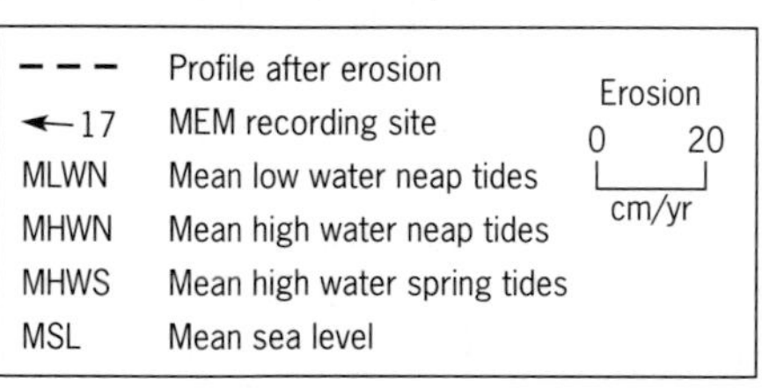

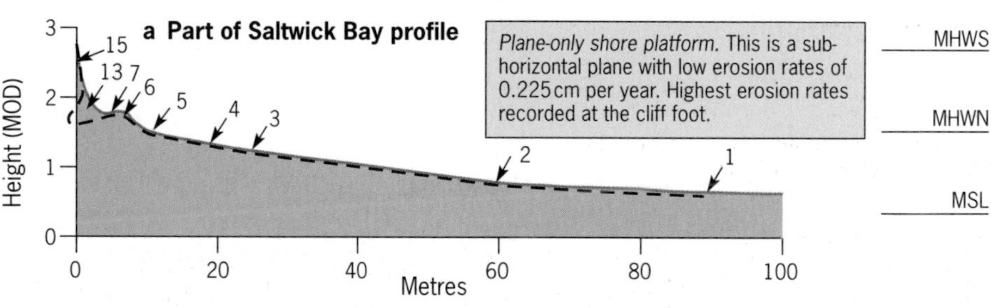

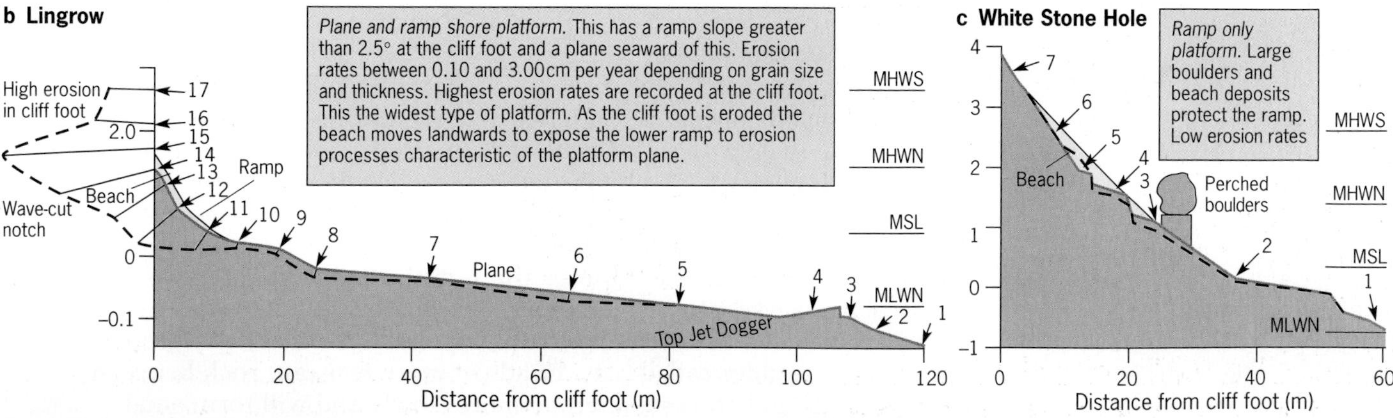

Figure 5.22 Shore platform morphology developed in uniform Lias clays (*Source:* Robinson, 1977)

Hard rock coasts: cliff foot processes and shore platforms in north east Yorkshire, England

?

18 Study Figure 5.20.

a What is the modal angle of the platform plane and ramp?
b Describe the fieldwork techniques you could use to produce the data shown.

19 Use Figure 5.21 to describe the influence of geological factors on the morphology of the shore platforms. Refer to bedding planes, relative resistance of rocks to erosion and rock dip.

20 By the use of annotated diagrams and notes, describe and explain what would be likely to happen to platform morphology following these changes in local environmental conditions:
a A sudden input of material from the upper cliff to a plane-only platform.
b A rapid reduction of the beach sediment store on a plane and ramp platform.

Where geological conditions are relatively uniform, e.g. the massive Lias Clay, the volume and calibre of beach sediment supply was found to be the key to platform morphology. Plane-only platforms were likely where there was little or no beach sediment store (Fig. 5.22a). Quarrying was dominant and created a low-altitude cliff foot (about 2 m OD). Cliff retreat and platform extension were slow. At sites with a moderate beach sediment supply, with both quarrying and corrasion active, plane and ramp platforms developed (Fig. 5.22b). Greater cliff-foot height and more rapid retreat allowed the steep ramp component to develop. Yet at sites with a deep or coarse beach sediment store, the cliff foot and ramp are, for long spells, protected. Seaward of the beach store, however, the exposed platform is slowly eroded and lowered, finally beneath low water mark (platforms are intertidal landforms) creating a ramp-only platform (Fig. 5.22c).

5.3 Differential erosion

Geological factors are important in coastal erosional landform development at a variety of scales. The resistance of different types of rock and structural weaknesses such as faults, joints and bedding planes account for the great variety in the morphology of cliffs and the form of the coastline.

At a medium and large scale, resistant rock outcrops will form areas of higher relief with cliffs and headlands, and outcrops of weaker rocks will form lowland areas with bays and inlets. The Pembrokeshire National Park coastline in south-west Wales clearly shows the importance of rock type, structure and differential erosion on the planform of the coast (Fig. 5.23). The coastline in plan view varies with the relationship between the line of the coastline and the geological outcrops and structures. Notice that between St David's Head and Abereiddy Bay the rocks are parallel to the coastline (concordant). The coastline is quite different from the areas where the rocks and structures are at right angles to the coastline (discordant), e.g. between Ramsey Sound and St David's Head. Remember that the relief and the structure of the coastline will result in variations in wave type, exposure and fetch (see section 4.2) which work with the varying geological structures to influence the coastline. Headlands at all scales will result in wave refraction and a concentration of wave energy at the headland.

At a small scale within the same rock outcrop, geological weaknesses, e.g. joints, bedding planes and fault lines, are eroded more quickly, and if the rock has the internal strength to support the opening, *sea caves* will develop. In some caves vertical shafts may extend to the ground surface to form a *blowhole*. Air and water are forced through the hole with an explosive force by breaking waves, causing large pressure changes in the cave which aid further erosion. Other joints and fault lines may be differentially eroded to form narrow gullies called *geos*. Bands of more resistant rock between weaker joints and cracks will erode more slowly and will form small headlands. Differential erosion of these may result in the formation of a *sea*

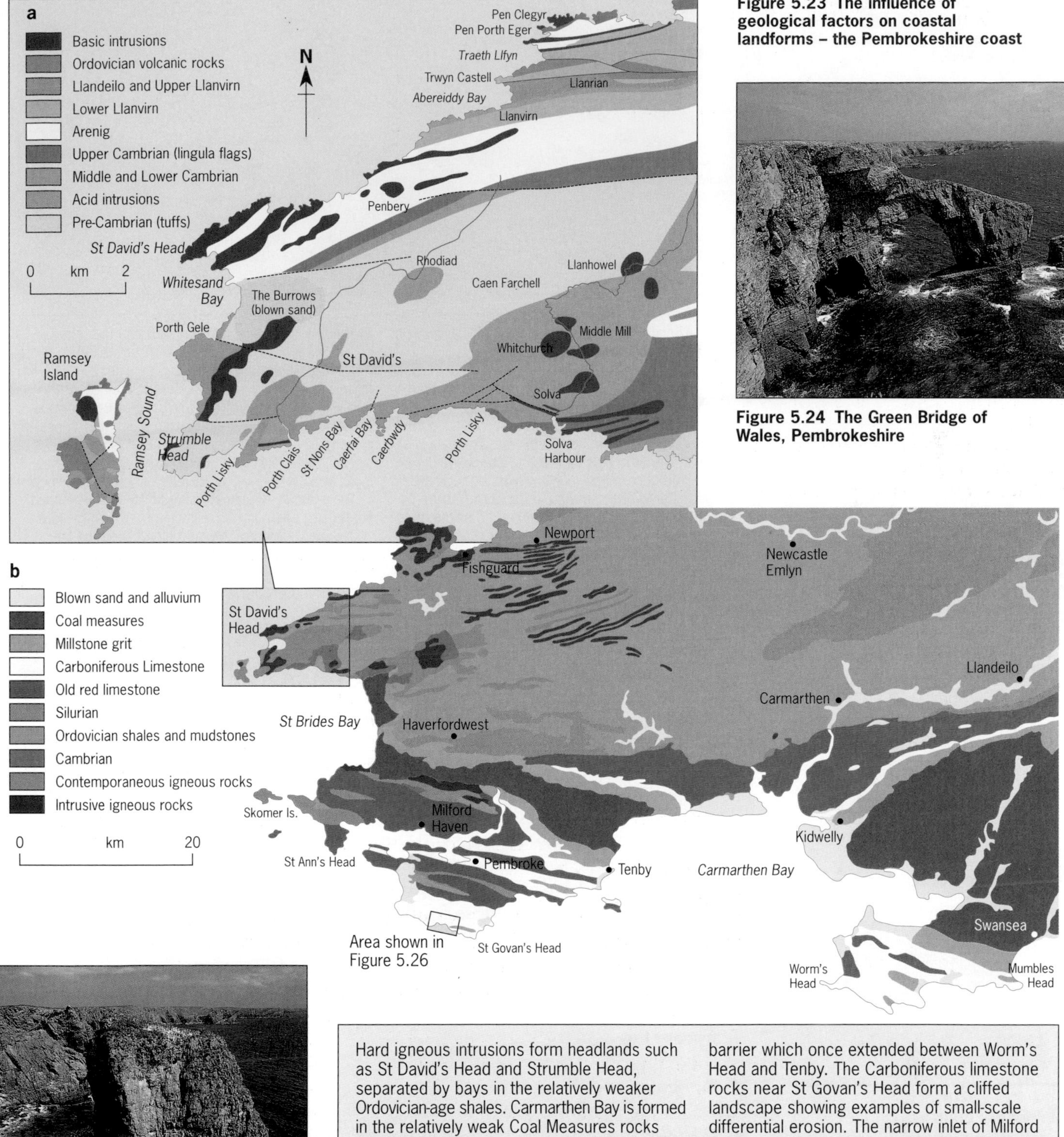

Figure 5.23 The influence of geological factors on coastal landforms – the Pembrokeshire coast

Hard igneous intrusions form headlands such as St David's Head and Strumble Head, separated by bays in the relatively weaker Ordovician-age shales. Carmarthen Bay is formed in the relatively weak Coal Measures rocks (shales and sandstones) once the erosion had breached the harder carboniferous limestone barrier which once extended between Worm's Head and Tenby. The Carboniferous limestone rocks near St Govan's Head form a cliffed landscape showing examples of small-scale differential erosion. The narrow inlet of Milford Haven is a drowned river valley or ria formed by the post-glacial rise in sea level.

Figure 5.24 The Green Bridge of Wales, Pembrokeshire

Figure 5.25 Elegug Stacks (Stack Rocks), Pembrokeshire

arch (Fig. 5.24). Collapse of the arch leaves upstanding blocks called *stacks* (Fig. 5.25) as the rest of the cliff face retreats. These will eventually be removed by erosion. The Carboniferous Limestone outcrop near St Govan's Head (Fig. 5.23) clearly illustrates the influence of small-scale differential erosion (Fig. 5.26).

Figure 5.26 Erosional landforms produced by the differential erosion of Carboniferous Limestone, south-west Dyfed, Wales

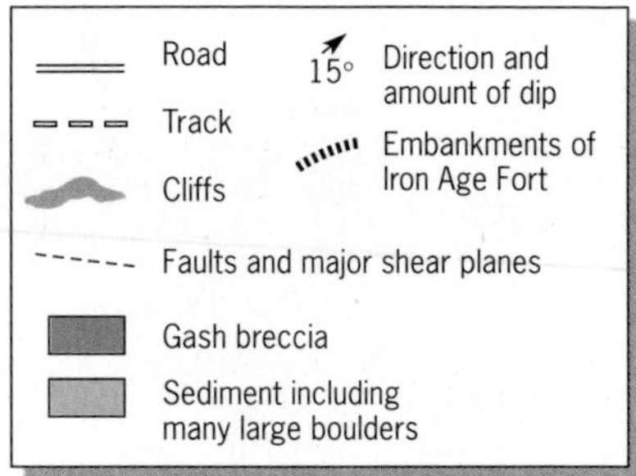

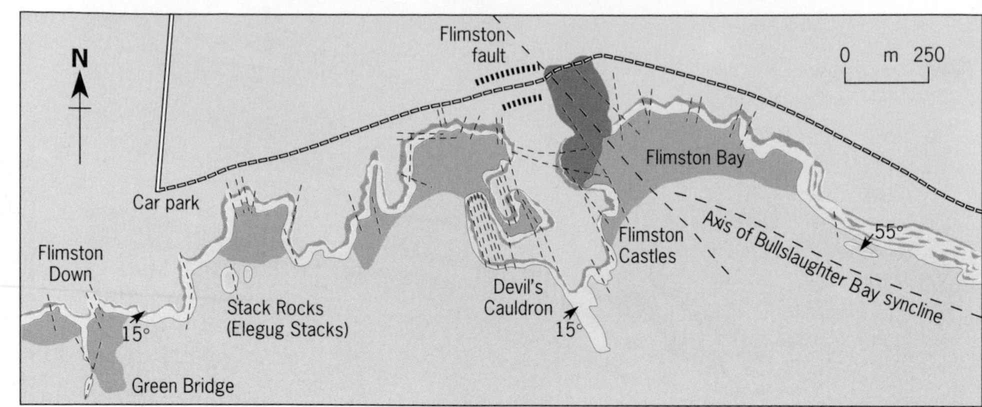

The Carboniferous Limestone dips at 12° landwards near Flimston Down forming steep, near vertical cliffs up to 50m high. Active erosion is taking place as shown by many rockfalls at the cliff base and overhangs of limestone blocks. To the east of the Flimston fault the limestones are more thinly bedded and dip steeply seawards at 55° into Bullslaughter syncline. The Gash breccia is due to intense fracturing of the limestone during earth movements. There are many erosional features in the area formed by the differential erosion and solution of the limestone along fault lines. The Green Bridge is a sea arch formed by caves on opposite sides of the small headland joining up. The arch is 24m high and spans 20m. The Stack Rocks are two sea-stacks which formed by the erosion of an old headland. The highest stack is 36m. To the east of the Stack Rocks cliffs are many caves, arches and geos where differential erosion has exploited fault lines or the weaker Gash breccia. The Devil's Cauldron peninsula is formed of a highly faulted mass of limestone projecting 200m from the cliffline. It has many caves, some with shafts extending to the surface to form blow-holes. The cauldron itself was formed by the collapse of former caves to form an enclosed shaft 45m deep and 55m in diameter. The cauldron is being extended by erosion along the fault line. Eventually it will collapse to form a series of stacks and arches. Caves and arches are also developed in the Gash breccia to the north of Flimston Castles.

5.4 Landforms of coastal deposition

Sediment ouputs from erosional landforms such as cliffs and shore platforms are transported downdrift by waves and currents. If they are not removed offshore they become inputs to depositional landforms. Additional sediment supply may be delivered by rivers and nearshore sandbanks (see section 4.4). The main depositional landform types are shoreline beaches, detached beaches (e.g. spits), sand-dunes and saltmarshes. Each landform acts as a dynamic sediment store within a sediment cell system (see Chapter 4, pp.69–70), and has a sediment budget, i.e. the net balance between sediment gain (input) and loss (output) over specified units of time.

Beaches

Beaches are the most widespread depositional landform. They are complex and varied stores of sediment which exhibit a wide range of minor landform features (Fig. 5.27).

21 Study Figure 5.28. Draw another copy of the diagram to show how human activity has changed the beach store at Barton on Sea. Show the changes by increasing or decreasing the size of your input/output arrows.

22 What will happen to the size of the beach store?

Beaches as sediment stores

Beaches are dynamic stores of sand and larger-sized sediment particles within the major sediment cells along the coastline. Fine sediment inputs of silt and clay do not settle in a beach environment and are quickly carried offshore or to areas of calmer water before being deposited. The crucial understandings we need for an individual beach are its sediment budget and the role that it plays as a sediment store within a major sediment subcell (see Chapter 4, p.69). The variables and processes involved in the calculation of this budget are given for Barton on Sea in Christchurch Bay, Hampshire (Fig. 5.28). Figure 4.16 on page 70 shows the main sediment sinks and movements within this subcell and how they affect the beach store.

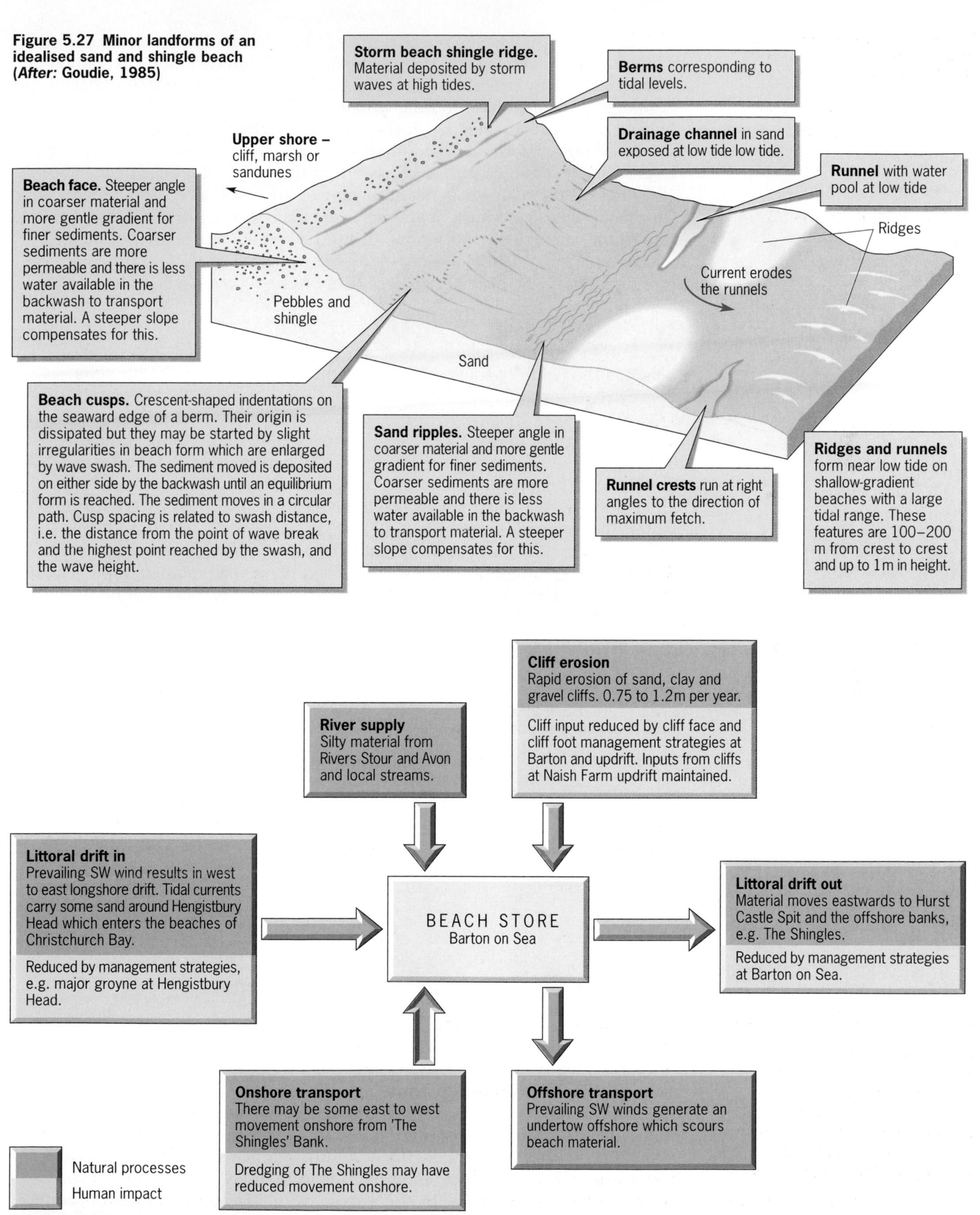

Storm beach shingle ridge. Material deposited by storm waves at high tides.
Berms corresponding to tidal levels.
Upper shore – cliff, marsh or sandunes
Drainage channel in sand exposed at low tide low tide.
Runnel with water pool at low tide
Beach face. Steeper angle in coarser material and more gentle gradient for finer sediments. Coarser sediments are more permeable and there is less water available in the backwash to transport material. A steeper slope compensates for this.
Ridges
Current erodes the runnels
Pebbles and shingle
Sand
Beach cusps. Crescent-shaped indentations on the seaward edge of a berm. Their origin is dissipated but they may be started by slight irregularities in beach form which are enlarged by wave swash. The sediment moved is deposited on either side by the backwash until an equilibrium form is reached. The sediment moves in a circular path. Cusp spacing is related to swash distance, i.e. the distance from the point of wave break and the highest point reached by the swash, and the wave height.
Sand ripples. Steeper angle in coarser material and more gentle gradient for finer sediments. Coarser sediments are more permeable and there is less water available in the backwash to transport material. A steeper slope compensates for this.
Runnel crests run at right angles to the direction of maximum fetch.
Ridges and runnels form near low tide on shallow-gradient beaches with a large tidal range. These features are 100–200 m from crest to crest and up to 1 m in height.
Cliff erosion
Rapid erosion of sand, clay and gravel cliffs. 0.75 to 1.2m per year.
Cliff input reduced by cliff face and cliff foot management strategies at Barton and updrift. Inputs from cliffs at Naish Farm updrift maintained.
River supply
Silty material from Rivers Stour and Avon and local streams.
Littoral drift in
Prevailing SW wind results in west to east longshore drift. Tidal currents carry some sand around Hengistbury Head which enters the beaches of Christchurch Bay.
Reduced by management strategies, e.g. major groyne at Hengistbury Head.
BEACH STORE
Barton on Sea
Littoral drift out
Material moves eastwards to Hurst Castle Spit and the offshore banks, e.g. The Shingles.
Reduced by management strategies at Barton on Sea.
Onshore transport
There may be some east to west movement onshore from 'The Shingles' Bank.
Dredging of The Shingles may have reduced movement onshore.
Offshore transport
Prevailing SW winds generate an undertow offshore which scours beach material.
Natural processes
Human impact

Figure 5.27 Minor landforms of an idealised sand and shingle beach (*After:* Goudie, 1985)

Figure 5.28 The beach as a sediment store – the example of Barton on Sea, Christchurch Bay, Hampshire

The UK coastline has many examples of beaches and associated landforms which contain volumes and types of sediment which cannot be explained completely by the processes operating today. Much of our beach material orginates from the offshore zone which was dry land during the Pleistocene. These areas were affected by glacial and periglacial processes which produced large amounts of sediment. This was moved landwards as the sea level rose (the Flandrian transgresssion) at the end of the Pleistocene about 10 000 BP. Many of our beach landforms are the result of coastal processes reworking this sediment. The implications for beach sediment budgets are important, since the greater inputs from these relict stores are not being maintained today. If the outputs increase as a result of natural processes or human activity, the beach store is unlikely to be replenished.

Beach morphology

The cross-section morphology (surface profile and internal structure) of a beach is the result of the interaction of three variables: wave character, energy rhythms and longshore drift.

Wave type and energy

The wave type and energy are key factors in determining the sediment which arrives and remains deposited on a beach. In a high-energy wave environment, finer sand particles will be easily transported by the turbulent water and only coarser pebbles will remain to form the beach. In lower-energy beach environments only sand-sized particles will be transported to the beach and remain in place as beach sediment. The relationship between wave type and beach form is important. A plunging breaker will be more erosive of material than a spilling breaker (refer back to Fig. 4.5). Thus we can make a general statement: a beach adopts a long-term profile which is in equilibrium with the prevailing energy conditions, and so two types of beach can be identified: dissipative and reflective (Table 5.1).

Table 5.1 Beach characteristics (*After:* Vines and Spencer, 1995)

Characteristics	Reflective beach	Dissipative beach
Breaker type	Plunging	Spilling
Wave breaker height (m)	<1	>2.5
Surf zone width (m)	<10	100–>1000
Angle of breaker approach to the shore	Strong oblique	Shore normal
Swash period (s)	5–10	Up to 60
Wave-driven currents	Shore parallel	Rip cells – circular
Rip currents	Absent or weak	Strong and persistent
Nearshore bars	Absent	Multiple
Beach slope (°)	>3°	<1°
Beach profile	Concave	Rectilinear
Common sediment size	Coarse sand and gravel	Silt to fine sand
Alongshore grading	Common	Rare
Sediment transport dominant directions	Alongshore	Onshore–offshore
Beach permeability	High	Low

Seasonality and the sweep zone

Although one type of wave may predominate, most beaches receive both plunging and spilling waves, and the beach profile constantly adjusts in the short term to the changing energy conditions. Since storm conditions are more common in the winter months in Britain, there is a tendency for a winter scouring of material by the erosive plunging breakers and a summer accretion of material by spilling breakers. This produces a variation in beach height and profile called a **sweep zone** (Fig. 5.29). For instance, the sweep

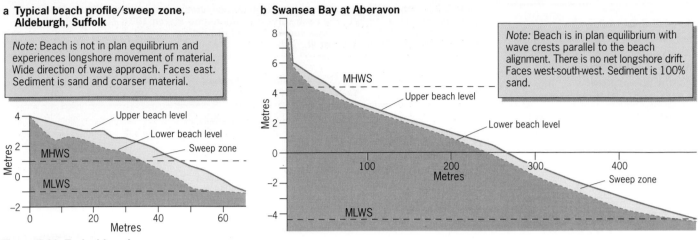

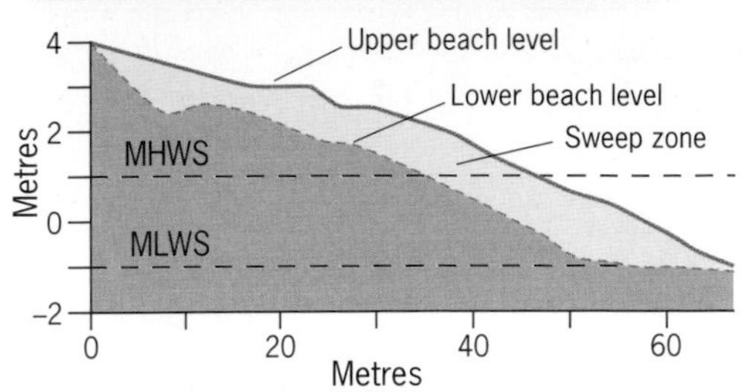
a Typical beach profile/sweep zone, Aldeburgh, Suffolk
Note: Beach is not in plan equilibrium and experiences longshore movement of material. Wide direction of wave approach. Faces east. Sediment is sand and coarser material.
Upper beach level
Lower beach level
Sweep zone
MHWS
MLWS
Metres
Metres

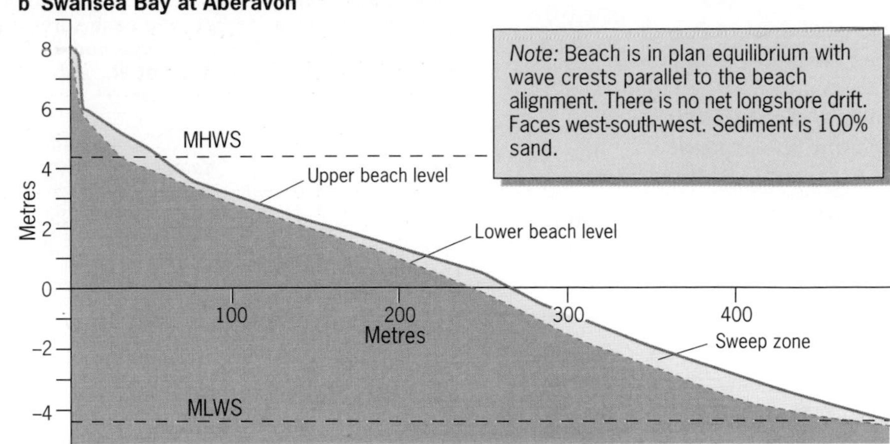
b Swansea Bay at Aberavon
Note: Beach is in plan equilibrium with wave crests parallel to the beach alignment. There is no net longshore drift. Faces west-south-west. Sediment is 100% sand.
MHWS
Upper beach level
Lower beach level
Sweep zone
MLWS
Metres
Metres

Figure 5.29 Typical beach profile/sweep zone (*Source:* Carr et al., 1982)

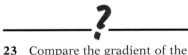

zone effect on the Holderness coastline results in a 2 m lower beach profile at the cliff foot during winter, allowing increased erosion rates as higher wave energy reaches the cliff foot.

Seasonal variations in beach profile have been identified from other parts of the world; for example, along the Kerala coast of south-west India the beach profiles are monsoon controlled, with cycles of erosion during the monsoon and accretion during non-monsoon months.

Some researchers suggest that the sweep zone concept is an over-simplified view of many beaches. It is the change in wave energy events and the degree of variability between the seasons which is the key (Tables 5.2 and 5.3). On the Forcados Beach, Nigeria, there is a net loss of material in the wet season and a net gain during the dry season, but erosion and deposition occur throughout the year. The frequency and magnitude of change was greatest in the wet season. In the high-energy wave environment

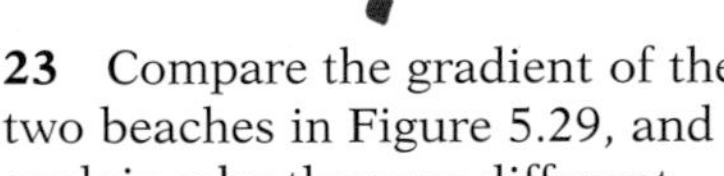

23 Compare the gradient of the two beaches in Figure 5.29, and explain why they are different.

24 Describe the changes in profile resulting from the sweep zone processes.

25 Study Tables 5.2 and 5.3.
a Complete the monthly change and cumulative change data for Aldeburgh to Southwold beaches.
b Plot the monthly change and cumulative change data for the two areas on to separate graphs. Use your graphs to answer the following, giving precise evidence for your answers:
i Are beach profiles eroded in winter (October–April) and accreted in summer (May–September)?
ii Are profile changes most pronounced in the winter months?
iii What happened to overall beach volume over the study periods?

Table 5.2 Swansea Bay: volume changes for the upper and lower beach with zero OD taken as MW (mid-water) level. Net monthly and cumulative volumes (m^3) are shown for the period from October 1975 onwards. (All 11 profiles are included; each is regarded as 1 m wide.) Sections extend down to low-water mark at time of survey. (*After:* Carr et al., 1982)

Date	Year	Above MW (upper beach) (m^3)	MW to LW (lower beach) (m^3)	Monthly change (m^3)	Cumulative change (m^3)
October–November	1975	35	119	154	154
November–December		74	3	77	231
December–January		–191	–82	–273	–42
January–February	1976	60	83	143	101
February–March		–94	–57	–151	–50
March–April		114	–30	84	34
April–May		–13	–125	–138	–104
May–June		157	35	192	88
June–July		–11	25	14	102
July–August		67	–16	51	153
August–September		2	–49	–47	106
September–October		–49	–56	–105	1
October–November		58	140	198	199
November–December		–256	–6	–266	–67
December–January	1977	89	9	98	31
January–February		–153	–55	–208	–177
February–March		–1	–23	–24	–201
March–April		–122	202	80	–121
April–May		107	–61	46	–75

26 Use Table 5.2 and appropriate techniques to investigate whether the upper and lower beach at Swansea Bay respond in the same way to erosion and accretion events.

27 What long-term changes in beach profiles at the two locations might you expect as a result of longshore drift? Explain your answer.

28 Which beach profile is most reflective and which is most dissipative of wave energy?

Table 5.3 Aldeburgh–Southwold: volumes of accretion/erosion, resultant and cumulative volumes (m^3) recorded between each survey for the period March 1978–May 1979. (All 10 profiles are included: each is regarded as 1 m wide.) Sections extend down to low-water mark at time of survey. (*After:* Carr et al., 1982)

Date	Year	Accretion+ (m^3)	Erosion— (m^3)	Monthly change (m^3)	Cumulative change (m^3)
March–April	1978	60.4	68.9	−8.5	−8.5
April–May		63.2	57.8	+5.4	−3.1
May–June		43.9	46.6		
June–July		52.7	23.5		
July–August		33.5	81.1		
August–September		71.4	26.1		
September–October		47.3	64.0		
October–November		51.8	122.2		
November–January	1979	106.5	82.0		
January–February		155.0	115.3		
February–March		81.3	76.3		
March–April		53.4	42.8		
April–May		24.9	56.0		

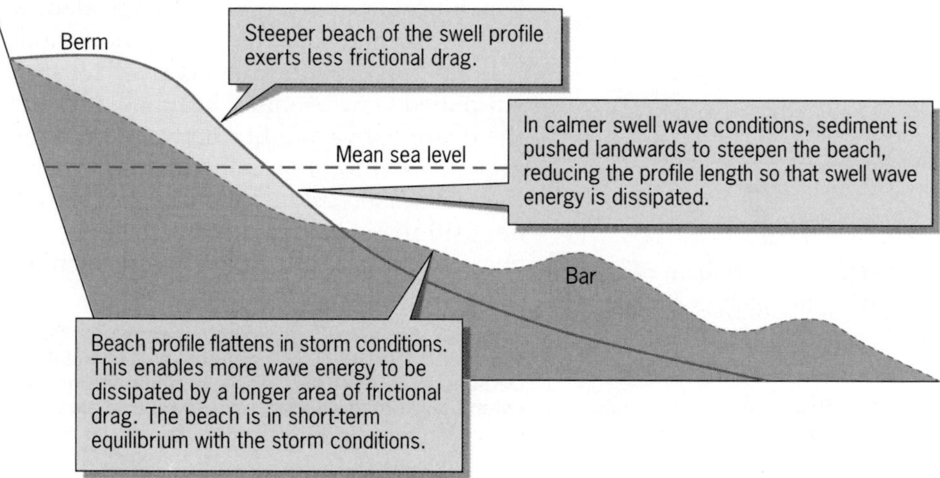

Figure 5.30 Short-term changes in beach profile – the steep beach profile characteristics of swell waves contrasted with the shallow profile of storm waves (*Source:* MAFF, 1993)

of the south coast of New South Wales, Australia, storm waves rapidly erode beach material to the offshore zone to form surf zone bars. The material is then moved gradually onshore during calm conditions.

The beach is the area over which wave energy is dissipated. The longer the beach profile, the more frictional drag is exerted and thus there is a greater dissipation of wave energy (Fig. 5.30). Short-term changes in beach profile occur in response to changing wave energy conditions, so that wave energy is more efficiently dissipated. There are onshore and offshore sediment movements with these short-term changes. However, the sediment remains in the nearshore zone and is not likely to be lost in the long term from the beach system.

Longshore inputs and outputs

In areas of oblique wave approach, longshore movement of sediment will cause beaches to be built up by net deposition or eroded by net outward movement of sediment, depending upon the beach location in the sediment cell. For example, in the Start Bay barrier beach, there is evidence of longer-term change in beach profiles as a result of longshore sediment redistribution (see Case Study, pp.96–99).

29 Describe an appropriate sampling technique which could have been employed at the 15 sediment sampling sites to obtain the shingle samples from which the sediment characteristics A, B and C in Table 5.4 were derived.

30a On three separate sheets of 2 mm graph paper, construct scatter graphs from Table 5.4 showing how the sediment characteristics vary with *distance* along the beach. Sketch in the best-fit trend lines for each of the graphical plots.
b Comment on the apparent trends, indicating how far they might be expected, and suggest reasons for any anomalies.

Some beaches show a grading of sediment size along a beach (Table 5.4). Although finer material may be moved offshore, coarse sand and shingle may show longshore grading. Assuming that wave energy inputs are equal along the beach, the finer shingle particles are likely to be carried further by longshore currents and become increasingly sorted and rounded as they move. Sand-calibre particles will be carried even further along the beaches. In reality, beaches show more complex patterns because of variations in wave energy and new sediment inputs, for example from cliff falls or river mouths.

Table 5.4 Sediment characteristics.
A large shingle storm beach occurs on the coast of Somerset aligned SW to NE (see Fig. 5.31). The main ridge increases in height from SW to NE. The shingle appears to decrease in size and become better rounded from SW to NE, suggesting a net transport in this direction. The following data were collected to ascertain the degree to which sediment characteristics reflect this expected direction of transport.

Distance (in metres from SW limit of beach)	Shingle line sampling site (see map)	Height of storm beach (metres above low-water mark)	Mean shingle diameter (Geometric mean size in cm) (A)	Sorting index (B)	Roundness index (C)
60	1	5.5	8.4	1.72	18
180	2	7.5	8.5	1.77	20
300	3	8.0	7.7	1.63	26
420	4	11.5	8.1	1.48	38
540	5	11.0	6.1	1.58	23
660	6	7.5	5.8	1.63	15
780	7	10.0	6.2	1.60	21
900	8	10.5	7.2	1.41	30
1020	9	10.0	7.5	1.36	27
1120	10	11.0	6.2	1.33	35
1220	11	14.0	6.5	1.35	44
1340	12	12.5	5.8	1.23	41
1460	13	18.0	4.8	1.18	62
1580	14	13.5	5.0	1.17	71
1700	15	15.0	5.8	1.25	65

A Mean shingle diameter: a simplified measure of average particle diameter.

B Sorting index: perfect sorting (i.e. all stones of the same volume) would give an index of 1.0: hence, the greater the index, the poorer the sorting.

C Roundness index: percentage of particles classified as rounded.

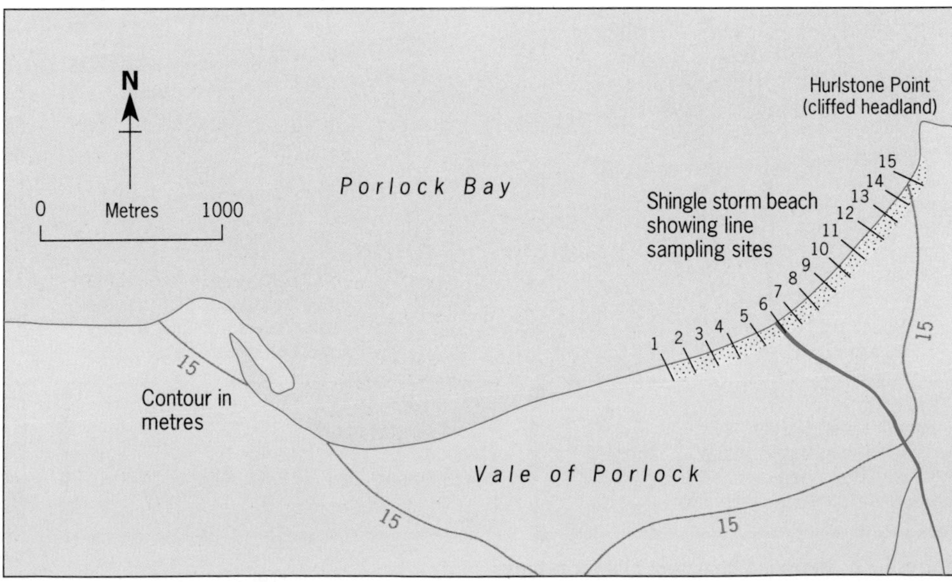

Figure 5.31 A section of the Somerset coast

Sediment type and size

Local conditions will determine the type (size, hardness and shape) of the sediments forming the beach: for example, the nature of the rock forming the cliffs and sediment inputs from river mouths. The size (calibre) of the sediment particles controls the slope of the beach. Shingle beaches have a steep profile with slope angles of up to 30°, whereas sandy beaches slope at about 1°. This is due to the angle of rest of these materials, and their permeability. Coarse sediments have a higher permeability than sandy sediments. As the swash moves up a shingle beach there is a large loss of water by percolation, reducing the effectiveness of the backwash. There is less energy to erode the coarse sediment, and the swash is relatively more important than the backwash, allowing the beach to be built up landwards. Fine-grained sandy beaches have lower rates of percolation and more powerful backwash, resulting in the removal of sediment and a decrease in beach gradient.

Beaches consisting of pebbles and sand often show cross-shore grading with finer materials towards the sea. Again, the reduced energy of the backwash is the key. Larger-sized sediment grains will settle before finer grains. The very coarse fragments forming the **storm beach** (Fig. 5.27) are thrown into position by very high-energy waves, and under normal conditions there is insufficient energy to move them.

?

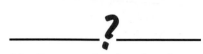

31 Compare the three beaches in Figure 5.32 in terms of • angle of the beach to the direction of dominant wave approach;
• sediment inputs and outputs;
• sediment movements alongshore and onshore/offshore.

32 For each beach type in Figure 5.32 suggest:
a What will happen during storm conditions.
b What will happen with wave crests approaching from a different direction.

The planform of beaches

The interactions between the amount of longshore movement of sediment, the prevailing wave conditions and sediment supply allow us to identify three main types of shoreline beaches in terms of their planform (Fig. 5.32).

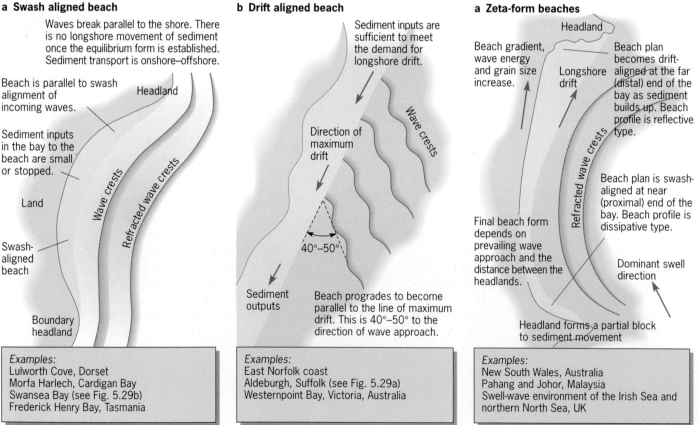

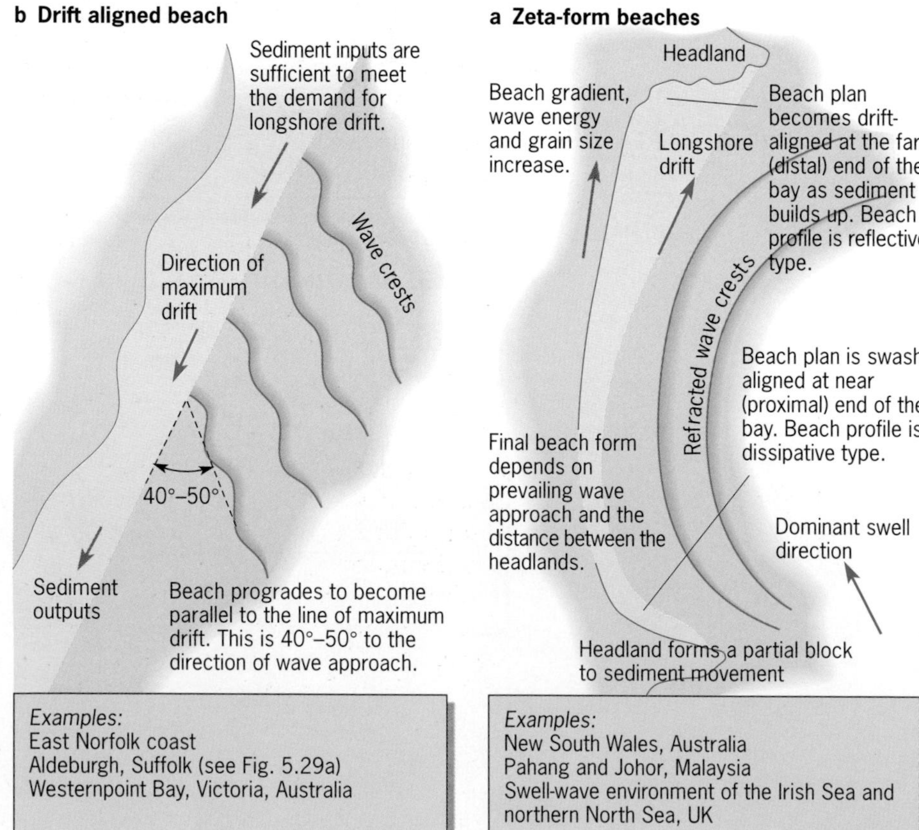

a Swash aligned beach
Waves break parallel to the shore. There is no longshore movement of sediment once the equilibrium form is established. Sediment transport is onshore–offshore.
Beach is parallel to swash alignment of incoming waves.
Headland
Sediment inputs in the bay to the beach are small or stopped.

Refracted wave crests
Swash-aligned beach
Boundary headland
Examples:
Lulworth Cove, Dorset
Morfa Harlech, Cardigan Bay
Swansea Bay (see Fig. 5.29b)
Frederick Henry Bay, Tasmania
b Drift aligned beach
Sediment inputs are sufficient to meet the demand for longshore drift.

Direction of maximum drift
40°–50°
Sediment outputs
Beach progrades to become parallel to the line of maximum drift. This is 40°–50° to the direction of wave approach.
Examples:
East Norfolk coast
Aldeburgh, Suffolk (see Fig. 5.29a)
Westernpoint Bay, Victoria, Australia
a Zeta-form beaches
Headland
Beach gradient, wave energy and grain size increase.
Longshore drift
Beach plan becomes drift-aligned at the far (distal) end of the bay as sediment builds up. Beach profile is reflective type.
Refracted wave crests
Beach plan is swash-aligned at near (proximal) end of the bay. Beach profile is dissipative type.
Final beach form depends on prevailing wave approach and the distance between the headlands.
Dominant swell direction
Headland forms a partial block to sediment movement
Examples:
New South Wales, Australia
Pahang and Johor, Malaysia
Swell-wave environment of the Irish Sea and northern North Sea, UK

Figure 5.32 Shoreline beaches: planform types

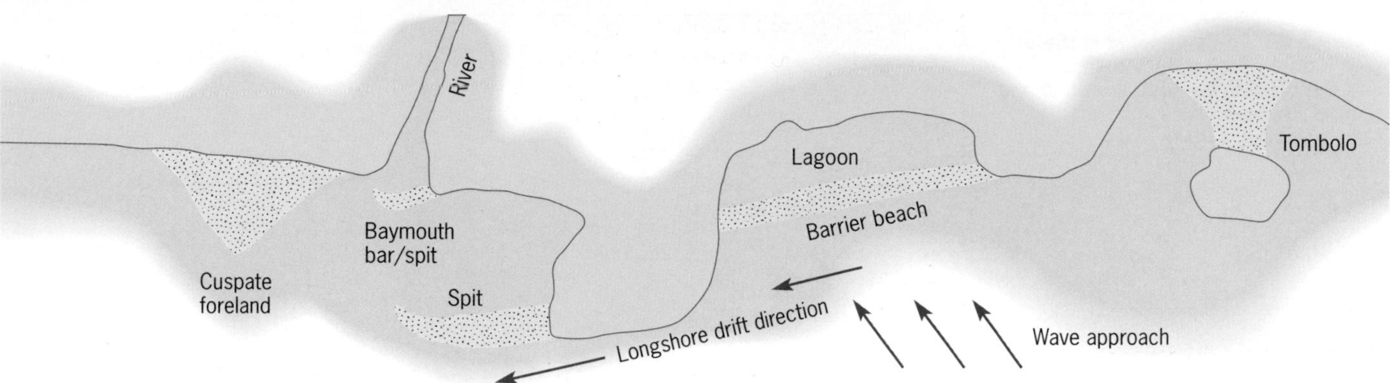

Figure 5.33 Diagrammatic representation of the main forms of detached beaches (the scale of these features varies greatly)

Detached beaches

Detached beaches do not always follow the direction of the coastline. There are a variety of types (Fig. 5.33) which vary enormously in scale and local detail. These beaches are shaped by the same processes which control shoreline beaches. However, high-energy events can break or breach parts of the detached beach, resulting in changes in morphology and therefore sediment movements until a new dynamic equilibrium is established. These cycles of breaching, rebuilding or reorientation are part of the natural evolution of these dynamic landforms. For example, Spurn Head spit (Fig. 5.3) has moved westwards since records began in Roman times by marine erosion of the east side and deposition on the west by wind and waves flowing over the spit. During high-energy storm events, the spit may be breached and then reform further to the west. Such a high-energy event occurred in February 1996 when a combination of gale-force north-easterly winds and unusually high spring tides almost completely breached the spit. About 250 m of the concrete causeway running down the spit were washed away, isolating the families who live there and crew the lifeboat station. A flexible roadway is to be installed so that it can be moved with the spit.

Spits

Spits and associated features form as a result of wave energy and longshore drift in areas where there is a change in the orientation of the coastline. Wave energy transports sediment along the shore by drifting and a spit builds out as a detached beach from the point where the coastline changes direction. In the UK they are mainly found in areas with a low tidal range (less than 3 m) since these areas are dominated by wave rather than tidal processes (Fig. 5.34).

Many spits are the 'end' depositonal feature in a sediment cell or subcell – for example, the cliffs and beaches of Christchurch Bay feed Hurst Castle spit in Hampshire, in East Anglia the Yarmouth beaches feed Orford Ness, and in east Yorkshire the Holderness cliffs and beaches feed Spurn Head. However, some spits are more complex than this. For example, at Christchurch Harbour to the west of Christchurch Bay (Fig. 4.15) there appear to be two spits extending into the bay, suggesting longshore drift in opposite directions either side of the bay. It is likely, however, that the more northern spit is in fact a remnant of a larger spit which once extended across Christchurch Harbour and was breached by a storm. The final form of a spit will reflect the balance between sediment supply, offshore gradients, tidal and river currents. The spit will extend outwards until the sediment supply is cut off or river and tidal currents become strong enough to prevent further deposition.

Spits extending across estuaries are also called *baymouth bars,* e.g. Dawlish Warren (Fig. 5.34). Some spits have developed sufficiently to divert the course of rivers reaching the sea, e.g. the River Ore has been diverted

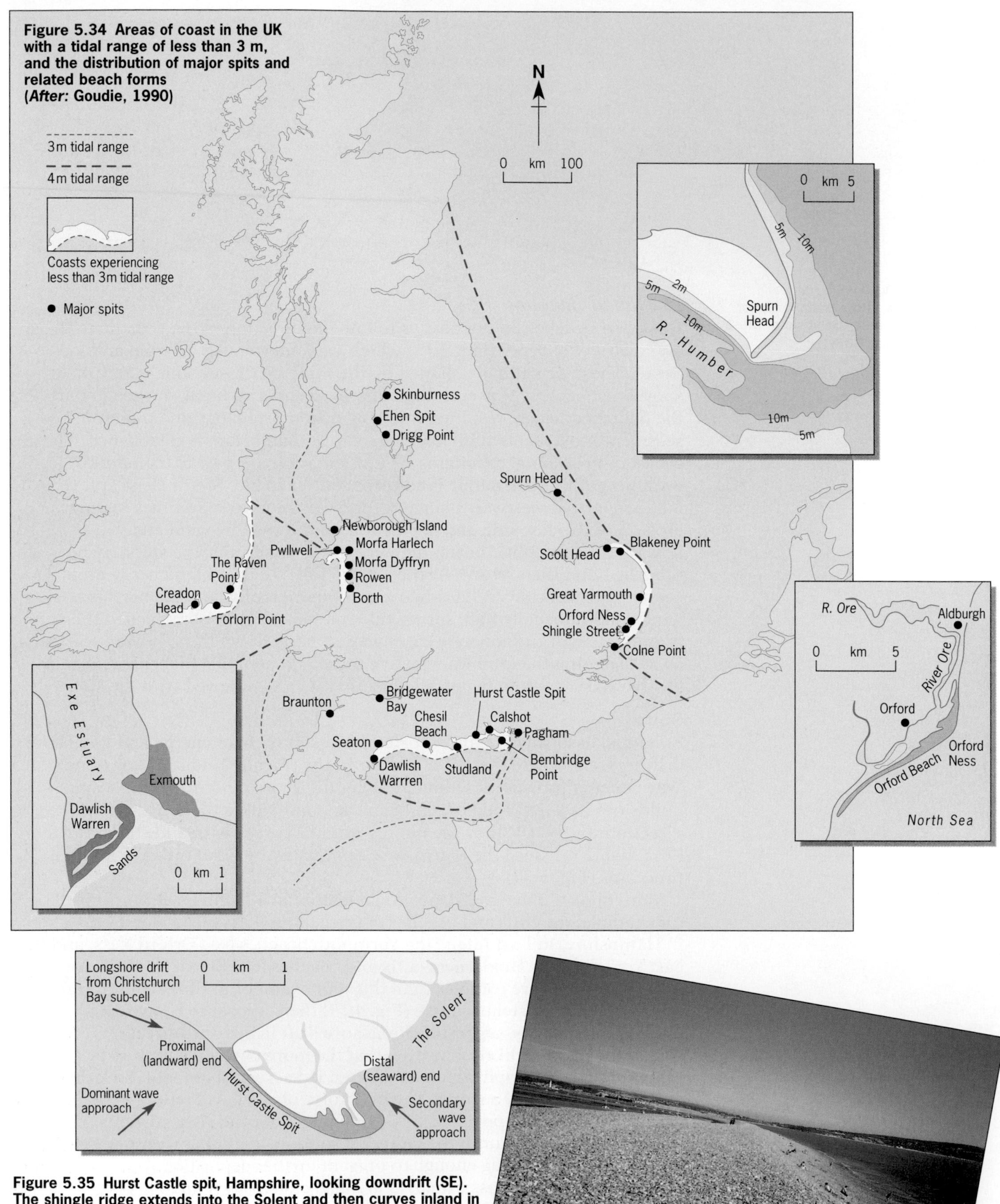

Figure 5.34 Areas of coast in the UK with a tidal range of less than 3 m, and the distribution of major spits and related beach forms (*After:* Goudie, 1990)

Figure 5.35 Hurst Castle spit, Hampshire, looking downdrift (SE). The shingle ridge extends into the Solent and then curves inland in the distance. The ridge is artificially raised and widened at this point to reduce the risk of overwashing and breaching.

almost 15 km to the south by the growing spit of Orford Ness (Fig. 5.34). More complex spits have recurves at their *distal* or seaward end, e.g. Hurst Castle spit (Fig. 5.35). These are thought to be due to longshore drift movements resulting from secondary wave directions and wave refraction around the distal end of the spit. Spits are composed of shingle and sand, although shingle is most common (Fig. 5.35) in the high-energy wave environments where spits occur. Much of the shingle forming the spits of the UK originates from the Flandrian transgression (see Chapter 7). For example, Blakeney Point in Norfolk is 97 per cent flint pebbles which could not be supplied by present-day erosion processes in the area. In addition, there is only a short distance for material to move by longshore drift to form the spit within this area. The main source of sediment supply to these beaches has therefore been removed and many are not actively being formed unless there is a suitable supply from cliff erosion or river sediments.

Barrier bars and beaches

Where a beach extends across an indentation of the coastline to join two headlands, the landform produced is called a *barrier beach*, e.g. Looe Bar in Cornwall and Start Bay, Devon (see the Case Study on p.96). If the beach is separated from the mainland, the feature is called a *barrier island*, e.g. Scolt Head Island, Norfolk. Barrier islands are very variable in scale and form, and are usually sand or shingle features backed by a landward wetland area or lagoon. They are common in areas with low tidal ranges, dominated by swell waves where the offshore coastline is gently sloping over wide continental shelves. Large-scale features are found worldwide – for example, the Netherlands coast, in North America, e.g. the South Texas coast and the barrier islands of the Carolinas, and the Lagos coastline of Nigeria. The origin of these features has been disputed. Some researchers consider them to be spits which have been broken through by storm waves and become detached from the mainland; others believe that they are sediment accumulations moved onshore by the Flandrian transgression. The third hypothesis is that barrier islands develop as a result of the drowning of sand-dune or large beach **berm** features. The explanation is likely to vary with different barrier islands.

Tombolos

A tombolo is a beach formed between an island and the mainland. Wave refraction around the island results in a wave-energy shadow where increased deposition occurs building up the tombolo. These may be covered by high tides, e.g. Lindisfarne in Northumberland and St Agnes in the Scilly Isles. The famous Chesil Beach in Dorset has a tombolo form since it links the mainland with the Isle of Portland. However, its origin is more complex. The feature probably developed as an offshore barrier beach which moved onshore with the Flandrian transgression. The Fleet lagoon developed between the barrier and the mainland.

Cuspate forelands

Cuspate forelands are triangular-shaped beach forms varying in scale from small (500 m^2) sand waves formed by wave currents, to the Capes of the Carolinas, USA, coast with each feature being 150–200 km^2 in size. Some researchers suggest that these may form in a similar way to tombolos, except that the sand and shingle accumulates in the wave shadow of submerged shoals rather than islands. However, in many examples there are no offshore shoals to support this hypothesis. There is little doubt that cuspate forelands result from the complex interaction of a number of variables. This complexity is illustrated by the history of the UK's largest example, Dungeness (Fig. 5.36).

?

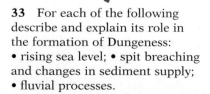

33 For each of the following describe and explain its role in the formation of Dungeness:
• rising sea level; • spit breaching and changes in sediment supply;
• fluvial processes.

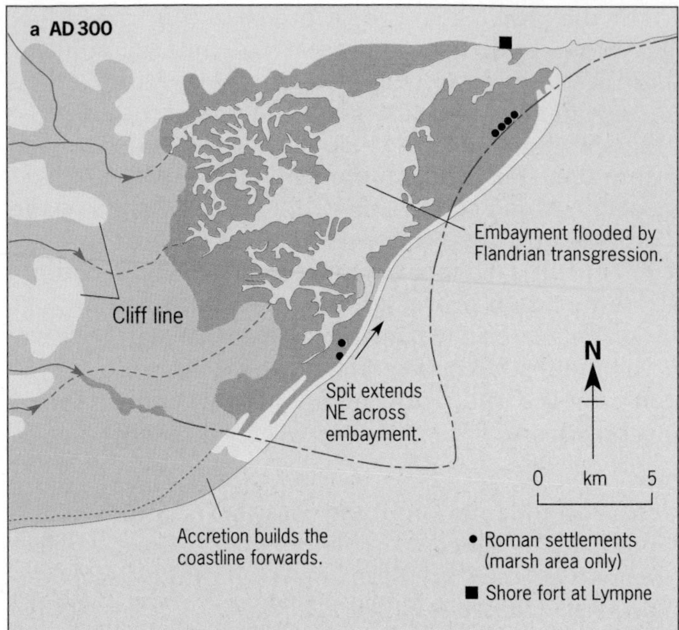

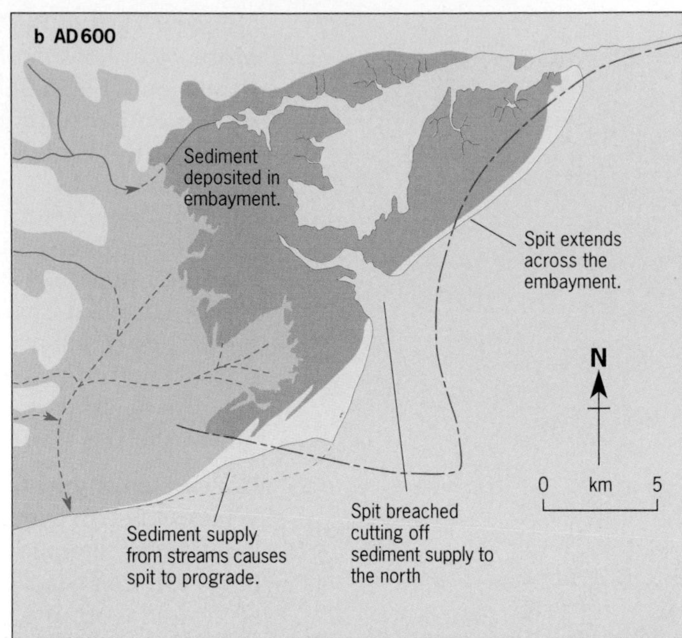

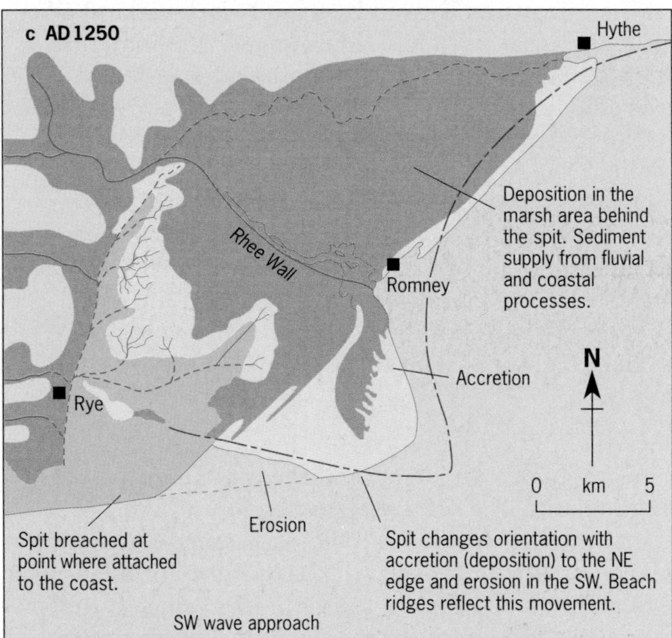

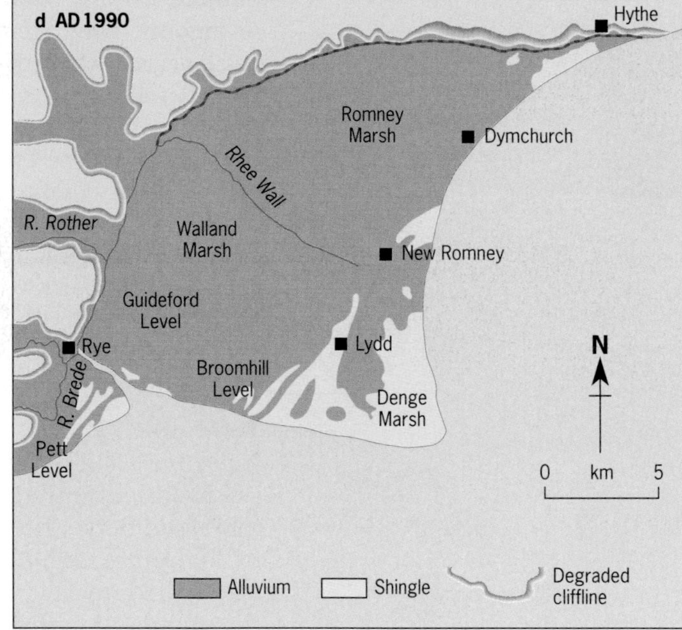

Figure 5.36 Stages in the evolution of the great cuspate foreland of Dungeness (the Romney Marsh region, Kent) since Roman times (*After:* Goudie, 1990)

The Start Bay barrier beach system

The Start Bay barrier beach system in the Slapton region of South Devon (Fig. 5.37) shows clearly how detached barrier beaches develop, the present-day processes acting upon the area, and how human activity interacts with the natural processes. Start Bay is a 60 km^2 zeta-curved bay with four main areas of deposition of different types of sediment (Fig. 5.37). The barrier beach is 9 km long, with cliffs and shore platforms at Pilchard Cove to the north and South Hallsands to the south. The barrier beach is divided into three subsystems by headlands formed by river valley spurs at Limpet Rocks and Tinsey Head.

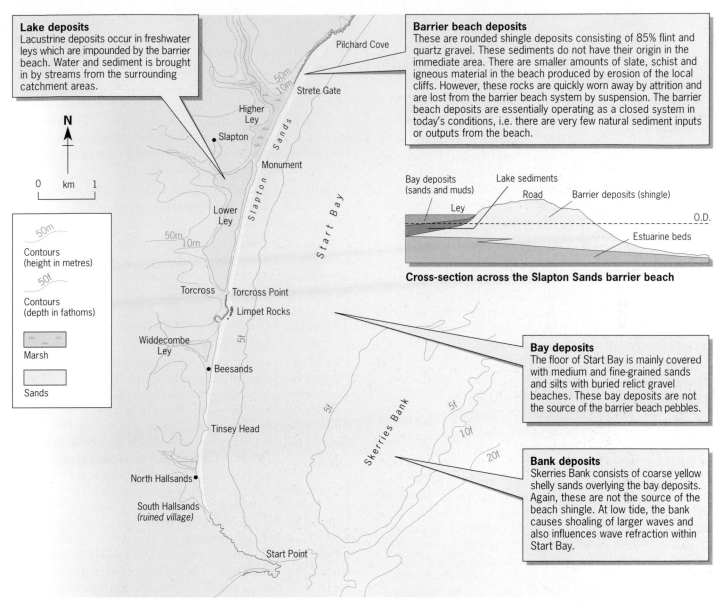

Lake deposits
Lacustrine deposits occur in freshwater leys which are impounded by the barrier beach. Water and sediment is brought in by streams from the surrounding catchment areas.

Barrier beach deposits
These are rounded shingle deposits consisting of 85% flint and quartz gravel. These sediments do not have their origin in the immediate area. There are smaller amounts of slate, schist and igneous material in the beach produced by erosion of the local cliffs. However, these rocks are quickly worn away by attrition and are lost from the barrier beach system by suspension. The barrier beach deposits are essentially operating as a closed system in today's conditions, i.e. there are very few natural sediment inputs or outputs from the beach.

Cross-section across the Slapton Sands barrier beach

Bay deposits
The floor of Start Bay is mainly covered with medium and fine-grained sands and silts with buried relict gravel beaches. These bay deposits are not the source of the barrier beach pebbles.

Bank deposits
Skerries Bank consists of coarse yellow shelly sands overlying the bay deposits. Again, these are not the source of the beach shingle. At low tide, the bank causes shoaling of larger waves and also influences wave refraction within Start Bay.

Figure 5.37 Start Bay and the types of sedimentary deposits

History of the coastline

Research suggests a four-phase evolution of the present-day coastline (Fig. 5.38).

1 During an interglacial, the sea level was about 7 m above present-day levels, with a shoreline along the western part of the present bay. This is marked by a relict cliff and raised interglacial shore platform along the landward edge of Slapton Ley.

2 During the Devensian glacial (see Chapter 7), sea level was lower than at present and the coastline was several kilometres to the east and south. Evidence for this is provided by river channels extending into Start Bay on the present-day seabed.

Figure 5.38
The Start Bay barrier beach

Start Bay

This lower sea level exposed flint-rich gravels, sandstones and pebble beds which provided the material for the present barrier beaches. Much of the flint in the barrier beach may have come from 40 km offshore.

3 As the ice sheets melted, sea level rose and the coastline moved towards its present position. As the sea advanced, from 10 000 BP, the flint-rich pebbles moved landwards in a series of barrier islands on top of estuary and lagoon sediments.

4 The rising sea level drowned the lower valleys of the rivers to form wide sea inlets called *rias*. The barrier beach was in its present position by 2800 BP, and the present system is essentially a relict, closed system. The barrier beach impounded the river water to form the present-day lakes or leys. These are freshwater but there is mixing with sea water.

The barrier beach system

The three barrier beach subsystems are in equilibrium with present-day prevailing wave conditions, but in the short term they are very dynamic in response to daily or seasonal wave conditions. The cliffed headlands which separate the subsystems are river valley spurs which have been cut back by marine erosion. Sediment moves between the three subsystems in both a northerly and southerly direction, depending upon wind and wave direction. In the past, this north and south longshore drift must have been in balance, but in recent decades wave activity is producing a dominantly northwards movement of material. In the north the barrier is higher and wider, and there is a net loss of material in the south; at Torcross, for example, the low-water beach width has declined from 100 m in the 1930s to 50 m in the 1980s.

The barrier beach profile shows changes, as evidenced by ten sites surveyed in the 1980s. The changes in beach morphology observed cannot be accounted for by longshore movements alone. All the sites showed a seasonal net gain or loss of material with the nearshore zone, balancing in the longer term. The beach dynamics can be explained in terms of four distinct wave environments.

1 Low wave height (< 0.3 m) and surging breakers generated by light or offshore winds. These act constructively to build up the beach producing high shingle levels, a steep intertidal profile, well-developed berms, well-sorted shingle surface, and a net onshore gain of beach material.

2 Plunging breakers and net offshore transport produced by moderate easterly winds. This lowers the intertidal beach, flattens the profile and destroys the berms.

3 High (4.5 m), steep, plunging breakers oblique to the shoreline produced by south and south-easterly gales. All beach profiles are lowered and flattened and there are large amounts of sediment transport to the nearshore zone. Shingle levels at Torcross dropped by 3 m under these conditions.

4 Spilling breakers produced by north-east to easterly gales which result in large amounts of beach accretion or deposition.

Thus, over the short timescale of individual events or cycles, transport between the beach and the nearshore zone becomes more important than longshore transport.

The impact of human activity upon natural processes and landforms

From the sixteenth century onwards the fishing villages of Hallsands, Beesands and Torcross developed along the coastline. Until the twentieth century, these settlements seem to have withstood major storms without much damage, but in 1917 South Hallsands was virtually destroyed by a storm. The village is built on the raised interglacial shore platform which before 1900 had the natural protection of the barrier beach. However, between 1894 and 1903, the shingle from the beach was lowered by 1.4 m. This may have been partly due to the increased northward drift of material observed this century, but a major factor must be the dredging of shingle between 1896 and 1901 to build the dockyards at Plymouth. Before the dredging, the shingle beach of the southern end of the barrier dissipated much of the wave energy. Despite the building of a sea wall, the 10 m waves of the 1917 storm overtopped the wall, attacked the shore platform and destroyed the village. The exposure of the shore platform has reflected wave energy and caused a further lowering of the shingle by 1.5 m by 1986. Today, the Hallsands Hotel and Mildmay Cottages are threatened by the increased cliff erosion resulting from the loss of beach material.

Beesands and Torcross are at the southern end of the sub-barrier systems, and are built on the barrier itself. The accelerated northward drift of beach material and the dredging mean that these villages, plus North Hallsands, are increasingly threatened. As the barrier system is essentially a relict feature and operates as a closed system, there is no natural source of new shingle material to replenish and nourish the barrier system. This has made the impacts of the Hallsands dredging particularly serious.

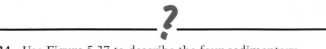

After storm damage in January 1979, the villagers campaigned for sea defences. In 1980 a concrete wave reflection wall fronted by a rock and concrete revetment was built at Torcross. At Beesands a rip-rap defence of limestone boulders placed on the beach surface was used as a defence. This had not taken into account the northward movement of beach material described above, and was quickly undermined by the erosion of the beach material by southerly gales. In 1992, a concrete wave return wall with steel piling fronted by rock revetment of Scandinavian gneiss replaced them.

Human activity has also affected the leys. The catchment area of the streams is mainly grassland used for livestock. Increased livestock numbers and grazing density have resulted in increased sediment yields from the catchments and increased rates of sedimentation in the leys: for example, in the Lower Ley the current rate of sediment influx is 8–12 mm per year compared with less than 2 mm before 1945.

Since the 1960s the leys have been increasingly prone to algal blooms due to the increasing export of nutrients from the catchments, presumably as a result of increased use of agricultural fertilisers.

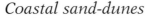

34 Use Figure 5.37 to describe the four sedimentary deposits in Start Bay.

35 Explain why the barrier beach operates as a closed system today.

36 Draw an annotated sketch map to describe and explain the short- and long-term changes in the barrier beach sediments and beach profile. Include details of different wind and wave characteristics and the effects they have.

37 What changes to the natural processes have been triggered by human activity, and why have control responses so far been unsuccessful?

Coastal sand-dunes

Sand-dunes are widespread depositional landforms which develop where there are strong onshore winds, and a low-gradient nearshore slope which provides large expanses of sand which dries out at low tide. These conditions occur on sandy spits at the mouths of estuaries, in bays, on indented coastlines, **prograding** coastlines with cuspate forelands, and on offshore islands and barrier beaches.

Sand moves from the intertidal area inland by the process of saltation (see section 9.3). The amount of sand moved will depend upon the wind velocity, the sand grain size and shape, and the length of time during which the sand surface is dry. The saltation process may be interrupted above the high-tide mark by obstacles such as seaweed, or even an empty drinks can! Sand is deposited in the lee of the obstacle. This deposition would go no further if it were not for the role of vegetation. Colonising plants such as sand twitch enable the deposition to continue by holding the dune together and building up as more and more sand is deposited. Eventually these small embryo dunes become larger and join up to form a sand-dune ridge.

Figure 5.39 Marram grass growing on a sand-dune

The key species in the next stage of development is marram grass (*Ammophila* spp). The marram grass helps to trap saltating sand and grows upwards with the dune (Fig. 5.39). A typical area of sand-dunes will show distinct changes in the landforms inland (Fig. 5.40). These changes result from a reduced supply of sand inland and increased shelter from the wind. The soil and vegetation show changes as well. The marram grass is not adapted to these more stable dunes and is replaced by lichens, mosses, grasses and shrubs. The amount of organic matter and the moisture content increase and the pH decreases as the beach shells are leached away. This inland change also shows how each dune ridge changes over time.

The most seaward dune ridges are the most mobile and unstable, and are easily modified by storm winds, very high tides and overwashing, or breaks in the marram cover. Once an area of dune is exposed to the wind, a large depression called a *blow-out* or deflation hollow forms in the dune crest

Oldest dune ridge

U-shaped or parabolic dunes

Dune slack

Dune slack

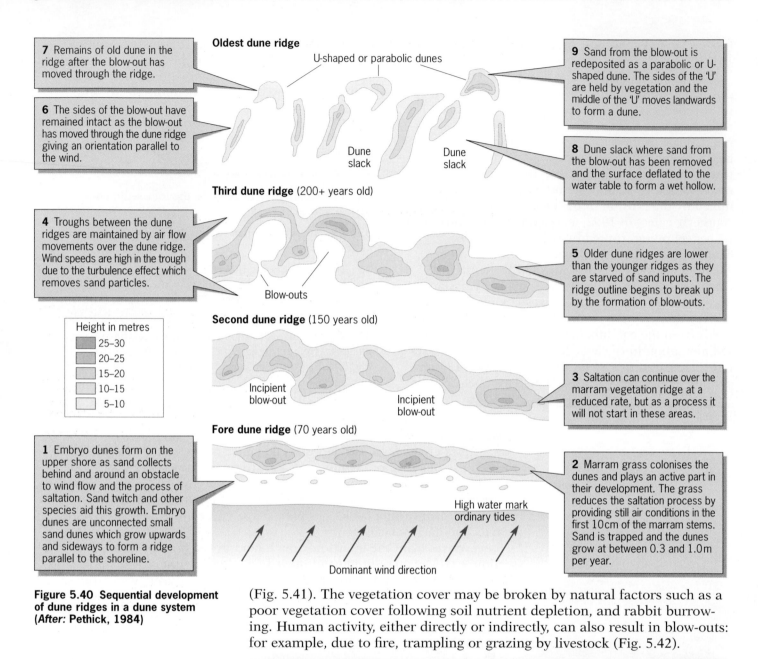

7 Remains of old dune in the ridge after the blow-out has moved through the ridge.

6 The sides of the blow-out have remained intact as the blow-out has moved through the dune ridge giving an orientation parallel to the wind.

9 Sand from the blow-out is redeposited as a parabolic or U-shaped dune. The sides of the 'U' are held by vegetation and the middle of the 'U' moves landwards to form a dune.

8 Dune slack where sand from the blow-out has been removed and the surface deflated to the water table to form a wet hollow.

Third dune ridge (200+ years old)

Blow-outs

4 Troughs between the dune ridges are maintained by air flow movements over the dune ridge. Wind speeds are high in the trough due to the turbulence effect which removes sand particles.

5 Older dune ridges are lower than the younger ridges as they are starved of sand inputs. The ridge outline begins to break up by the formation of blow-outs.

Second dune ridge (150 years old)

Incipient blow-out

Incipient blow-out

Height in metres
25–30
20–25
15–20
10–15
5–10

3 Saltation can continue over the marram vegetation ridge at a reduced rate, but as a process it will not start in these areas.

Fore dune ridge (70 years old)

High water mark ordinary tides

Dominant wind direction

1 Embryo dunes form on the upper shore as sand collects behind and around an obstacle to wind flow and the process of saltation. Sand twitch and other species aid this growth. Embryo dunes are unconnected small sand dunes which grow upwards and sideways to form a ridge parallel to the shoreline.

2 Marram grass colonises the dunes and plays an active part in their development. The grass reduces the saltation process by providing still air conditions in the first 10 cm of the marram stems. Sand is trapped and the dunes grow at between 0.3 and 1.0 m per year.

Figure 5.40 Sequential development of dune ridges in a dune system (*After:* Pethick, 1984)

(Fig. 5.41). The vegetation cover may be broken by natural factors such as a poor vegetation cover following soil nutrient depletion, and rabbit burrowing. Human activity, either directly or indirectly, can also result in blow-outs: for example, due to fire, trampling or grazing by livestock (Fig. 5.42).

?

38 Draw a cross-section of Figure 5.41. Add notes on changes in dune form, vegetation and wind.

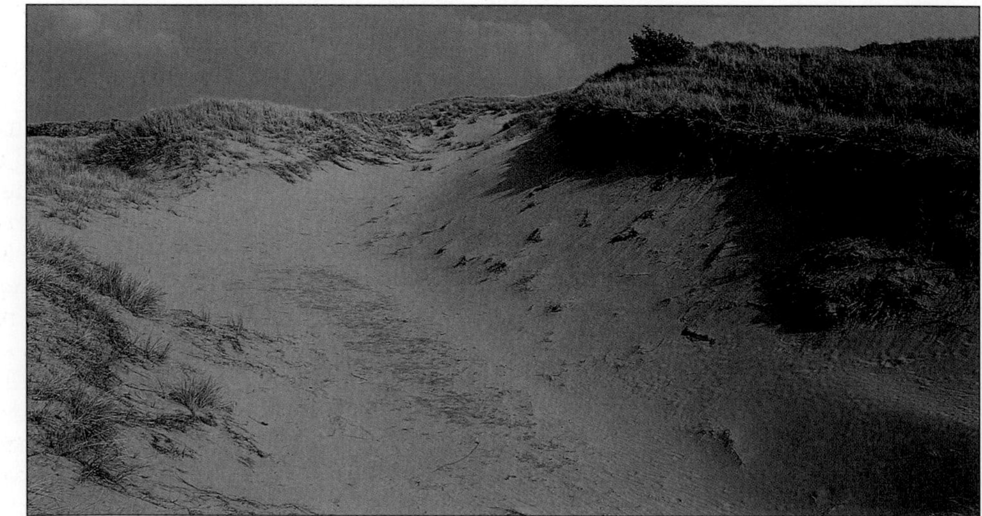

Figure 5.41 Deflation hollows in a sand-dune

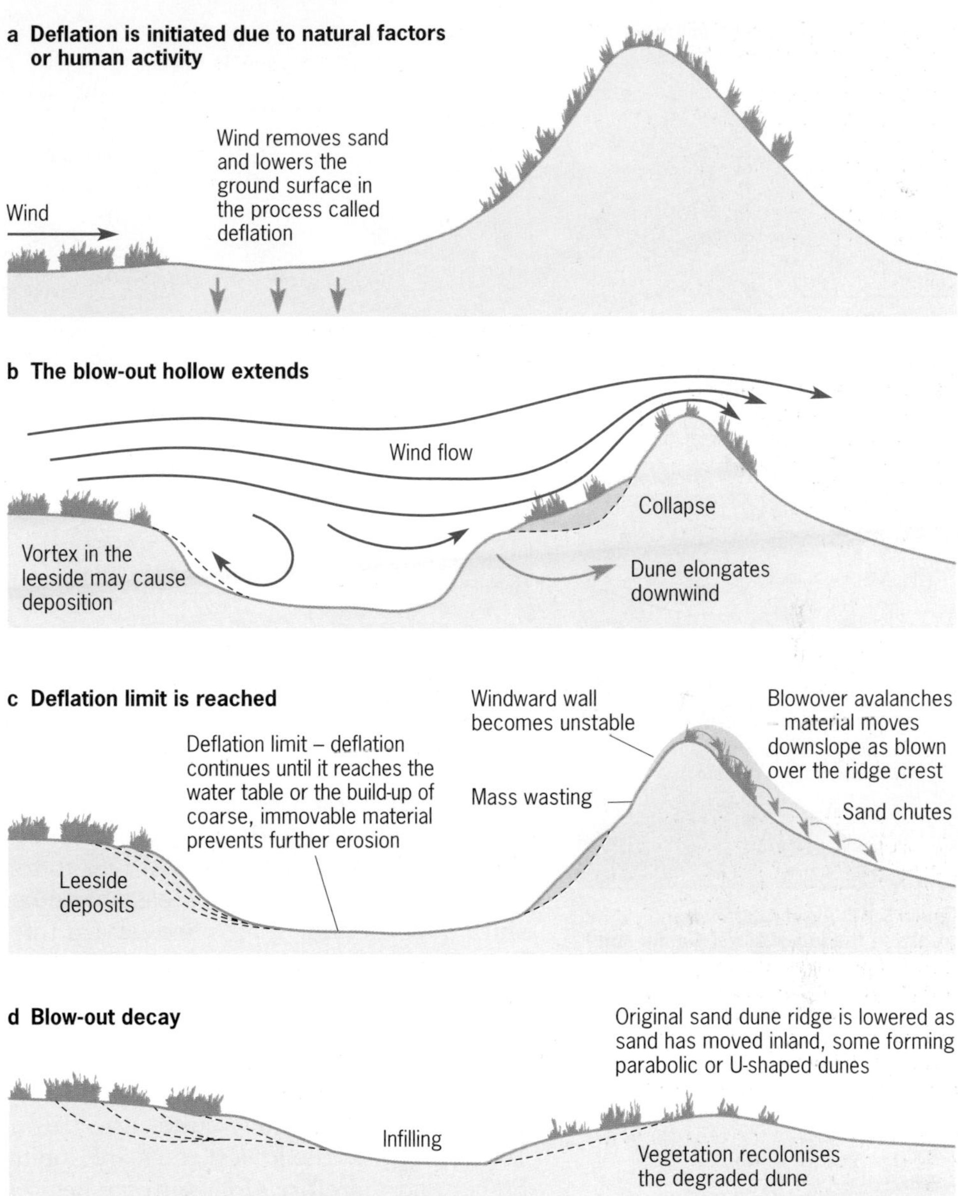
a Deflation is initiated due to natural factors or human activity
Wind
Wind removes sand and lowers the ground surface in the process called deflation
b The blow-out hollow extends
Wind flow
Collapse
Vortex in the leeside may cause deposition
Dune elongates downwind
c Deflation limit is reached
Deflation limit – deflation continues until it reaches the water table or the build-up of coarse, immovable material prevents further erosion
Windward wall becomes unstable
Mass wasting
Blowover avalanches – material moves downslope as blown over the ridge crest
Sand chutes
Leeside deposits
d Blow-out decay
Original sand dune ridge is lowered as sand has moved inland, some forming parabolic or U-shaped dunes
Infilling
Vegetation recolonises the degraded dune

Figure 5.42 Stages in blow-out formation (*After:* Carter, 1988)

5.5 Mudflats and saltmarshes

Mudflats and saltmarshes (Fig. 5.43) develop where wave energy is low due to the sheltering effect of estuaries, barriers and spits, and where there is fine sediment available. The main energy inputs to these environments are generated by tidal rhythms, and the resulting saltmarshes and mudflats are tidal landforms consisting of silt and clay grains. These fine particles are transported as suspended sediment which becomes an unvegetated mudflat or a vegetated saltmarsh. Fine-grained sediments are cohesive sediments which *flocculate* in seawater, i.e. they are attracted together to form larger particles which have a higher settling velocity than their individual grains would suggest. Once deposited they are not as easily entrained back into the water flow as sand particles (see the Hjulström curve on p.36). Most deposition occurs near high- and low-tide levels, when flow velocities are low. Deposition rates are highest on the higher mudflats where water velocities are lowest. The upper limit of deposition is the high-tide mark and deposition extends seawards from this point, building out the mudflat. The

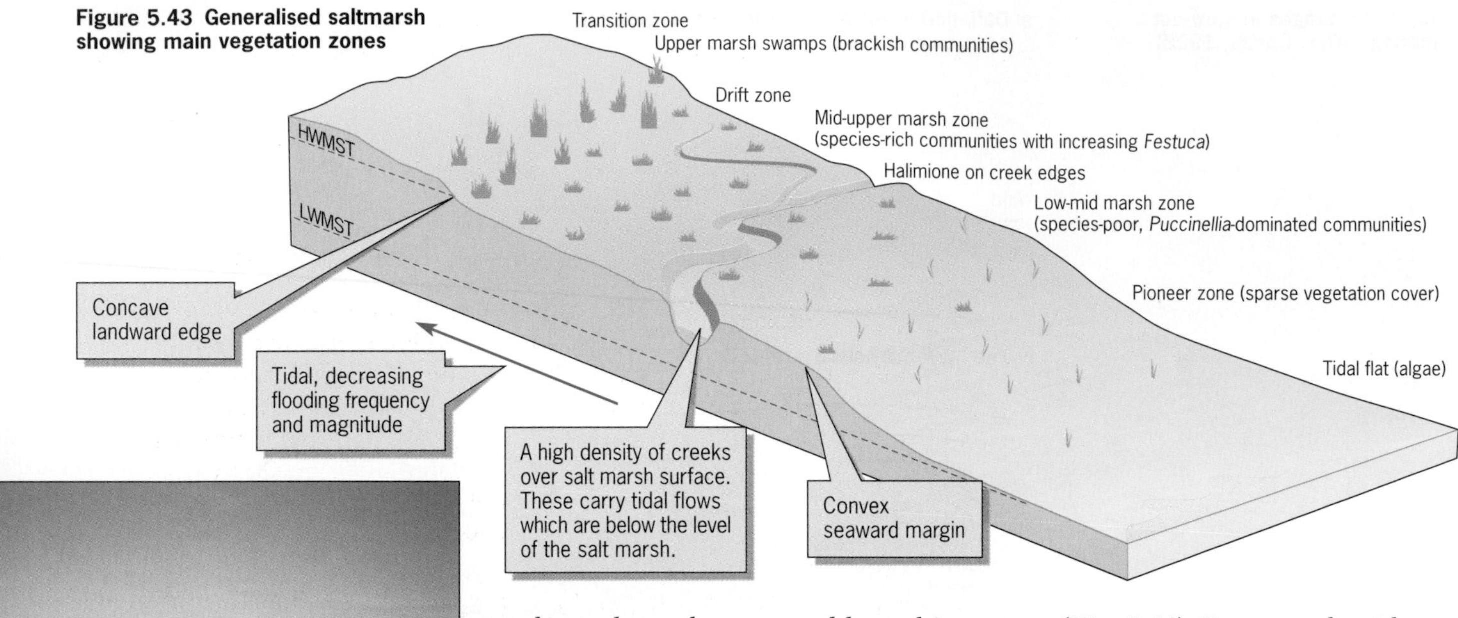

Figure 5.43 Generalised saltmarsh showing main vegetation zones

Figure 5.44 Algal mudflats and creeks in the lee of Hurst Castle spit. The shingle beach of the spit is in the foreground and the change to the higher vegetated saltmarshes can be seen in the background.

saltmarsh creek system adds to this process (Fig. 5.44). Between the tides, water flows fastest within the creeks, and as the tide rises the fast flow continues in the creeks. On the intervening areas the velocity is suddenly reduced as the creeks overspill with the approaching high tide and sediment is deposited at a faster rate. The creeks are areas of reduced deposition rather than being formed by erosion.

As the mudflat builds up, the frequency of tidal flooding falls and eventually vegetation can colonise the area to form a saltmarsh (Fig. 5.45). The depositional rates in these areas would be very small without the vegetation since only very fine grains are carried this far by the tides, and the period of low-velocity water flow is very small. The presence of the vegetation, e.g. pioneer species such as marsh samphire (*Salicornia* spp) and marsh cordgrass (*Spartina* spp), is crucial. The plants reduce current velocities and allow deposition to occur through more of the tidal cycle, not just at the lowest-velocity period of high tides.

The saltmarsh surface builds upwards at rates of between 0.001 cm and 10 cm per year; the lowest rates are due to the less-frequent flooding of the higher and therefore older saltmarshes. This aggradation of the saltmarsh

Figure 5.45 Middle-level saltmarsh community at Hurst Castle spit. The debris from the earlier high tides which flood this area can be seen.

Figure 5.46 Upper saltmarsh community at Keyhaven Nature Reserve in the lee of Hurst Castle spit. A drainage channel can be seen in the foreground and the flood protection dyke to the left. The dark green plant is *Spartina*. This stiff, tough grass with a near-horizontal leaf arrangement is very effective in trapping sediment.

surface is the net result of the deposition of material, minus erosion of the marsh during high-energy events, and compaction of the marsh surface (Fig. 5.46).

The tropical equivalent of this aggradational process is the mangrove coast. The raised root systems of the mangrove trees trap the sediment, and the leaf litter provides a significant organic input.

5.6 Deltas – coastal and fluvial interactions

'Deltas are formed where sediment-laden rivers flow into standing bodies of water' (Chorley et al., 1984). At such locations, velocity and available energy are reduced, and so is the capacity to transport load. This results in progressive deposition of sediment. In many ways the mechanics of delta formation are similar to those of alluvial fans (see p.43), but in this case most of the deposition takes place below the water surface: what we see is only the upper surface of a complex three-dimensional landform (Fig. 5.47).

Deltas are mainly formed in coastal locations, but they can occur wherever a river or stream enters a water-filled basin, e.g. a lake. Indeed, some of the great deltas of the world are *inland deltas,* for example the Niger Inland and Okavango deltas. Like saltmarshes, and mudflats in estuaries, the flat and constantly changing wetlands of deltas possess high ecological and conservation values, e.g. the Copper River delta of Alaska, which provides a seasonal home for up to 13 million migrating water birds, and high attractiveness for people because of their agricultural fertility, e.g. the Nile delta.

The principles of delta formation are similar in all locations. As we look at maps and photographs, two key questions arise:

1 Why is it that some rivers have deltas at their mouths and some do not?

2 Why do deltas vary so much in shape?

The main variables involved in answering these questions are set out in Figure 5.47.

Clearly the first variable we need to establish is the sediment input: how much sediment is being delivered and what is its character? Deltas are formed mainly of fluvial sediment inputs, whereas estuaries tend to be dominated by marine sediments. For instance, the Mississippi delivers approximately 450 million tonnes of sediment into its delta distributaries

Factors influencing delta form and formation

1 The volume and character of the river sediment load, especially the proportion of bedload to suspended load

2 The river hydrology, especially the volume and variations of the discharge

3 The relative densities of the river water and the body of water into which the river flows

4 The geometry of the coast, including the plan view and the offshore gradient

5 The strength and direction of the coastal processes, e.g. wave action; tidal scour; longshore currents

6 The tectonic stability of the coastal zone over time, i.e. whether the coast is subject to rises and falls in relation to sea-level

7 The climate and its effects upon amount and type of vegetation cover, and the growth of marine organisms

Figure 5.47 The delta formed by the River Mississippi as seen from Shuttle Challenger, 1985. This photograph shows clearly the pattern of sediment of this type of river delta. Between the kinks in the river, top left, is the town of Buras.

each year. Over 90 per cent of this is fine-grained silt and clay, much of which is transported in suspension. Once the sediment arrives at the coast, the offshore gradient becomes a crucial factor. A delta can form most easily if the sea bed slopes gently, although gently shelving coastal waters may be subject to strong waves and current action which may remove just as much sediment. For a delta to grow, more sediment must arrive (*input*) and be deposited (*store*) than is removed (*output*) by waves and currents.

As a delta progrades, i.e. builds out over time, each set of beds is progressively overlain by more recent deposits. This has led to the classification of delta deposits into three categories – topset, foreset and bottomset (Fig. 5.48). As the river's load is deposited, the main channel splits into many smaller channels called distributary channels. The internal structure of a delta is further complicated by the recurrent shifting of these distributary channels, by lateral movement, by abandonment and by building new channels. These shifts may destroy some earlier deposits as well as change the location of deposition and progradation over time. Where a river makes a fundamental shift or series of shifts of delta location and formation, it is known as *avulsion*. The best-documented example of large-scale avulsion is the bird's-foot-type delta of the Mississippi, where seven distinct lobes of delta progradation within the past 5000 years have been identified (Fig. 5.49). Thus, the main outlet today has been in use for less than 600 years. Even so, if engineers had not constructed a comprehensive flood control system through the delta, it is likely that the river would by now have reverted to an earlier course along the Atchafalaya channel, thereby moving the main discharge some 200 km to the west. This would be a shorter route by 160 km, resulting in changes upstream. Effectively this would represent a lowering of base level (see p.40), with rejuvenation of the channel and increased sediment production.

It is clear that, because of the number of variables involved, each delta is unique in detail (Fig. 5.47). Attempting a classification of delta types is, therefore, a difficult task. One useful classification uses the major processes controlling their development, i.e. fluvial, wave and tidal controls (Fig. 5.50).

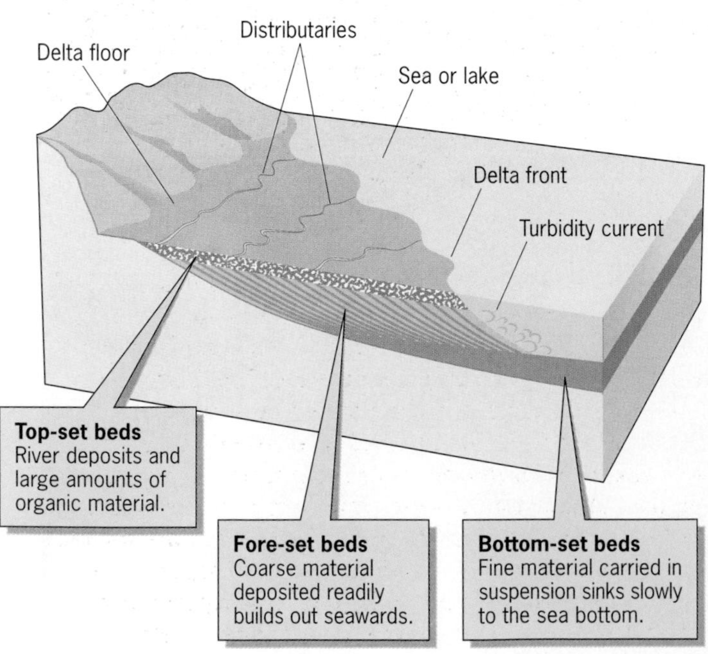

Figure 5.48 Delta structure

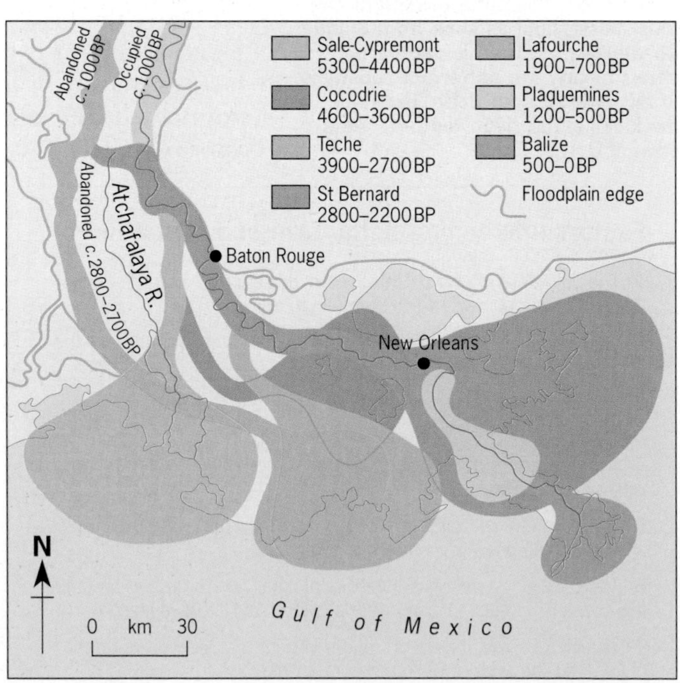

Figure 5.49 The evolution of the Mississippi delta (*Source:* Chorley et al., 1984)

I HIGH-DESTRUCTIVE

a Wave-influenced deltas Flattened planform as a result of vigorous sediment removal by longshore wave currents, e.g. Sao Francisco River, Brazil

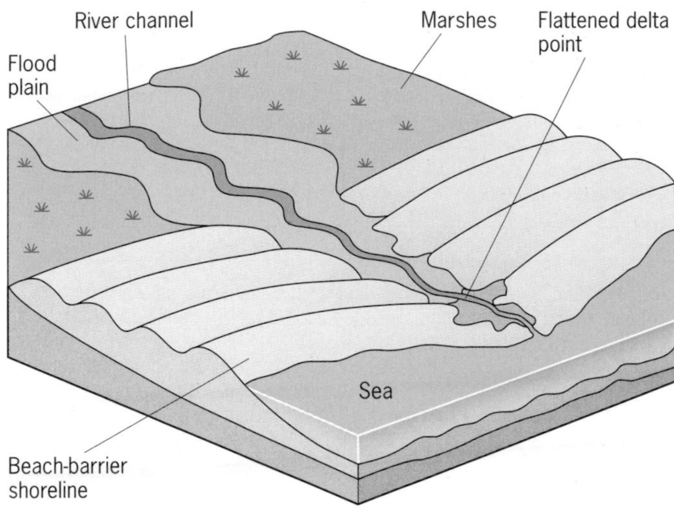

b Tide-influenced deltas Long, funnel-shaped distributaries kept open by strong tidal scour. In high latitudes, there may be frequent channel shifts, e.g. Copper River delta, Alaska, while in humid tropics, vigorous mangrove forest development may stabilise the channels, e.g. Mekong.

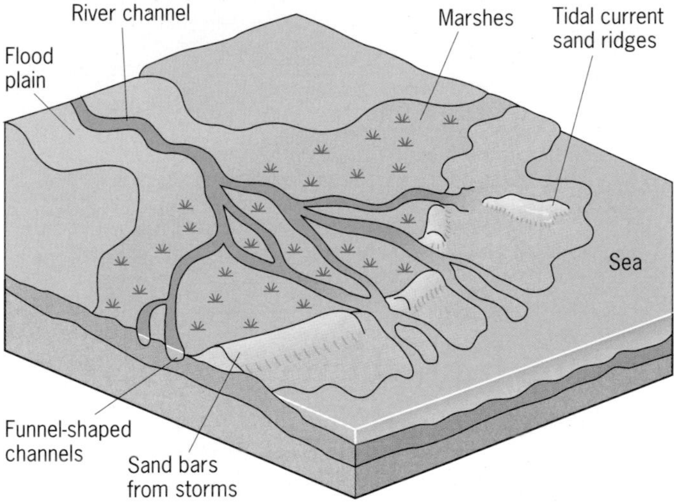

II HIGH-CONSTRUCTIVE

c Fan-shaped deltas Most likely to be formed where there is an ample supply of relatively coarse sediment. Distributary channels constantly shift, and so change the location of deposition, creating a compact fan, e.g. Nile, Rhône (note, both these examples occur in the Mediterranean – an almost tideless sea).

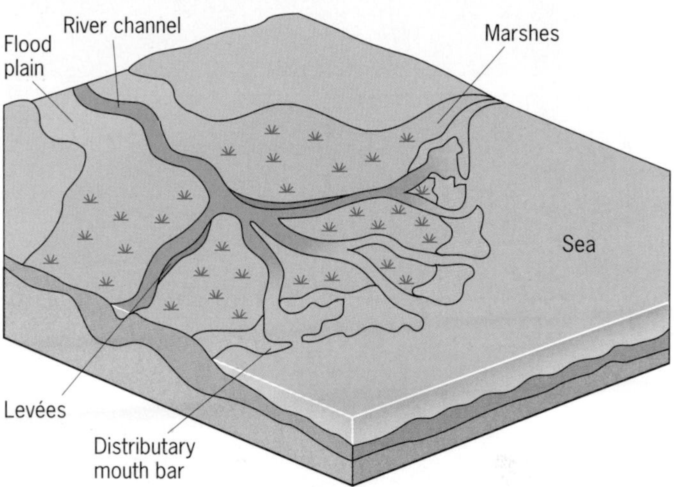

d Bird's-foot delta An elongated form which develops when a river delivers large volumes of fine sediment. The delta progrades around a recurrently shifting set of distributaries flanked by levées, e.g. Mississippi.

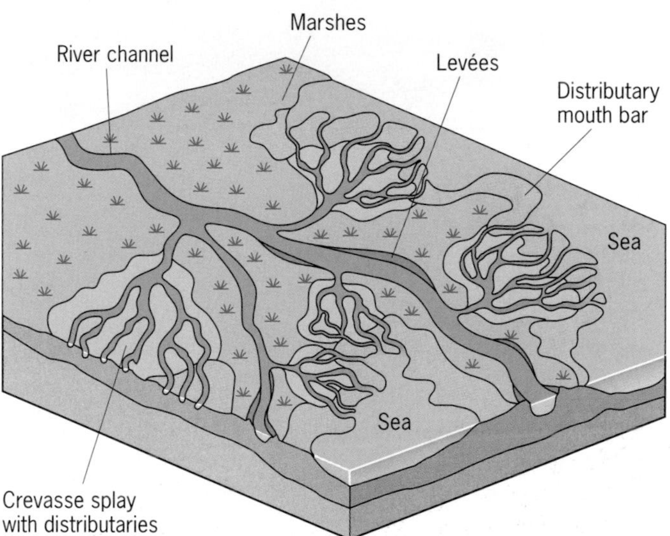

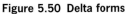

Figure 5.50 Delta forms

Deltas are susceptible to change by human activity, especially if the fluvial inputs change in terms of discharges of both water and sediment. River management schemes can change both of these inputs and thus affect the balance between the fluvial inputs and the marine outputs. The Ebro delta, Spain, has changed to a wave-dominated coastline as fluvial inputs have been reduced by 95 per cent since the 1960s as a result of the dams along the River Ebro. This has changed the dynamic equilibrium between river discharge and wave transport capacity, and wave transport now dominates. The central lobe of the delta has retreated 1500 m in 30 years at Cap Tortosa. Spits on the delta coastline have accreted due to longshore transport and deposition of the eroded material. The Nile delta, Egypt, has experienced similar change. Figure 5.51 shows how this arcuate (fan-shaped) delta landform has evolved over geological time and how it is expected to change in the future as a result of human activity.

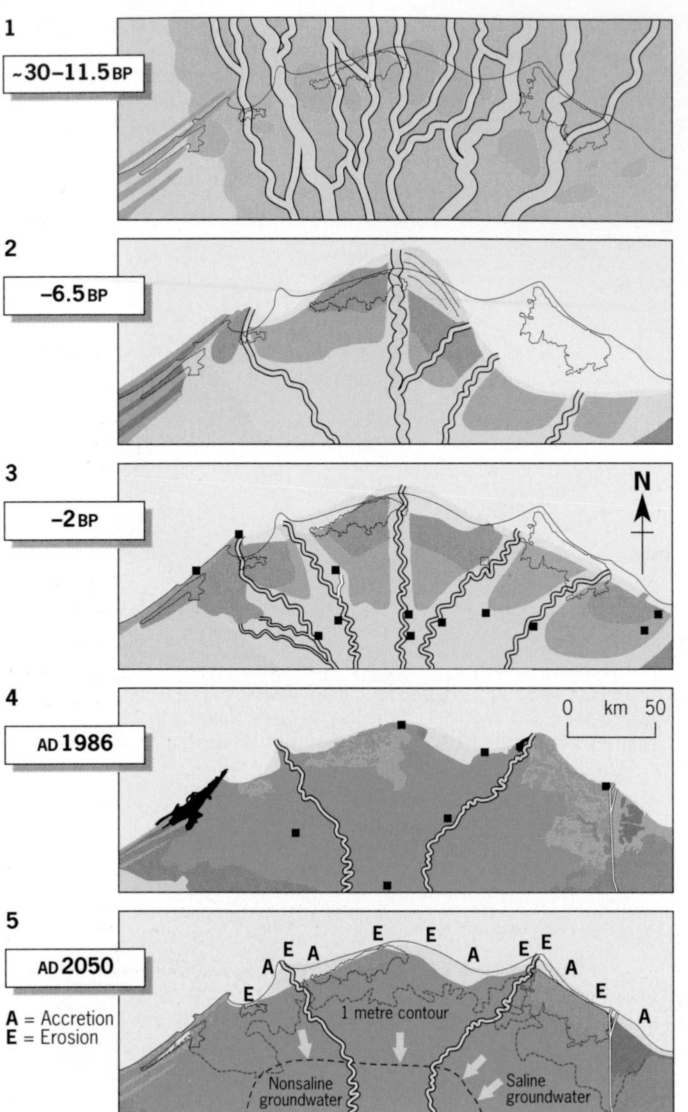

1 The region was an alluvial plain with seasonally active braided river channels. Sea level lower than today due to the Pleistocene ice age.

2 Sea level begins to rise moving the coast to the south. The alluvial plain deposits were reworked. By 6500 BP the modern delta had begun to form. Sea level was 10 m below present levels. The climate was wetter than today and large volumes of coarse sand were transported to the coast by the distributaries, especially the Sebennitic channel. The delta was a cuspate-shaped area dominated by fluvial deposition. Beach ridges accreted on the seaward edge of the delta.

3 The delta changed from a river-dominated system to a wave-dominated cuspate delta. The NE area prograded. Sea level 2 m below present but the coastline resembles the present-day outline. At least five distributary channels bring great volumes of sediment to the coast adding 1 mm of sediment to the delta surface each year. Human activity was starting to modify the delta by irrigation and wetland drainage projects. By AD 1000 only two distributaries remain – the Damietta and Rosetta.

4 98% of sediment now trapped behind dams, especially the Aswan Dam. Increased coastal erosion and marine flooding of the northern delta plain. Lagoons and coastal fish populations decline. Water flow into the delta is now agricultural runoff and municipal/industrial waste. SE winds result in coastal currents moving W to E. Delta now has a negative sediment budget. Sea level rising. The delta area is 66% of Egypt's habitable land.

5 Predictions. Sea level will rise 1 mm per year at least and delta is subsiding between 1–5 mm per year. By the year 2050 sea level rise in the NE delta will be between 12.5–30 cms. Promontories eroded and lagoons infilled on seaward edge by overwashing. Some local accretion, e.g. near Rosetta and east of Damietta. Water table will rise relatively as delta subsides but is not built up by sediment. Pumping of groundwater for irrigation will cause the movement of saline groundwater inland resulting in a decline in agricultural productivity.

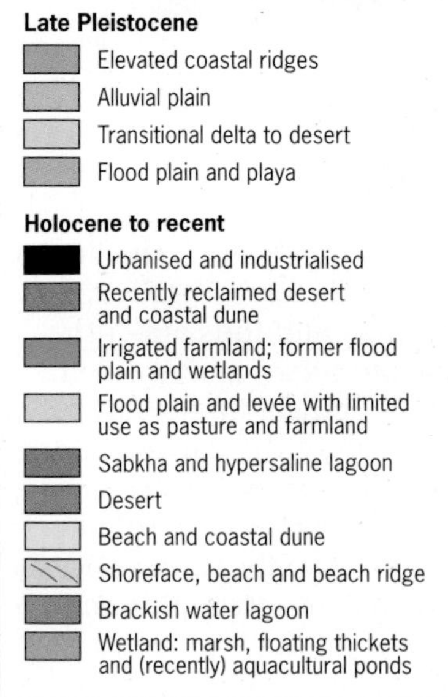

Figure 5.51 The evolution of the Nile delta

5.7 Landforms resulting from changing sea levels

Sea level has varied considerably over geological timescales, particularly during the Pleistocene when the sea level fell and rose several times as a result of glacio-eustatic change. It reached its present position about 7000 years ago as a result of the Flandrian transgression. This rising sea level resulted in drowned river valleys (rias), and glaciated valleys (fiords), increased accretion in the lower river valley floodplains and submerged forests, e.g. at Borth in Dyfed, Wales. Most estuaries result from this post-glacial sea-level rise.

However, the processes involved are not as simple as they may first appear. The Flandrian transgression may have raised the sea level, but we must also consider what was happening to the land surface as well. In areas of little earth crust movement or tectonic activity, the result would be a relatively simple drowning of the land, with lowland areas forming inlets and bays, and higher land forming headland and islands. However, many land surfaces are also moving as well. A sinking landmass would have the same net effect as a rising sea level, and both happening together makes the overall subsidence twice as big. Remember from earlier sections that this is the situation in south-east England.

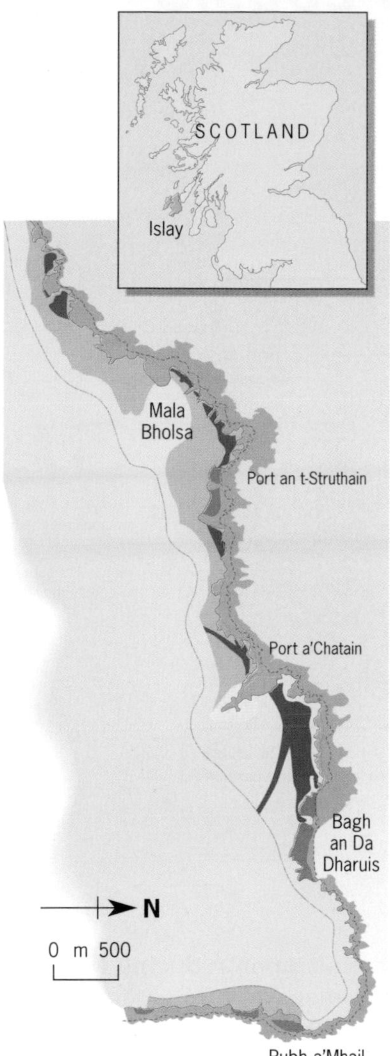

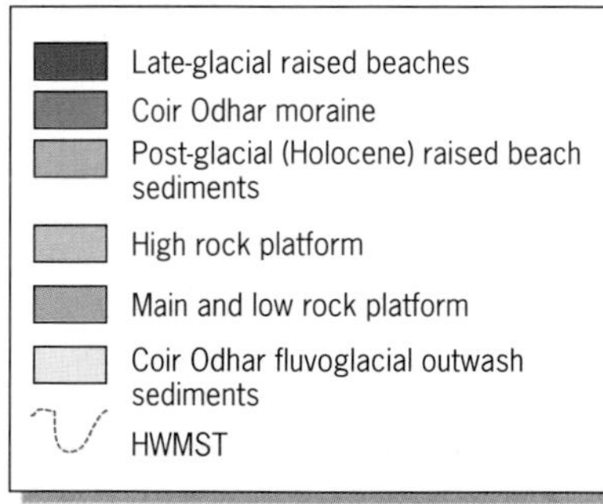

Figure 5.52 The geomorphology of northern Islay, showing the distribution of the principal raised shoreline features

Many land surfaces are rising relative to sea level, particularly as a result of the process known as *isostatic recoil*. During glaciation, the weight of the ice mass depresses the earth's crust beneath. The amount of that depression is related to the thickness of the ice mass. Once the ice has melted, the land begins to rise again at a rate determined by the ice thickness. So when we look at former glaciated areas, we must explain landforms along the coast by rising post-glacial sea levels and rising land as well. If these happened at the same rate, then the shoreline would remain in the same place. In most locations, however, the isostatic recoil is faster than the sea-level rise. The landforms produced are a complex set of erosional and depositional landforms which vary over short distances due to differing rates of isostatic recoil and the nature of the former shoreline. Just to complicate the picture further, remember that the Pleistocene consisted of several glacials and interglacials which have differing extents (see Chapter 7). To explain these landforms fully, geomorphologists must also consider the coastal landforms existing before the Pleistocene and how glacial and periglacial processes have modified these! The result is a complex set of coastal landforms above the present sea level, such as the raised shore platforms, cliffs and beaches around the coast of Scotland, Wales and other previously glaciated areas. For example, northern Islay, Scotland (Fig. 5.52), has a high rock platform with a fossil cliff line at its seaward edge and a lower platform beneath dated as the late Pleistocene (Loch Lomond stadial) shoreline when the sea level was 50 m below present. The height of the platform today depends upon how much isostatic recoil has occurred at any particular place.

Summary

- Coastal erosional and depositional processes produce distinctive landforms: However, erosional and depositional processes can occur in all areas at different times.
- Cliff morphology results from the interaction of a number of processes related to marine, subaerial and geological factors which vary in importance from place to place.
- Cliff evolution is episodic and cyclical.
- Shore platforms are erosional landforms developed in the intertidal zone. Their morphology is determined by the interaction of marine and subaerial processes and lithology and structure
- Geological factors, e.g. lithology, joints, bedding planes and faults, influence coastal morphology at a variety of scales as a result of differential erosion.
- Depositional landforms are dynamic stores of sediment. Beaches are stores of sand and larger-sized particles. Finer sediments form saltmarshes and mudflats. Sand particles may move inland to form sand-dune systems.
- Beaches are the most widespread depositional landform. Their morphology results from the interaction of wave type and energy, sediment supply and longshore inputs/outputs. Human activity is increasingly important in beach dynamics.
- Sand-dunes develop in areas of strong on-shore winds and low-gradient nearshore slopes. Wind and biological processes combine to form dune systems.
- Mudflats and saltmarshes develop in areas of low wave and high tidal energy. Tidal and biological processes play an important role in their development.
- Deltas are dynamic landforms resulting from the interaction of marine and fluvial processes. For a delta (store) to form, the fluvial inputs must exceed the outputs by marine erosion.
- Relative changes of sea level result in distinctive landforms – rias, fiords, estuaries, and raised cliffs and beaches.

6 Managing Britain's coasts

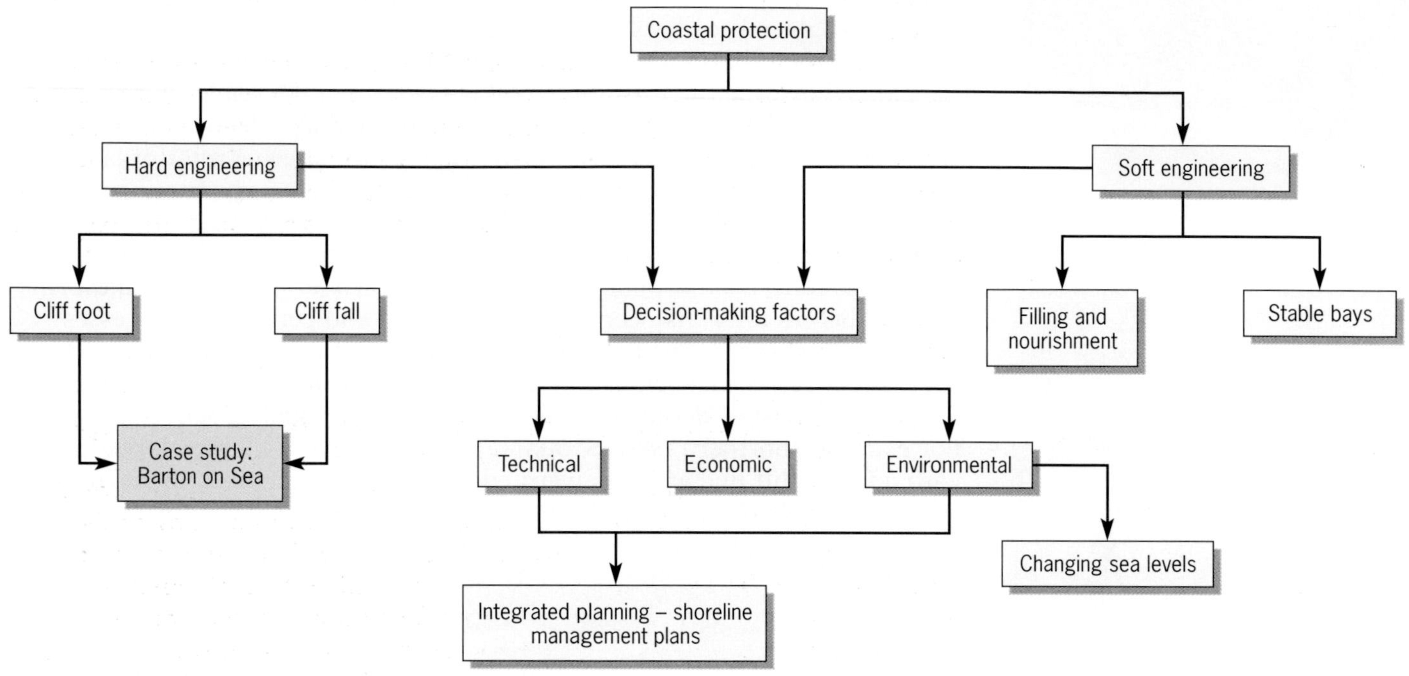

6.1 Introduction

Coastal management is wide-ranging, and may focus upon reducing the impact of coastal processes and landform development on human activity, protecting coastal ecosystems, and managing human use of the coastal zone. There are many organisations involved in managing the coastal zone (Table 6.1), and there is no single government department with responsibility for the coast. As a result, management to date has been rather piecemeal, involving the problems of eroding coastlines, visitor pressure and the protection of coastal ecosystems.

In this chapter we shall study the way in which people attempt to manage coastal processes and landforms and how this human activity can have unintentional impacts. People have attempted to manage coastal processes for hundreds of years, particularly since the end of the nineteenth century. Coastal defence is a term covering measures against coastal erosion (coastal protection) and flooding (flood defence). Our concern here is coastal protection.

Many cliffed coastlines are developed for residential, industrial and commerical use. Cliff-top homes and hotels have always been popular for the wonderful views that they provide. Erosion, however, causes conflict with human activity because people often expect the coastline to be static rather than the dynamic system that it is. Fast rates of cliff recession will often threaten buildings that were originally built well away from the cliff line. Although lives are rarely threatened by cliff erosion, the loss of people's homes and livelihoods, and the financial loss of commercial and industrial buildings and other cliff-top land uses, can be very high. It is in these circumstances that coastal protection measures may be used to defend the human developments along the eroding coastline. Defence measures are

Table 6.1 Organisations involved in managing Britain's coastline

- Local maritime councils
- County councils
- National Rivers Authority
- National Trust (the Enterprise Neptune appeal aims to protect 650 km of threatened coastline)
- English Nature (Sites of Special Scientific Interest (SSIs), Areas of Outstanding Natural Beauty)
- National Nature Reserves
- National Park Authorities
- Countryside Commission (Heritage Coasts in areas of special scenic attraction under heavy visitor pressure)
- Ministry of Agriculture, Fisheries and Food (MAFF)

used along eroding sedimentary coastlines where beaches of all types are being eroded at increasing rates due to natural processes through rising sea levels or land subsidence or an increase in stormy conditions, or changes in sediment supply, or because of human activity changing sediment budgets and amounts of longshore drift. Beaches are the key amenity in holiday resorts, and they provide valuable protection from wave energy and coastal flooding.

As a result of the Coast Protection Act 1949, responsibility for coastal protection in the UK lies with 121 local district councils whose boundaries include the coastline (called maritime councils). Management of coastal flooding is the responsibility of the National Rivers Authority. All coastal defence measures require planning permission and consultation before they can be implemented. In England, 570 km, about 20 per cent of the coastline, is protected against erosion.

6.2 Coastal protection

Hard engineering

Hard engineering is the use of structures which aim to resist the energy of waves and tides. These engineering structures have to withstand huge variations in energy conditions from small waves to extreme storm events, and as a result they need expensive maintenance. However, many of these structures have been very successful and have allowed large-scale human development of the coastline or have protected existing developments. Nevertheless, these schemes can have significant impacts upon natural processes and coastal ecosystems. They result in a narrower shoreline, reduce habitats, and impede sediment movement between the land and the sea. This effect has been called 'coastal squeeze' by English Nature.

Cliff erosion can be reduced by locating structures at the foot of the cliff on the upper or lower shoreline. These *cliff-foot strategies* aim to manage erosion by reducing wave erosion or providing a protective beach, which dissipate wave energy due to their permeability, and the mobile sediment of sand and pebbles which provide a huge surface area for energy loss by friction. Thus some of these cliff-foot strategies are used along depositional landforms to increase rates of sedimentation (Fig. 6.1), e.g. where beaches have declined because of updrift management, or sand and gravel extraction from the beach, or offshore dredging which affects the nearshore sediment budget. Where mass movement processes on the cliff are particularly rapid, *cliff-face strategies* may be used to reduce the subaerial processes.

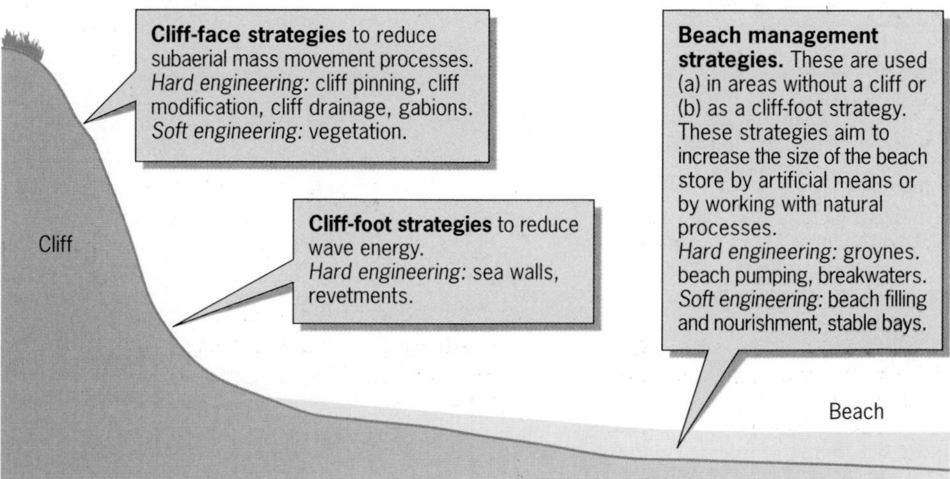

Figure 6.1 Coastal protection strategies

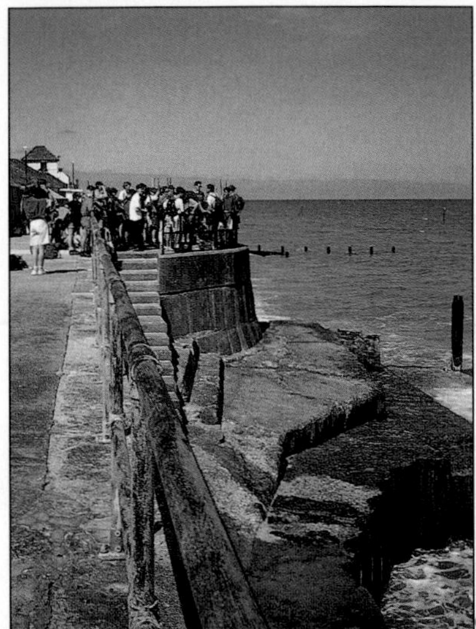

Figure 6.2 The tiered sea wall at Sheringham, Norfolk, before the recent renovations. This type of wall incorporates a promenade for holidaymakers.

Cliff-foot strategies

SEA WALLS

Sea walls aim to prevent erosion by providing a barrier (Fig. 6.2) which reflects wave energy rather like the standing wave clapotis effect of a cliff in deep water (Fig. 6.3a). Older sea walls are vertical in design and reflect much of the wave energy. Although this was the original aim, the reflected energy is able to pick up sediment from the beach in front of the sea wall and transport it into the offshore zone. This *beach scouring* can be great enough to undermine the sea wall foundations and result in expensive maintenance. As the beach levels fall, the shape of the beach changes to become less dissipative of wave energy and more reflective. A sea wall built at Porthcawl in South Wales in 1906 caused beach levels to be scoured by an average of 0.04 m per year; a new sea wall had to built in 1934 because the first sea wall was being undermined. This structure was built in front of the old sea wall and the gap between was infilled. The beach scouring caused a fall of 3 m in beach levels over 75 years, and by 1986 the new wall was threatening to collapse. Sea walls now tend to be curved in order to dissipate rather than reflect wave energy (Fig. 6.3b, c).

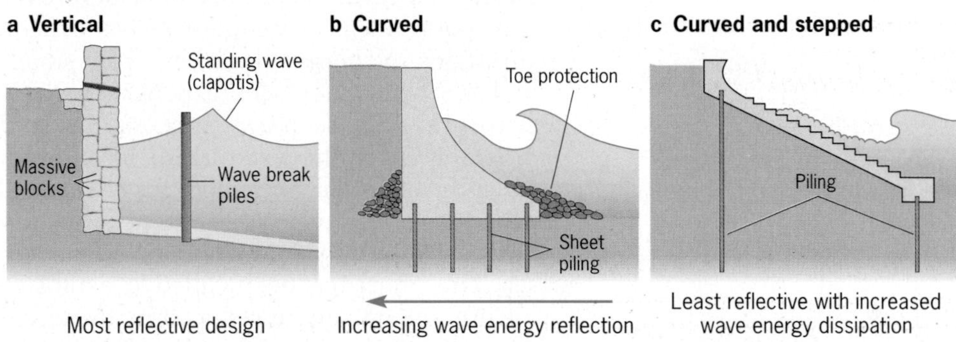

Figure 6.3 Types of sea wall

Sea walls are very expensive structures, costing from £2000 per metre for a simple structure up to £5000 per metre for a two-level wall with a promenade in a seaside town. Some 90 per cent of sea walls require expensive maintenance within ten years of being built, and 96 per cent suffer damage over a 30-year period. Although they are effective defences against erosion and flooding, in addition to the financial costs involved there are other problems resulting from their construction, some of which can only be solved by further engineering structures (Fig. 6.4).

?

1 Refer back to Figure 4.1. Describe and explain the problems resulting from the building of the sea wall at Mappleton.

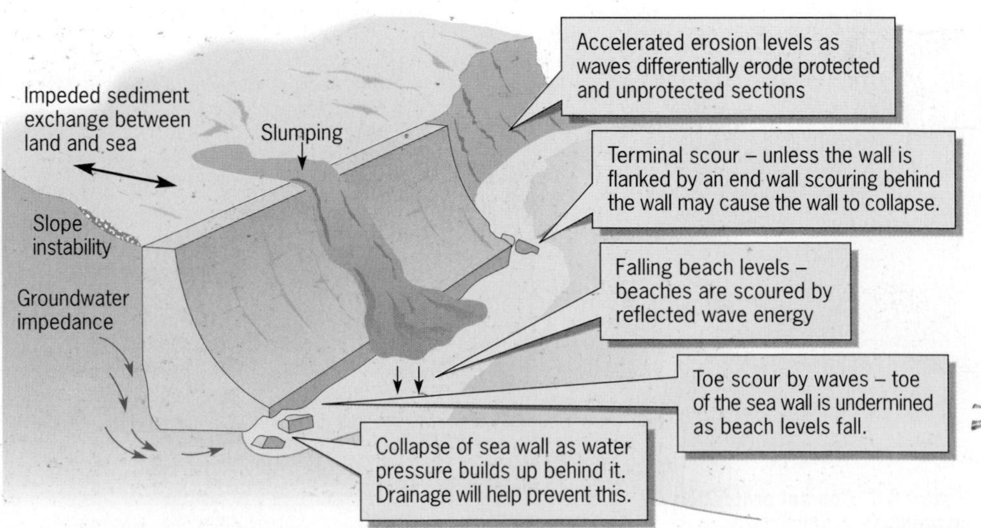

Figure 6.4 Some problems related to sea walls (*After:* Carter, 1988)

Figure 6.5 Wooden revetments at West Runton, Norfolk

REVETMENTS

Revetments are a cheaper alternative to sea walls. They consist of timber, steel or concrete posts driven into the beach with connecting timber or a rock boulder infill. Timber revetments (Fig. 6.5) aim to dissipate wave energy rather than reflect it. The waves pass through the structure but have reduced erosive energy at the cliff foot. They are permeable and allow sediment to pass through them, which means that the beach can build up behind the revetment and provide further protection for the cliff foot. However, time has shown that timber revetments do reflect some wave energy and beach scouring on their seaward side can occur. Although timber revetments are relatively cheap (£1200 per metre), they also have a relatively short life since the timbers can be quickly damaged by storm wave action and the foundations can be undermined by beach scouring.

Block revetments have a concrete or rock boulder infill between concrete or timber posts, or they are made up of a sloping rock armour (rip-rap) as at Barton on Sea (see case study, p.117). These aim to reflect or dissipate wave energy depending upon whether the structure is vertical like a sea wall or sloping boulders.

?

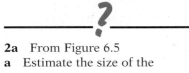

2a From Figure 6.5

a Estimate the size of the revetment structure.

b What evidence is there from the photo that the revetments have helped beach material to build up behind the revetment?

c Comment upon the visual impact of the revetments on this area.

GROYNES

Groynes are common along the coastline of the UK. They aim to increase the natural rate of sedimentation and thus build up a protective beach at the cliff foot. They are also used in any location to retain a beach on sedimentary coastlines such as a spit (to reduce the chances of breaching), as protection from coastal flooding, or for recreational purposes. The groyne's structure, made of timber, steel or concrete (Fig. 6.6), is built perpendicular to the shoreline to stop or reduce sediment movement by longshore drift. Groynes aim to achieve a balanced sediment sub-cell between the groynes which is in equilibrium with the dominant waves (Fig. 6.7). The groynes interrupt and reduce longshore currents, and as deposition occurs the shoreline reorientates towards the wave front, reducing the angle of wave approach and therefore the current velocity. This results in further deposition and the formation of groyne bays. The length and spacing of groynes is important if this is to be achieved (Fig. 6.8): for example, at Heacham north beach to the south of Hunstanton, three groynes are being shortened because they are currently diverting material drifting southwards into the offshore zone.

Groynes which are sized and spaced in balance with coastal processes of the area work very well, and a groyne field can be 100 per cent efficient in

Figure 6.6 A concrete groyne at Sheringham East, Norfolk

Figure 6.7 A wooden groyne field

Figure 6.8 **Plan view of a groyne field, showing groyne functions and the importance of groyne spacing**

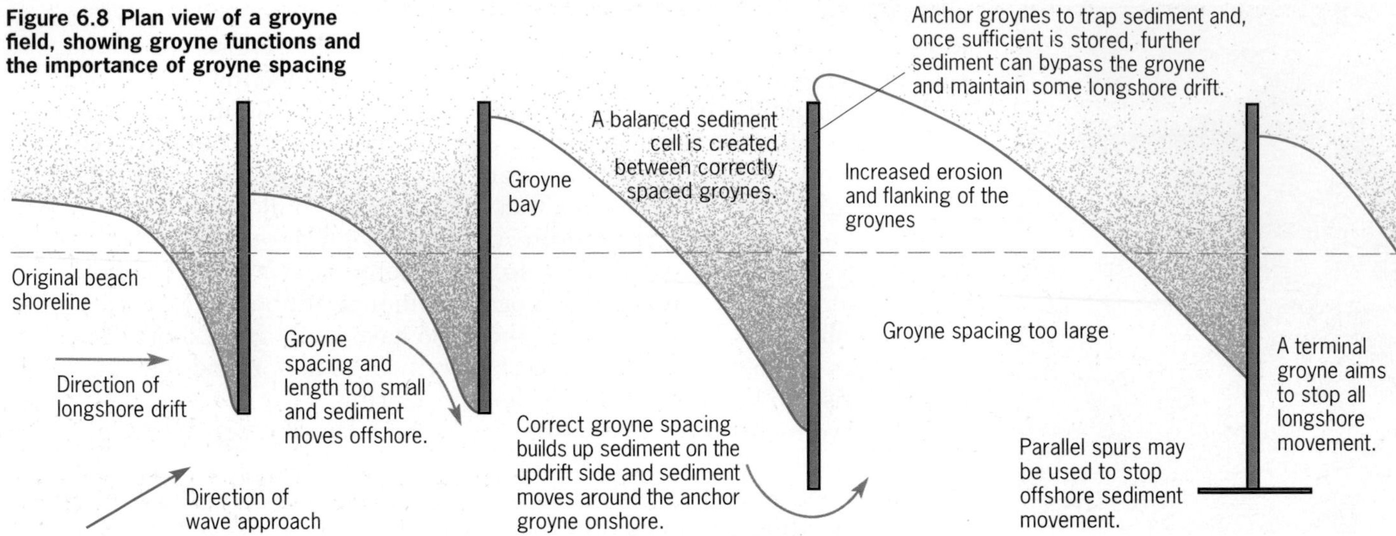
Anchor groynes to trap sediment and, once sufficient is stored, further sediment can bypass the groyne and maintain some longshore drift.
A balanced sediment cell is created between correctly spaced groynes.
Groyne bay
Increased erosion and flanking of the groynes
Original beach shoreline
Groyne spacing too large
Groyne spacing and length too small and sediment moves offshore.
Direction of longshore drift
A terminal groyne aims to stop all longshore movement.
Correct groyne spacing builds up sediment on the updrift side and sediment moves around the anchor groyne onshore.
Parallel spurs may be used to stop offshore sediment movement.
Direction of wave approach

?

3 Use Figure 6.8 to explain why the correct spacing of groynes is important if they are to be a successful management strategy.

stopping longshore drift. This efficiency can result in major problems for beach systems downdrift of the groyne(s), however. The sediment input is reduced or cut off by the groynes, but the outputs by longshore drift continue. This reduction in beach material has serious consequences for seaside resorts or for cliffs that may have a reduced protective beach at their foot and are likely to experience increased erosion rates (see, for example, the Naish Farm cliffs case study in Chapter 5). During storms or high tides, groynes act like headlands and refract wave energy. This alters local sediment patterns, and the groyne structure becomes the focus for rip-current formation. These currents move sediment offshore so that they are lost from the beach system (at least temporarily).

BEACH PUMPING

As with groynes, beach pumping aims to increase natural rates of sedimentation. Shingle beaches are highly permeable, so the energy from wave backwash is low as the water soaks into the beach. On sandy beaches, however, the closer packing of the particles means that the beach is less permeable and the size of the sand sediment particles results in some transport into the offshore zone. By installing plastic drainage pipes into the beach at 2–3 m below the surface, the backwash energy is reduced as more water soaks into the beach, backwash scour is reduced, and sediment build-up on the beach increased.

BREAKWATERS

Breakwaters in the wave zone aim to break waves further offshore, or create new wave refraction patterns. The structures need to be capable of withstanding direct wave forces and are therefore expensive to build. In recent years, rubber tyres and oil drums have been used in efforts to find effective cheaper alternatives. The most common design of a breakwater (Fig. 6.9a) is a core of loose fill overlain by a veneer of large interlocking rip-rap. These dissipate and refract wave energy before it reaches the shore.

Breakwaters are also used to increase sedimentation. A series of shore-parallel breakwaters create 'wave shadows' which can be humanly infilled with sediment or allowed to infill naturally as deposition rates increase in the calmer, shallower water (Fig. 6.9b).

The design of offshore breakwaters must take into account local conditions such as sediment movement patterns, wave approaches, offshore relief, and wave return intervals. Without knowledge and modelling of local conditions the structures can be short-lived.

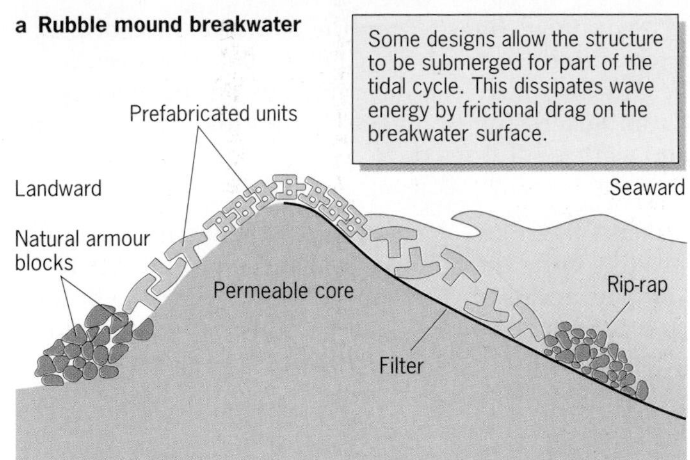

Figure 6.9a Cross-section of a rubble-mound breakwater, showing the types of material which might be used in its construction (*Source:* Carter, 1988)

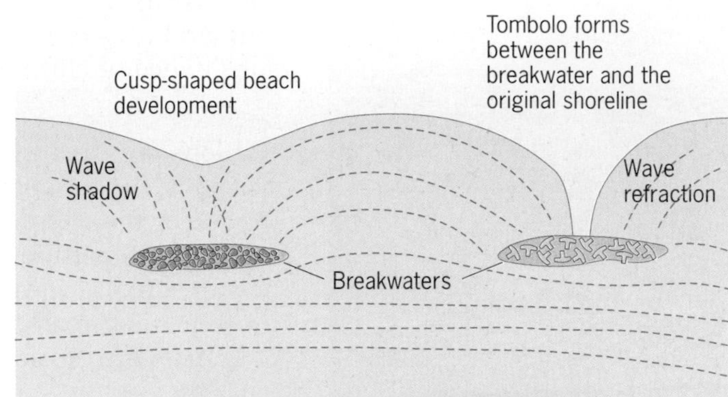

Figure 6.9b Plan view of the effects of periodic offshore breakwaters, promoting wave refraction sedimentation (*Source:* Carter, 1988)

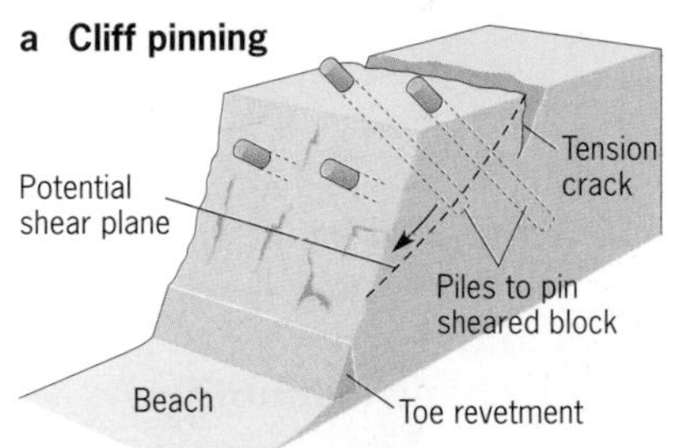

b Cliff modification

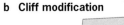

Original cliff

Lower cliff height

Reduce cliff face angle

Shear plane

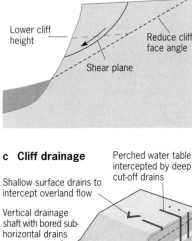

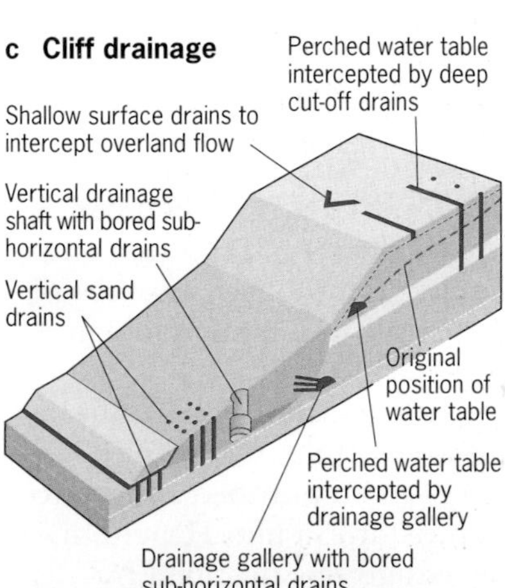

Figure 6.10 Cliff-strengthening measures (*Source:* McKirdy, 1992)

There may be impacts on local ecosystems (for example, upper-shore ecosystems may be affected by reduced wave splash), and the structures themselves are visually intrusive, though there may also be some environmental opportunities. The increased sedimentation creates new ecosystems, and the structures themselves provide habitats for marine invertebrates and plants.

Proposals for breakwaters in England include the Holderness coastline of Yorkshire (Fig. 4.1), and between Happisburgh and Winterton in east Norfolk.

Cliff-face strategies

Cliff-face management strategies aim to reduce mass-movement processes, particularly on very susceptible soft-rock cliffs or cliffs of glacial material. They aim to increase the stability of the cliff either by increasing the strength of the cliff materials or reducing the deforming forces. They are often used along with cliff-foot strategies.

CLIFF PINNING

The cliff is pinned through the likely shear planes in the cliff (Fig. 6.10a).

CLIFF MODIFICATION

If the height and slope angle of the cliff are decreased then there is a reduced chance of mass movements occurring (Fig. 6.10b). This needs to be planned with care. At Llantwit Major on the Glamorgan coast of the Bristol Channel, rockfalls from cliffs in horizontally bedded limestones (relatively strong) and shale (relatively weak) were causing a danger to recreational users. The upper cliff was blasted in 1969 to reduce its slope angle, and the blast material used to stabilise the cliff toe as toe armouring. Nine years later the toe of material was less than had been predicted because of an underestimation of the outputs due to wave attrition and longshore drift. The upper cliff was still prone to rockfalls because the blasting had weakened the rock and allowed increased weathering. The interbedding of soft and weak rocks were unstable even with a reduced angle. The scheme was not an erosion control one, but it does show how such schemes need to take into account the processes acting on the cliff and at the cliff foot.

CLIFF DRAINAGE

Cliff drainage is the cliff-face method of control which is most widely used in the UK, particularly for cliffs with a high clay content. Pore water

Figure 6.11 Gabion boxes at the base of the cliff face, Overstrand, Norfolk

pressures can be reduced by drainage lines within the cliff face, field drains, gravel trenches and by intercepting overland flow (Fig. 6.10c). However, if successful, changes in the cliff hydrology can have impacts on the ecology and land uses of the cliff top. Successful drainage schemes may also result in subsidence of cliff-top land as the cliff dries out.

GABIONS

Gabions are steel wire baskets (about 2 m × 1 m in size) filled with rock and placed at the foot of an eroding cliff. They aim to provide some stability to the toe of the cliff and help cliff drainage. They are relatively cheap to build (about £350 per metre), but cannot be used in situations where they would be subjected to wave erosion. They are usually used where ground-water movements threaten another protection structure, e.g. a sea wall as at Overstrand in Norfolk (Fig. 6.11).

VEGETATION

Grading and terracing of cliff faces can increase slope stability. At Bournemouth the soft Eocene-age sand and clay cliffs have been planted with privet hedges and shrubs, and seeded with grasses and other groundcover plants. Vegetation alone is unlikely to be effective in stopping major cliff failure but it can be helpful in the long-term stabilisation of a regraded slope.

Soft engineering

These solutions aim to work with rather than against natural processes and provide a more environmentally friendly approach to the management of coastal erosion and flooding (Fig. 6.12). They do not aim to stop coastal erosion completely but to work with it to achieve a dynamic equilibrium where erosion is reduced.

There are two main methods to reduce coastal erosion problems.

Beach filling and nourishment

These are the import of foreign or borrowed material. Beach filling is the dumping of a large amount of material. Beach nourishment is the gradual feeding of beach material over a period of time. These are attractive protection measures since they do not involve costly engineering structures. The compatibility of the imported material with the natural sediment is important. If the imported material is too small it will quickly be removed by wave action. If the material is too coarse, then supratidal ecosystems may be changed and sand-dune formation reduced.

Beach nourishment and filling works by building up a beach which can dissipate more wave energy (see section 5.4) than that which already exists. This applies where the natural beach sediment has been depleted by artificial causes, as with updrift protection measures such as groynes.

The major problem with this method is the source of the imported sediment. Most comes from relatively cheap offshore dredging, which can leave holes in the sea bed capable of affecting wave refraction patterns, and may impede nearshore–offshore sediment exchanges. Land-based sources of sediment pose problems of cost and the environmental problems associated with quarrying.

Beach nourishment schemes are being implemented increasingly in the UK, for example at Sheringham, Norfolk, and between Mablethorpe and Skegness in Lincolnshire where sea-dredged sand is to be pumped ashore to nourish the 24-km length of coastline. Between Hunstanton and Heacham, Norfolk, the beach below the sea wall built 40 years ago has dropped by 2.5 m. As well as repairs to the sea wall, there are plans to recharge a 7-km length of beach. The Sand Bay scheme in Avon is one of the earliest beach

Study Figure 6.12.

4 Why does the UK need to improve its sea defences?

5 Why are soft approaches considered a better alternative to coastal defence than hard engineering approaches?

6 What is strategic retreat and in what coastal environments can it be used?

Taking the soft option

Nuala Moran on protecting the UK coastline

Britain's coastline is under threat. Not only do many of the country's sea defences date from before 1939, but sea levels are rising through geological changes and global warming.

To find better ways of protecting Britain's coastline from both erosion and flooding, the national Coastal Research Facility has been set up. This large tank – built to model the interactions of sea and shore – is the main element of a five-year, £5 m coastal engineering programme funded by the Science and Engineering Research Council.

The tank, a joint venture with the contract group HR Wallingford in Oxfordshire, will be able to model 800m stretches of coastline. Paul Meakin, secretary of SERC's environmental civil engineering committee, says one of the main results will be improved software for civil engineers who design and build defences.

Britain has more than 17 000 miles of coastline, subject to the constant attrition of the sea. It is difficult to get a true picture of the scale of the threat because the last survey of flood and coastal defences was completed in 1981, although the Ministry of Agriculture, Fisheries and Food expects to finish its national survey of coastal defences by September.

With sea levels forecast to rise from 4 mm a year on the Northumbrian coast to 6mm in the south and east, there is no doubt that the UK needs to improve its sea defences. But there is no consensus among researchers, engineers and funding bodies about how to do so. It is already well recognised that 'hard' defences will not solve Britain's serious coastal erosion problem. 'In fact sea walls do nothing to stop erosion – they protect the land behind them but overall they reflect the energy and cause the sea to take more material away,' says Dick Thomas, consulting engineer at Posford Duvivier, maritime engineering consultants.

He says rough, sloping sea walls are now acknowledged as the best way of absorbing rather than reflecting the sea's energy, and walls of this type have been built at Mablethorpe and Skegness in Lincolnshire.

SERC is also investigating alternative 'soft' defences. 'The aim is to find ways of placing sand and shingle beaches and make sure they stay there,' says Michael Owen, chairman of the Coastal Defence Research Committee, a unit of SERC's environmental committee.

Beaches are often preferred to sea walls because they can absorb all the wave energy. But they do take up more room. A typical sea wall 5m high may need a strip 10m wide. A shingle beach 5m in height would typically have a slope of 1:8, and a sand beach a slope of 1:20.

The difficulty of designing beaches is to predict what slope is needed for different waves and tidal situations.

Beaches are also subject to drift and so need frequent maintenance. Despite this, they are cheaper than walls, which typically cost around £5000 per metre. For example, Owen estimates that in the long run the shingle beach constructed at Seaford, near Newhaven, will cost 75 per cent of building a sea wall. But in common with sea walls, beaches can have knock-on effects. It is recognised that raising the height of the sand beach at Bournemouth to cut down erosion has resulted in neighbouring beaches being deprived of their natural source of recharge.

Another problem with building soft defences is finding the right materials. Shingle is best because it absorbs most wave energy, but good shingle is difficult to find, Owen points out.

The Construction Industry Research and Information Association is about to start a £500000 project on coastline protection and beach management which will investigate this problem. 'The aim is to survey resources that for one reason or another are not currently exploited, ' says Judy Payne, research manager in water engineering at CIRIA.

Another approach, which according to Chris Birks, head of flood defence at the National Rivers Authority, is 'relatively novel in the UK', is the construction of artificial reefs to protect the coast by reducing wave energy. The NRA is constructing reefs 250m off the north Norfolk coast consisting of layers of rock held in place with mesh.

The trend in coastal defences is towards soft engineering, Birks says. 'This means working with the coastal processes and using nature against nature.'

The most fashionable example is strategic retreat. This means allowing the sea to inundate existing sea walls and create a salt marsh, building a new sea wall on higher ground.

The approach is favoured by the conservation body English Nature because it will create important wildlife habitats and by the ministry because it is much cheaper than building higher and higher sea walls.

No one advocates strategic retreat in areas where there is any risk to property, but there is a feeling that it is pointless to protect marginal farmland when farmers are being paid to take land out of production.

An experiment is taking place at Northey Island in Essex where the tide has been allowed to flood over the old sea wall to begin creating a salt marsh and a lower sea wall has been built on higher ground.

A 10 m stretch of salt marsh offers the same protection as 1 m on top of a sea wall. But Alan Gray, head of population ecology and genetics at the Institute of Terrestrial Ecology, says it is not just a matter of allowing land to be flooded by seawater.

'Salt marshes need to be designed just as much as seawalls and beaches are. It is essential to carry out tidal studies because there is a chance of altering the balance between the ebb and flow of the tide so that sediments are sucked out rather than being deposited.'

The life expectancy of a sea wall is 50 years – the same time it takes for a mudflat to develop into a saltmarsh. So a managed retreat scheme would have a much longer life than a hard defence.

However, Dick Thomas is sceptical about building soft defences to replace sea walls. 'Even if the sea wall has been undermined, with a bit of work it will be as good as new. If you go to a softer defence you are sometimes rejecting all existing assets.'

Figure 6.12 Newspaper report: Taking the soft option (*Source: Financial Times*, 3 June 1993)

nourishment schemes as part of a coastal flood defence strategy. The scheme shows how working with natural processes can be successful. The sediment was obtained from sources in the Bristol Channel. The scheme was completed in 1984 and renourishment was expected to be needed every ten years. To date there has been very little loss of beach sediment as the Sand Bay area is a closed sediment cell.

Stable bays

Bays act as discrete sediment cells which trap sediment to build up a beach. The shape of a bay lengthens the coastline, refracts waves, and therefore

Figure 6.13 Artificial headlands and bays (*After:* MAFF, 1993)

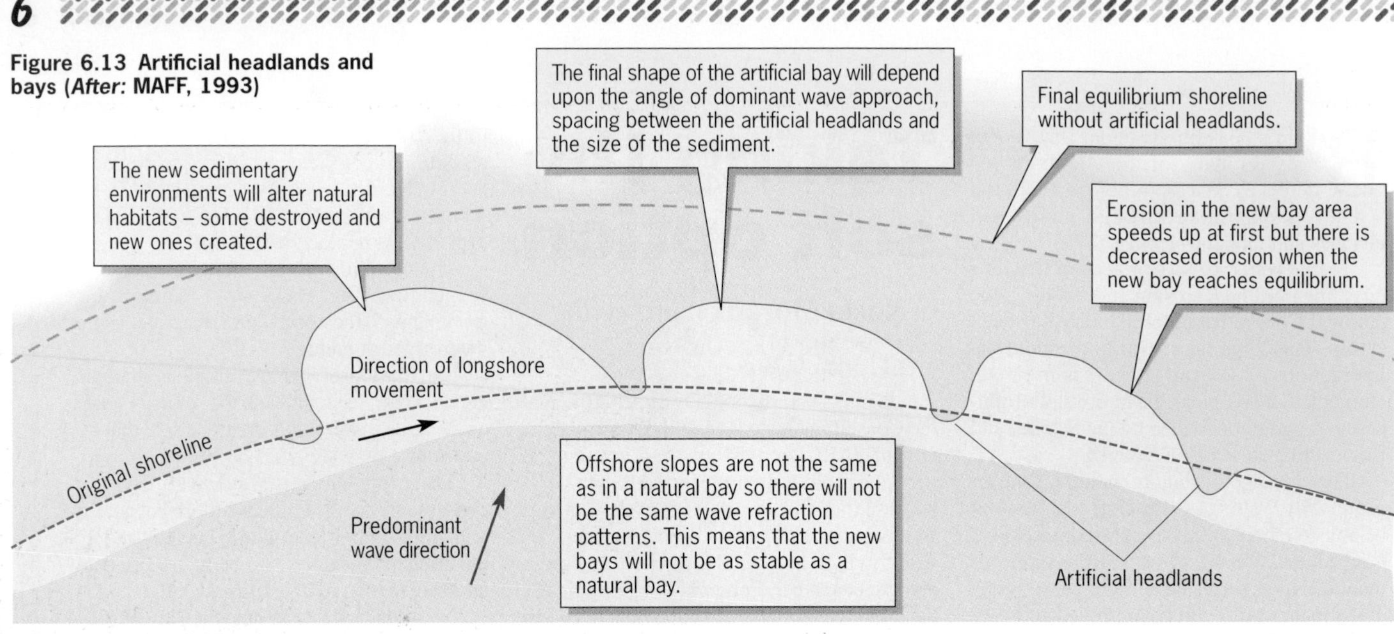
The final shape of the artificial bay will depend upon the angle of dominant wave approach, spacing between the artificial headlands and the size of the sediment.
Final equilibrium shoreline without artificial headlands.
The new sedimentary environments will alter natural habitats – some destroyed and new ones created.
Erosion in the new bay area speeds up at first but there is decreased erosion when the new bay reaches equilibrium.
Direction of longshore movement
Original shoreline
Predominant wave direction
Offshore slopes are not the same as in a natural bay so there will not be the same wave refraction patterns. This means that the new bays will not be as stable as a natural bay.
Artificial headlands

?

7 What are the implications for the Holderness coastline of allowing the equilibrium zeta-curve bay to develop (Fig. 6.14a)? Why might this be unacceptable?

8 Describe and explain how the 'reef' proposals work (Fig. 6.14b).

9 Suggest why some people argue that the coastal protection measures along the Holderness coastline made the 1996 breach of the Spurn Head spit more likely.

reduces the wave energy per unit area of coastline. A relatively new approach to coastal protection is to try to create a bay which will trap sediment and reduce wave energy in the same way as a natural bay. Artificial bays (Fig. 6.13) are created by building artificial headlands as strongpoints, or by building offshore breakwaters. Remember the reef of tyres proposals for the Holderness coastline (Fig. 4.1). In this technique the line between soft and hard engineering is blurred, but the technique aims to work with natural processes once the artifical hardpoints have been created by engineering techniques (Fig. 6.14). There are some major environmental impacts as a result of the creation of a new depositional environment. Currently, artificial bays are only considered when more expensive techniques are not economically viable, and after there has been study and modelling of natural processes in the area.

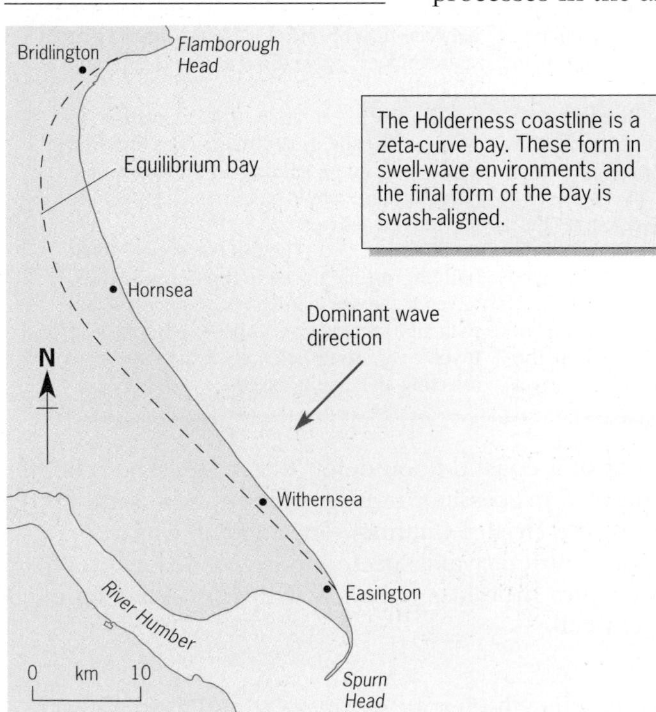
Bridlington
Flamborough Head
Equilibrium bay
The Holderness coastline is a zeta-curve bay. These form in swell-wave environments and the final form of the bay is swash-aligned.
Hornsea
Dominant wave direction
N
Withernsea
Easington
River Humber
0 km 10
Spurn Head

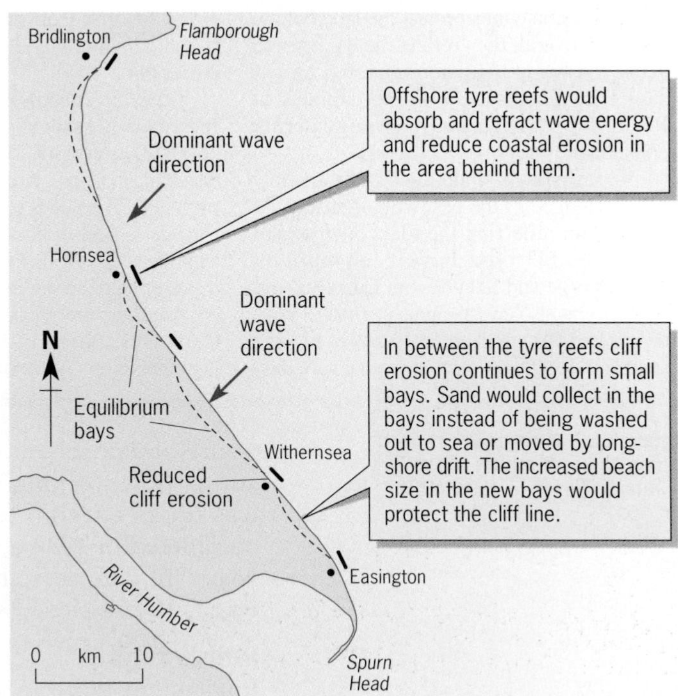
Bridlington
Flamborough Head
Offshore tyre reefs would absorb and refract wave energy and reduce coastal erosion in the area behind them.
Dominant wave direction
Hornsea
Dominant wave direction
N
In between the tyre reefs cliff erosion continues to form small bays. Sand would collect in the bays instead of being washed out to sea or moved by long-shore drift. The increased beach size in the new bays would protect the cliff line.
Equilibrium bays
Withernsea
Reduced cliff erosion
Easington
River Humber
0 km 10
Spurn Head

Figure 6.14 The Holderness coastline: equilibrium forms

Managing cliff erosion at Barton on Sea, Hampshire, England

The 30-m-high cliffs at Barton on Sea are in a heavily developed area in the centre of Christchurch Bay, Hampshire. The complex subaerial processes operating on the cliffs are shown by the unmanaged section of the cliffs at Naish Farm 1.2 km to the west (see the case study on p.77). The effective environment for wave erosion within Christchurch Bay (see Fig. 4.15) means that there has been a long history of coastal protection measures. At Naish Farm, the cliff-top land has not been heavily developed and is used for wooden chalet and caravan holiday accomodation. At Barton, however, there has been intensive urban development in a 2-km-deep strip extending from Barton inland to New Milton. Erosion rates at Barton were 0.5 m per year until the 1950s, but, as at Naish, the rates increased following the building of the major groyne at Hengistbury Head and defences at Christchurch. The built-up area of Barton was therefore under increased threat from erosion. The first defences at Barton were a timber groyne system, built in the 1930s. By 1960 these were in disrepair and the cliff line was virtually unprotected against recession rates of 1 m per year. The cliff top was 10 m from a group of shops, flats and cafés. Since 1964 protection work and emergency measures have been continual. In order to be effective, both cliff-foot erosion by wave activity and the cliff-face mass movement processes needed to be addressed.

Cliff-foot strategies

Between 1966 and 1968, 1800 m of the cliff foot was protected by timber groynes and revetments of wooden piles filled with rock (Fig. 6.15). During February 1974, severe storms resulted in waves breaching the revetment and damaging 200 m of the defences. Since 1972, there has been gradual replacement of the wooden groynes with seven rock strongpoints between 300 m and 500 m apart (Fig. 6.16). These aim to control longshore movement of sediment in a similar way to the wooden groynes, and they have proved to be effective in this as well as having a stronger structure to resist storm waves. The old timber revetments were replaced by a 1.8-km-long rock revetment in 1991 (Fig. 6.17) at a cost of £4.5 million, with a design life of 30 years (Fig. 6.18).

The increased rates of erosion at Naish and Barton had been monitored by the New Forest District Council. It was hoped that by converting the most westerly wooden groyne at Barton into a rock strongpoint a stable bay would form between this and the Chewton Bunny outfall (see Fig 5.10), this structure acting as the second strongpoint to provide some protection to the Naish cliffs. Unfortunately, this scheme had limited success. Plans for a further

Figure 6.16 A rock strongpoint at Barton on Sea, Hampshire, looking south into Christchurch Bay. The build-up of beach material on the updrift (west) side of the strongpoint is clearly visible. These rock structures have been more successful at retaining beach material than the 1960s wooden groynes that they have replaced.

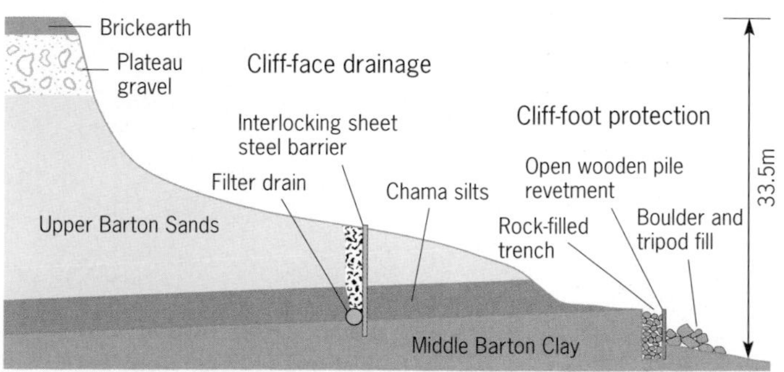

Figure 6.15 Cross-section of coastal defence works at Barton on Sea (*Source:* Clark et al., 1976)

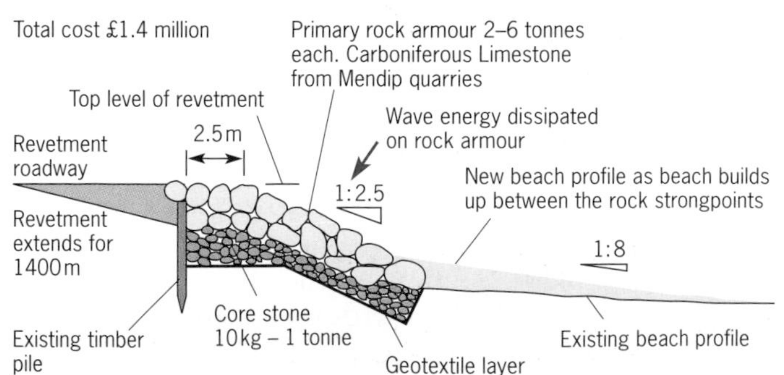

Figure 6.17 Typical cross-section through a revetment (*Source:* New Forest District Council)

Barton on Sea

Figure 6.18 The rip-rap of the rock revetment at the cliff foot, Barton on Sea, with the newly constructed revetment roadway behind. The high-energy waves breaking on the shore are noticeable as this photograph was taken on a hot, calm July day.

Figure 6.19 Installing cliff-face drainage pipes at East Barton, July 1995. The lines of the pipes are visible on the cliff face by the lines of stones. To the right of the working tunnel a drainage pipe can be clearly seen (*above right*).

three strongpoints and rock revetments at Naish were suggested but a cost-benefit ratio of 6:1 meant that the scheme was not economically viable (see p.120). This contrasts sharply with the more favourable cost-benefit ratios at Barton which have allowed the major defence works to proceed.

Cliff-face strategies

The first engineering works (1964–8) at Barton comprised a cliff drainage scheme with an interlocking sheet-steel barrier to intercept the water movement though the cliff which produces much of the mass movement (Fig. 6.15). However, in 1974, major mass movements occurred. The storm which had damaged the revetments saturated the undercliff. Following this, during November 1974, there was a period of heavy rainfall. The resulting cliff failure damaged 200 m of the steel barrier and pushed some of it from its vertical position to near horizontal. A toe of landslip material extended 10 m across the beach, and the upper cliff moved by 3 m as it settled behind the damaged steel piling.

By 1995 much of the original drainage work had been replaced (Fig. 6.19), with underground channels in a fan shape which drain water into a drain at the cliff foot (Fig. 6.20). Sections of the cliff have been regraded and strengthened with rock armouring (Fig. 6.21). The free face of the upper cliff will continue to retreat by weathering and mass movement until it reaches a stable angle (Fig. 6.22).

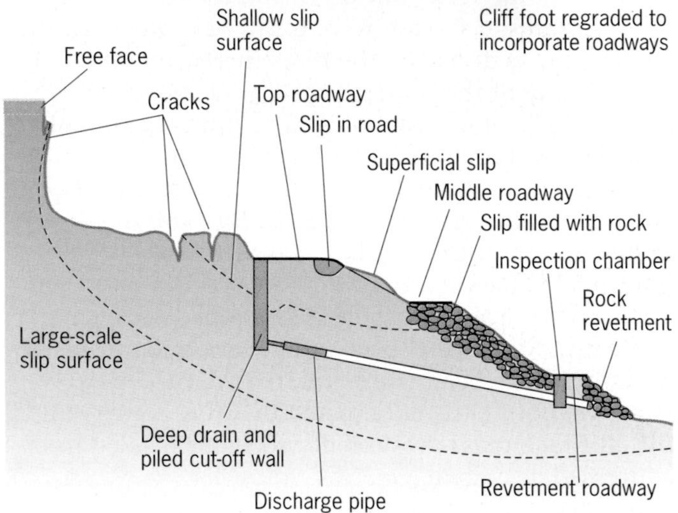

Figure 6.20 Cliff cross-section showing typical drainage and slip-surface details (1995 scheme) (*Source:* New Forest District Council)

Figure 6.21 An engineered landscape: the boulder revetment, revetment roadway, upper revetment and rock armouring of the lower cliff face at Barton on Sea, Hampshire, looking east

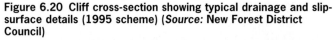

Figure 6.22 Barton on Sea, Hampshire, looking east towards Hurst Castle spit and the Isle of Wight. The upper cliff shows continued weathering and will retreat until it reaches a stable angle. The cliff-foot revetments and rock strongpoints are visible, with sand building up on the updrift (west) side of the rock strongpoints.

Local issues

At Barton on Sea, the protection strategies employed since the 1960s have been generally successful at reducing cliff erosion, although this has involved large capital investment. Between 1965 and 1987, £5.4 million (at 1987 prices) was spent on the schemes and emergency works, and a further £4.5 million on the works since 1987. The cliff and beach area has been visually transformed in the bid to stop the erosion. The local residents have been actively involved in the issue, although there has been conflict between those directly affected by the cliff erosion and those not. The defence measures in the 1960s had helped to instil a perhaps false sense of security at Barton, with new building developments taking place and some residents extending their properties.

At Naish, the proposed works were opposed by the Nature Conservancy Council (English Nature) since Christchurch Bay has been designated an SSSI. The geological formations and fossils in the cliffs are of scientific importance. Any defence works at Naish would obscure the cliff face and stop the production of new exposures by erosion, as has been the case at Barton. A further consideration is the consequences for the Barton cliffs and other areas downdrift to the east, such as Milford and Hurst Castle spit (see Fig. 4.15 and p.94), if there was reduced sediment input to the system by reducing erosion at Naish.

Hurst Castle spit to the east of Christchurch Bay in Hampshire is downdrift of Barton and has shown an increased tendency for breaching under storm conditions than it has in the past. Before 1954 breaching was very rare, but since then there have been several breaches of the spit. The spit has also moved 100 m inland since 1898. Behind the spit are sheltered waters and boatyards as well as saltmarshes which are of ecological significance. The way people use landforms often does not allow for them to be dynamic systems which evolve and change. The shingle bank of the spit has been protected at the western (proximal) end since 1966 by a rock armour rip-rap (Fig. 6.23). In 1985–6 work to strengthen the spit continued with the widening and steepening of a 450 m length of the spit. The crest of the shingle ridge has been widened to 6 m with a 1:3 slope each side. Ironically, the increased rate of breaching may be the result of the updrift management aimed at reducing cliff erosion and retaining beaches in front of the built-up parts of Christchurch Bay. The sediment inputs from cliff erosion, and the longshore drift of beach material to the spit have both been reduced.

Figure 6.23 The proximal end of Hurst Castle spit looking north-west (updrift) into Christchurch Bay. The rock armour rip-rap was added in 1966 to prevent the spit from breaching at this point.

10 Use Figures 6.15 and 6.20 to compare the 1995 cliff drainage scheme with the 1964–68 scheme.

11 Describe the 1991 cliff-foot strategies at Barton using Figures 6.16, 6.17 and 6.18.

12 Comment upon the environmental impact of the scheme (use the photographs to help).

13 Are the rock strongpoints successful in retaining a beach and reducing longshore drift? (Use Figures 6.16 and Figure 6.22.)

14 What is the evidence from Figure 6.22 that the cliff-top scarp is still retreating?

15 Should planning permission have been given for new developments and extensions at Barton?

16 What is the function of regrading the cliff face and toe armouring?

What is the nature of the problem (e.g. rates of erosion)?

What alternative schemes are available to use?

What is the risk involved to people and property?

What will the scheme cost? Is it economically justifiable?

Who will pay? Is grant aid available from MAFF?

Is the scheme environmentally acceptable?

Does the area have an amenity value and will this be affected?

What are the views of local residents and council tax payers?

What land uses need to be protected?

How many people are affected by the erosion?

How quickly do we need to respond?

Who else needs to be consulted? What are their views?

Will there be impacts elsewhere along the coastline?

Is the scheme technically feasible?

What schemes already exist, what condition are they in, and how effective are they?

Figure 6.24 Questions for consideration when choosing coastal protection strategies

6.3 Choosing a management strategy

The maritime councils involved in managing the coastline are required to consult with other authorities and other interested parties such as MAFF, NRA, county councils, and conservancy and navigation authorities, as well as the public, before any schemes can be implemented. Some private landowners and organisations such as British Rail and the highways authority may also undertake coastal protection measures on their land.

Once a coastal erosion issue has become apparent along a section of coastline, there are many factors to be considered before deciding on the best way forward. There are three broad options available to decision-makers.

DO NOTHING

Take no action at all to maintain existing defences or to build new ones.

SUSTAIN

This involves work to maintain the current standard of protection by repairing engineering structures or adding sediment by nourishment.

CHANGE

This strategy includes new control measures to improve the performance of existing defences; constructing a new defence forward of the existing shoreline; a new defence behind the existing defences; or creating a tiered defence where the current position is maintained and a new defence created further inland.

The choice of the preferred option depends on a variety of factors (Fig. 6.24), the most important criteria being technical feasibility, economic viability, and the results of an environmental impacts assessment.

Technical feasibility

Whether a scheme is technically feasible is primarily the concern of coastal engineers. Not all schemes will be appropriate to an area, for they will depend on the complex natural processes operating locally. The need for a detailed knowledge of coastal processes and continued monitoring is important in these decisions. Other factors such as local geology are also important: for example, at Barton on Sea, a sea wall was not a technically feasible scheme due to the nature of the Barton Clay and its lack of suitability to form the foundations of a wall.

Economic viability

Cost-benefit analysis (CBA) is a widely used technique to analyse a project. The aim of CBA is to ensure that finite financial resources are allocated for the maximum benefit of people. The technique aims to balance the costs of a scheme, e.g. the costs of building and maintaining any hard-engineering structures, over their expected useful life (the design life). The benefits of a scheme would include the value of the buildings and infrastructure which would be protected from erosion if the project takes place. The end result is a cost-benefit ratio which is used to make the final decision on the most appropriate strategy. If the cost-benefit ratio is 1:1, the cost and benefits are equal and there would be a financial gain from the project equal to the expenditure. A cost-benefit ratio of 1:0.8 would mean that the benefits are less than the financial outlay and the scheme is therefore not economically viable. In these circumstances the 'do nothing' option is the likely result. Maritime councils are unlikely to receive grant aid from the government with a negative cost-benefit ratio. Where benefits exceed the costs, e.g. a cost-benefit ratio of 1:1.3, the scheme is more economically acceptable and the project is more likely to be approved provided that other considerations support the scheme.

CBA is becoming more sophisticated as a technique, but there are problems with its use. These stem from the difficulty of giving financial values to all the factors which need to be considered. It is difficult to assign monetary values to environmental impacts, in particular, although there has been some progress in this area of CBA. For example, if the erosion of beach material is adding to cliff erosion problems, the costs to the area from the loss of visitors can be included in the analysis. This can be significant for a holiday resort area and helps to put a financial value on the beach as an environmental asset through its amenity value. It is more difficult to put a value on human anxiety and suffering. Although CBA can value a house which may be lost through cliff recession, it cannot so easily put a value on that house as a home.

The Fairlight Cove scheme in East Sussex illustrates some of the controversy which can surround coastal protection schemes and CBA. The scheme consists of a 50-m-long rock revetment structure at the base of the soft clay and mudstone cliffs. This aims to reduce the wave undercutting of a soft band of clay near the base of the cliff. The project was completed in 1990 at a cost of £2.5 million, supported by a 70 per cent grant from MAFF. The balance was paid by East Sussex County Council and Rother District Council. Originally the CBA came to the conclusion that a coastal protection scheme was not economically viable, and Rother District Council adopted the 'do nothing' option. However, the local residents were unhappy with this decision and formed the Fairlight Coastal Protection Society (FCPS), which appointed its own consultants to consider the costs and benefits. There was evidence that the rate of cliff erosion was increasing due to a loss of updrift material. This was thought to have been caused by updrift coastal protection works and harbour works at Hastings 5 km away. Although this updrift loss has been reduced by installing a gate in the terminal groyne at Hastings and beach recharge schemes, the rate of cliff erosion was crucial in the outcome of the CBA. Faster rates of erosion would mean that more houses would be lost or made uninhabitable over the design life of the project. These would therefore need to be included in the CBA figures. The FCPS estimated that 57 houses would be lost over the next 100 years without protection measures. Although the scheme is estimated as having a design life of 75 years, the 100-year figure was used because a significant number of houses would have been threatened in the future just beyond the projected 75-year cliff line. Table 6.2 shows the results of the CBA produced by the FCPS based upon erosion rates for a 7-year (1981–88) and 15-year (1973–88) period. Since erosion rates are increasing, the average rate for the 7-year period is higher than that for the 15-year and give a higher benefit from the protection scheme.

Further controversy surrounded the pricing of houses and inflation over the project life (Table 6.2), plus the pricing of the loss of roads and services such as sewers, water and electricity. These would be damaged before some of the houses and so would need replacing, including a new sewage pumping station in 15–20 years, a new electricity substation, and the creation of new access roads. These were added to the costs in the 'do nothing' option, but as benefits in the proposed protection scheme. As a result, the need for coastal protection measures was reappraised and Rother District Council resubmitted plans for grant aid which were accepted by MAFF.

CBA may help to justify the financial outlay on a coastal protection scheme and help councils to receive grant aid from the government.

Table 6.2 Summary of the benefit-cost analysis for the Fairlight Cove coast protection scheme (*Source:* Penning-Rowsell et al., *The Economics of Coastal Management – a manual of benefit assessment techniques*, Belhaven Press, 1992)

Real house price inflation rate projected over scheme life	Sum of benefits (75 year scheme life)	Cost-benefit ratio costs = 1
7-year erosion lines (based on 1981–88 higher erosion rates)		
0.0	£2 163 488	1:1.26
1.0	£2 473 622	1:1.44
2.5	£3 179 135	1:1.84
3.0	£3 511 122	1:2.04
15-year erosion lines (based on 1973–88 lower erosion rates)		
0.0	£1 841 819	1:1.07
1.0	£2 125 444	1:1.23
2.5	£2 805 813	1:1.63
3.0	£3 142 284	1:1.82

Notes: The FCPS argued that house price increases due to inflation should be incorporated in the CBA. The impacts of possible inflation are shown for 0%, 10%, 25% and 30% over the design life of the scheme.

However, a large amount of money still has to be found from local funds provided by local taxpayers. This can raise important issues at the local level. Not all residents may be in favour of their tax money being used to protect a relatively small number of residents and businesses. Although the cliff-top residents at Fairlight Cove were able to prove their case financially, the local councils still had to provide £0.75 million from taxpayers most of whom are not affected by coastal erosion.

The nature of the land to be protected is crucial in the outcome of a CBA (see Fig. 6.25). In recent years, these economic criteria have become even more important. The result is that only relatively high-density housing areas, and commercial and industrial properties, are likely to receive protection measures, especially for the more expensive techniques such as sea walls.

Environmental considerations

The environmental impacts of the main types of protection works have been discussed. These are important considerations in deciding on the type of scheme or whether a scheme should take place at all. Of particular concern are the impacts on sediment movement, especially downdrift impacts and land–coast exchanges if cliff erosion is reduced. These have been illustrated by many of the examples in this book. Sediment exchanges can be affected over a large area within a coastal sediment cell and offshore. Time has shown that these impacts can be very severe and result in the need for even more protection measures along the coastline where there may not previously have been a problem.

Figure 6.25 A landslide at Overstrand, Norfolk, which undermined the cliff road and the bungalows behind it. The slide began in January 1990, and caused a cliff retreat of 85 m in three years before being managed by a barrage and stabilisation.

The on-site impacts are also significant and can influence the outcome of the decision-making process. At Fairlight Cove, the cliffs are a geological SSSI. Any scheme which completely stopped cliff erosion would have been opposed since the rock exposures needed to be maintained for their geological value. The scheme that was implemented does not obscure the cliff face and allows continued weathering and marine erosion at a reduced rate. At the hamlet of Easton Bavents, near Southwold in Suffolk, a proposal for a sea wall was stopped in 1990 following objections from the Countryside Commission which claimed that the sea wall would damage one of the finest stretches of unspoilt coastline. It argued that the environmental

?

17 Should compensation be used as a management technique?

damage was not justified to protect nine houses, and that the people should be compensated for the loss of their homes. Unfortunately, there is no provision for compensating people who lose their homes as a result of cliff erosion, and many householders find it difficult to insure their property as cliff erosion proceeds.

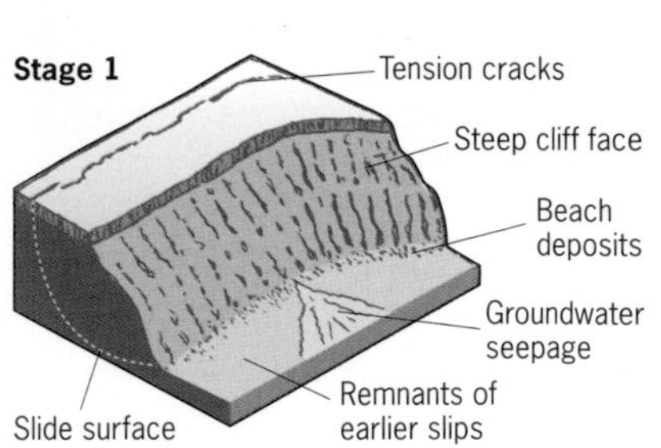

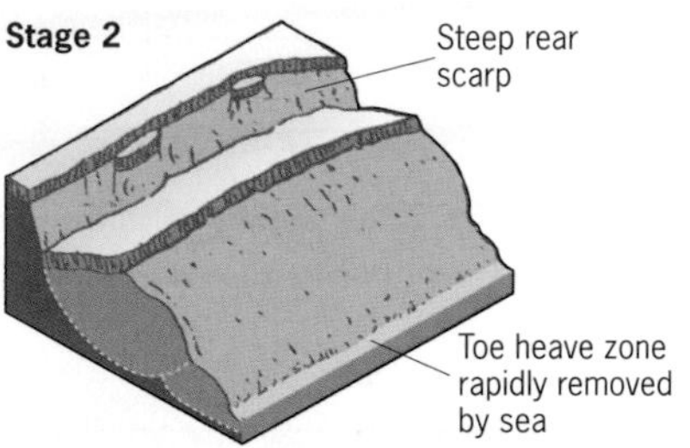

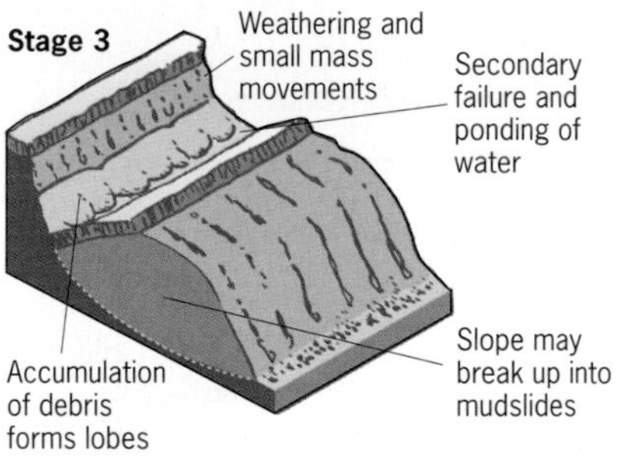

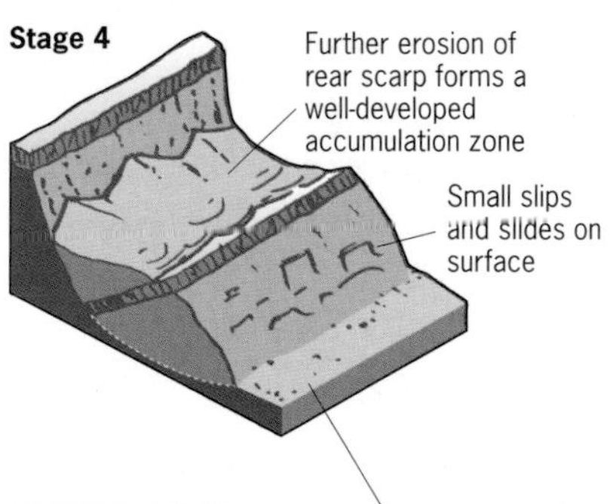

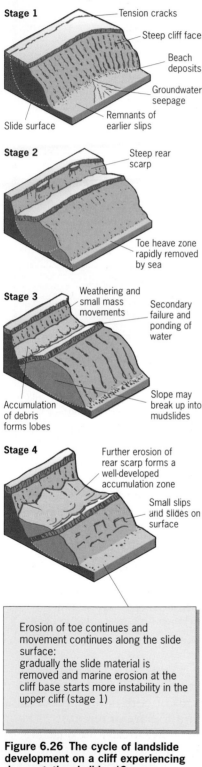

Erosion of toe continues and movement continues along the slide surface:
gradually the slide material is removed and marine erosion at the cliff base starts more instability in the upper cliff (stage 1)

Figure 6.26 The cycle of landslide development on a cliff experiencing deep rotational slides (*Source:* Richards and Lorriman, 1987)

6.4 Coastal management and the future

Rising sea levels

Global sea level reached its present position during the last 3000 years, and the eustatic rise is approximately 1mm per year. This small rise means that natural processes can maintain an equilibrium by moving sand and shingle to maintain the most energy-efficient beach profile or by adjusting cliff-recession rates so that shoreline platforms remain at a relatively constant level with wave height. However, a sudden rise of sea level as a result of human-induced global warming means that natural processes will not have time to make adjustments to the new conditions. Deeper water at the foot of cliffs will result in greater wave energy to increase cliff-foot erosion. Material delivered to the cliff foot by mass-movement processes will be more quickly removed, resulting in a shorter period of toe protection and therefore an increase in the rate of the cliff erosion cycle (see Fig. 6.26). The consequences will be most significant in areas of soft-rock and glacial cliffs in the more densely populated south and east of England. The result will be a need for even more expensive protection measures in the future. The strategies of coastal planners are having to be modified in the light of these possible changes, and due to the increasing geomorphological, economic and environmental arguments against continued spending on defences. More councils are considering 'do nothing' or 'fall-back' strategies which restrict planning permission for any developments in areas of erosion threat. This allows more money to be available for protection in areas of more intensive development.

Shoreline management plans (SMPs)

Much coastal management in the UK to date has had a piecemeal approach by maritime councils acting for the benefit of their own areas. A greater understanding of coastal processes and the impacts of schemes on other areas of coastline has led to changes in the way the coastline is being managed. In the late 1980s, groups of local authorities organised themselves into coastal defence groups covering most of the coastline and have been working together to try to co-ordinate their management strategies. In 1995, MAFF encouraged the development of a more integrated approach to coastal management in the form of Shoreline Management Plans (SMPs) (Fig. 6.27). These are defined as 'a document which sets out a strategy for coastal defence for a specified length of coast taking account of natural coastal processes and human and other environmental influences and needs' (MAFF, 1995). SMPs are based upon the 11 major sediment cells making up the coastline of England and Wales (see Fig. 4.16). The key here is that the SMPs are based on units and boundaries based upon natural processes rather than the administrative boundaries of individual councils. Although sub-cells may initially form a more practical level for producing an SMP, the aim is that all the maritime councils and other authorities and interested bodies within the major sediment cells should work together to manage the coastline. This should help to reduce downdrift impacts and produce a more coherent approach to coastal management. SMPs are voluntary strategy plans along with Coastal Zone Management Plans which consider wider coastal issues.

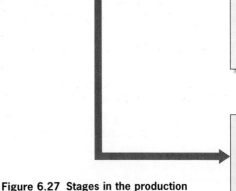

The aim of a SMP is to provide the basis for sustainable coastal defence policies within a sediment cell and to set objectives for the future management of the coastline.

Stage 1 Data collection and analysis

Collect and analyse data regarding:

a Coastal processes – e.g. physical characteristics; present-day processes; rates of erosion and deposition; areas at risk from erosion and/or flooding; offshore characteristics; historical and future evolution of the coastline; and identification of any gaps in knowledge.
b Coastal defences – ownership; existing defences and their location, type, condition and effectiveness.
c Land use and the human/built environment – e.g. commerce, ports, aggregate extraction, residential, industrial and commercial land uses, infrastructure and recreation.
d Natural environment – areas of conservation, ecosystem distribution, areas of biological, geological, geomorphological and landscape interest.

Where possible, data should be mapped for use in a Geographical Information System.

Stage 2 Set management objectives for the plan

Some objectives may apply to the whole sediment cell and others may be more specific. These objectives should form the basis for the future appraisal and development of strategic defence options.

Consultation
Consultation at all stages with all those with an interest in the area so that links are established; objectives for the coast are identified and possible areas of conflict addressed; information is gathered on coastal processes, coastal defence and the natural human environment. Agreement is reached regarding the SMP.

Stage 3 Plan preparation

a Define management units within the sediment cell/subcell
b Appraise strategic defence options
Those available are:
- do nothing
- hold existing defence line by maintaining or changing the standard of protection
- advance the existing defence line
- retreat the existing defence line.

The option(s) chosen must be considered in terms of its effect on adjacent management units and the sediment cell as a whole. They must be based on the objectives for the plan area and each option must be evaluated in economic, engineering and environmental terms.
c Compile the plan
The preferred strategic defence options which have been identified for each management unit are presented as a single document – the SMP. Individual operating authorities, e.g. marine councils, will develop strategy plans covering the management units within the area of responsibility. This will include a detailed assessment of coastal defence option(s) chosen for each management unit.

Stage 4 Monitoring and review

The SMPs are working documents and need to include details of on-going and future monitoring requirements. The plan needs to be reviewed at regular intervals and should include within it a timetable for updates. This review will and should consider any integration with other coastal initiatives.

Figure 6.27 Stages in the production of a Shoreline Management Plan

One of the stages in the preparation of an SMP (Fig. 6.27) is to identify management units within the sediment cell. A management unit is 'a length of shoreline with coherent characteristics in terms of both natural processes and land use' (MAFF, 1995). These units are likely to have different characteristics when any schemes are appraised economically for coastal defence measures: for example, a town is a single management unit and will have more favourable cost-benefit ratios for expensive schemes than rural areas (Fig. 6.28).

Coastal defence options for each management unit should be identified. The options available are:

- do nothing
- hold existing defence line by maintaining or changing the standard of protection.
- advance the existing defence line
- retreat the existing defence line.

Each option should be considered in relation to its effect on adjacent management units and the sediment cell as a whole. They must be based on the objectives of the Shoreline Management Plan (Fig. 6.27)

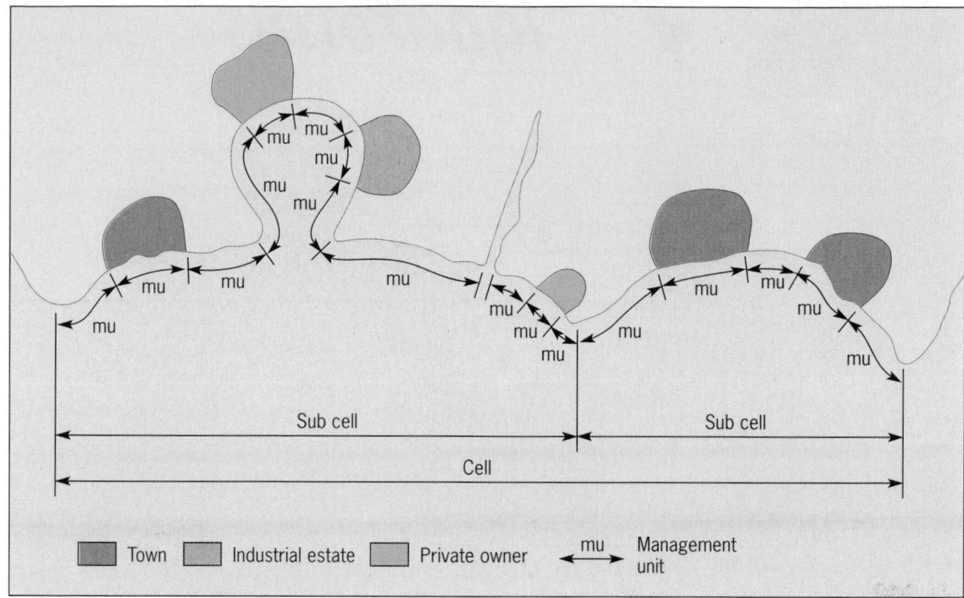

Figure 6.28 Sediment cells and management units (*Source:* MAFF, 1995)

18 For an area of coastline with which you are familar, e.g. from fieldwork or Ordnance Survey maps, divide the coastline into distinct management units. Justify your decisions.

19 Essay. Continued spending on hard-engineering coastal defences cannot be justified in geomorphological, economic or environmental terms. Discuss the validity of this statement with reference to examples you have studied.

Summary

- Coastal management is the responsibility of a range of organisations.
- Human activity conflicts with coastal processes due to 'static' human uses of dynamic systems
- People attempt to manage eroding coastlines with a range of coastal protection measures. These include hard and soft engineering strategies which aim to reduce cliff-foot and/or cliff-face processes.
- The choice of management strategy depends upon a range of factors, especially technical feasibility, economic viability and environmental considerations. CBA is a major technique used to assess economic viability.
- Coastal management strategies are likely to have to be modified if sea levels rise at faster rates due to global warming.
- Shoreline Management Plans are an attempt to integrate coastal management strategies and the different organisations involved, based upon coastal processes in major sediment cells.

7 Glacial processes and landforms

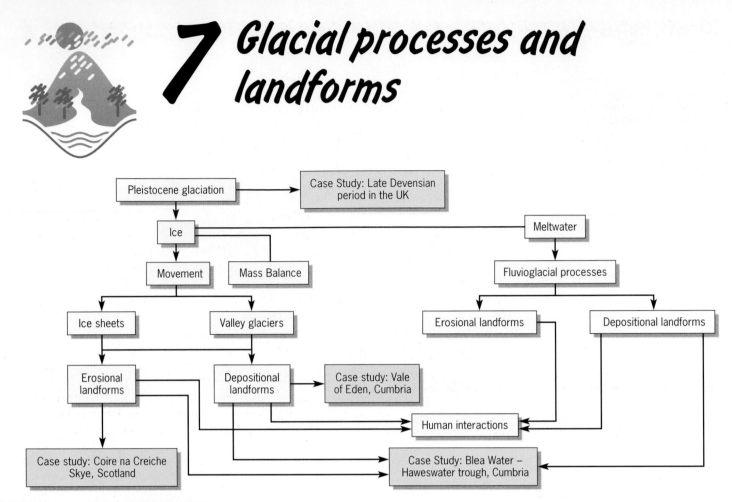

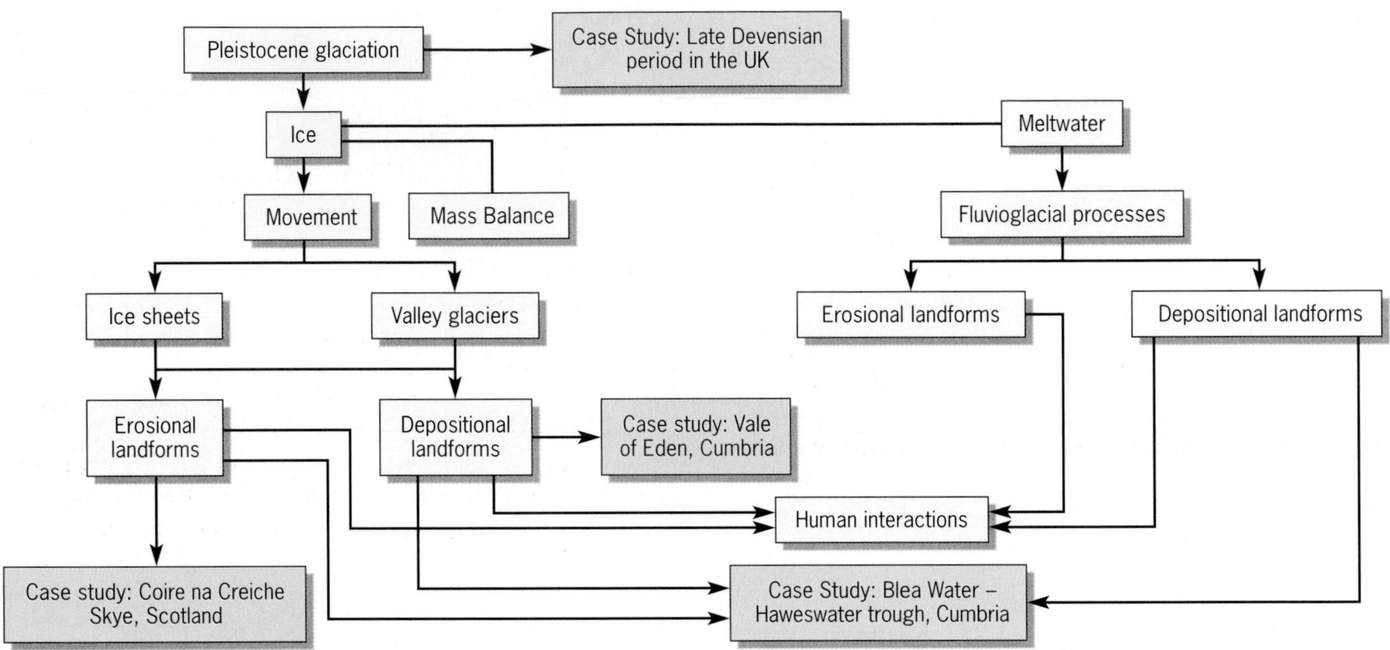

Figure 7.1 Yosemite Valley, California, USA, a fully developed glacial trough. The waterfall, Bridal Veil Falls, spills from a hanging valley carved by a small glacier which fed the main valley glacier.

7.1 Introduction

Ice is a powerful agent of erosion–transportation–deposition, capable of creating spectacular landforms (Fig. 7.1). However, the landforms which make up glaciated landscapes are not the work of ice alone. If you visit a glacier during the summer, you will see numerous streams, fed by melting snow and ice, running across the ice and emerging from the glacier margins (see Fig. 7.7) These seasonal *meltwater streams* produce a range of erosional and depositional *fluvioglacial landforms*. Thus, we use 'glacial' as a general term covering landforms resulting from the work of ice and meltwater, with 'fluvioglacial' referring to a subset of features formed by glacial meltwater. Note, however, that 'glacial' is also used as a noun, i.e. a 'glacial', meaning an episode of glacier advance – so take care!

As you follow the story of glacial landforms through this chapter, keep two key understandings in mind: first, a glacier is a *system* (Fig. 7.2); second, a glaciated landscape will be made up of a variety of glacial and fluvioglacial landforms.

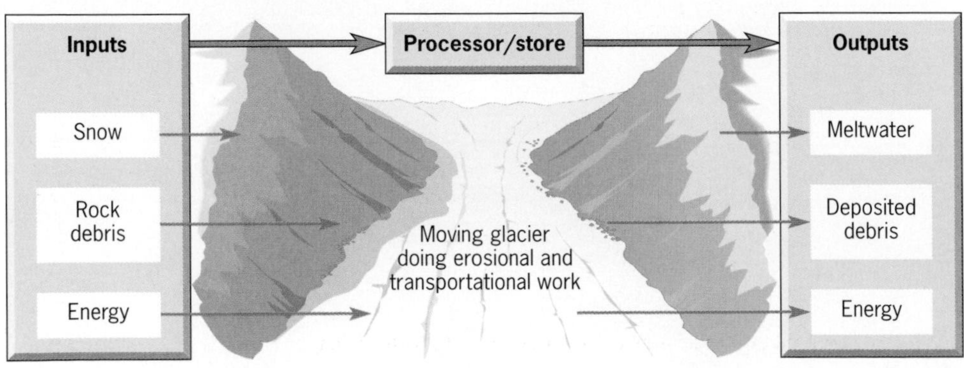

Figure 7.2 A glacier as a system

7.2 The Pleistocene glaciation

The glacial landforms we see in the landscape today are the outcome of the *Pleistocene glaciation* when ice sheets advanced well beyond their present-day limits. This began around 2.4 million years BP (Before the Present), as global temperatures fell. Across the British Isles, the Pleistocene epoch lasted about two million years and ended around 10 000 BP, when the last permanent ice disappeared. Yet ice still covers approximately 10 per cent of the earth's surface.

Three key characteristics of the Pleistocene glaciation provide a useful basis for our understanding of glacial landforms. First, this was not a single 'Ice Age', with one steady advance of the ice sheets, followed by a retreat. Within the two million or so years, temperature fluctuations were large enough to trigger a number of ice advances and recessions. The timing and number of these advance–recession cycles varied from region to region, but all experienced a series of glacial episodes, or *glacials*, separated by warmer *interglacials*. For example, Jones and Keen (1993) identify seven glacial–interglacial cycles for the British Isles (Table 7.1).

Second, the extent of ice advance during each glacial was different (Fig. 7.3). Third, within each major glacial, there were complex, often localised fluctuations. These relatively short-lived pulses of ice advance are known as *stadials*, and the warmer periods of recession as *interstadials*. For example, the most recent stadial in the British Isles, known as the Loch Lomond Stadial, was completed in only 1000 years (see Case Study: The Late Devensian).

Table 7.1: The Pleistocene sequence in the British Isles (*Adapted from:* Jones and Keen, 1993)

Approximate date of commencement (years BP)	Epoch	Glacial	Interglacial
10 000	Holocene		
115 000	Late Pleistocene	Devensian	
			Ipswichian
		Wolstonian	
300 000			Hoxnian
500 000	Middle Pleistocene	Anglian	
			Cromerian
		Beestonian	
			Pastonian
1 million		Pre-Pastonian	
			Bramertonian
	Lower Pleistocene	Baventian	
			Antian
		Thurnian	
2 million			Ludhamian

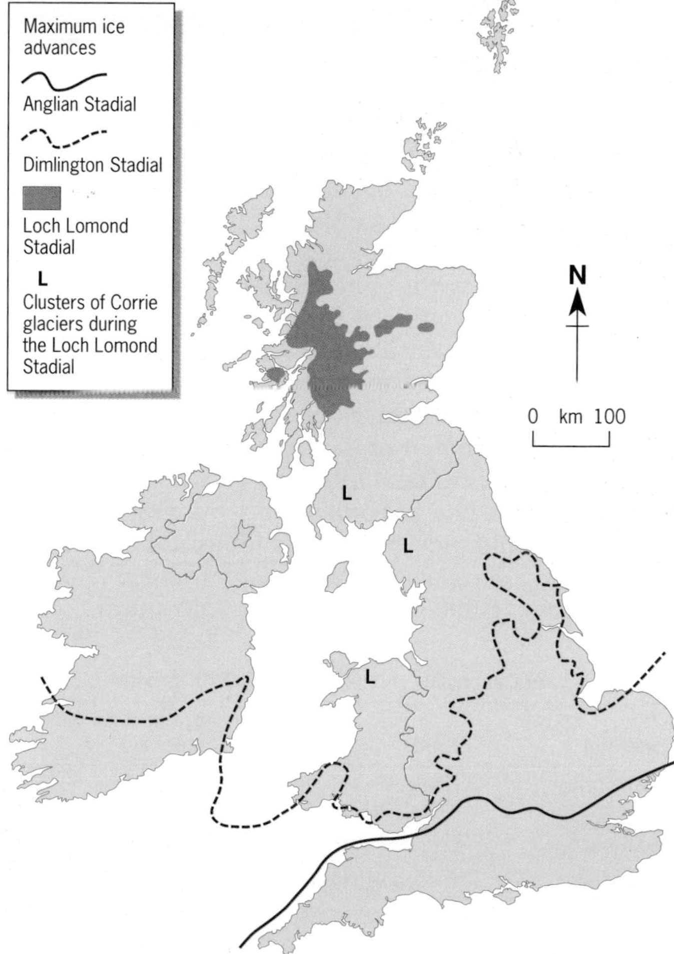

Figure 7.3 The limits of three glacial advances in the British Isles

One important outcome of these characteristics is that any specific landform is the result of work done during one or more glacial–interglacial episodes. For example, major troughs such as the Yosemite Valley in the USA (Fig. 7.1) have been progressively enlarged during several ice advances. Thick sheets of glacial deposits such as those which cover much of East Anglia (up to 143 m thick) and the English Midlands have been built up in layers. Earlier materials would be modified during interglacials and subsequent glacier readvances.

It is important to remember that periglacial conditions and processes would be found across a broad zone beyond the ice fronts. This zone would shift as the ice fronts advanced and retreated (see Chapter 8 for coverage of periglacial landforms). Periglacial processes modified pre-existing glacial and fluvioglacial landforms.

Remember, too, that in deglaciated regions such as the British Isles, we are studying relict or 'fossil' landforms. However, visit a glacial region today, e.g. the Swiss Alps, and you can see what the UK would have been like thousands of years ago. Comparison of the landforms left by glaciation with current processes allows us to make two general statements:

1 Medium- and large-scale erosional landforms are likely to have been influenced by successive glacial advances.

2 The form of depositional features tends to result from conditions and processes at work during the most recent glacial–interglacial cycle.

The Late Devensian: The most recent glaciation of the British Isles

The Devensian period was triggered by a rapid drop in average temperatures around 115 000 BP. Climatic fluctuations caused at least three pulses of ice advance and recession during this period. The most extensive ice advance occurred during the *Late Devensian*, between 26 000 and 10 000 BP (Table 7.2). Notice that even within this relatively short time frame, climatic conditions fluctuated sufficiently to divide the Late Devensian into two stadials of ice advance, separated by an interstadial.

During the Dimlington Stadial ice spread across more than 50 per cent of the British Isles (Fig. 7.4). Figure 7.5 shows in detail the complex relationships between the ice and the underlying topography in eastern England. Across the land areas not covered by ice, many periglacial landforms developed.

Yet within 10 000 years of this ice maximum, permanent ice had almost disappeared during the Windermere Interstadial. An even more vivid illustration of the sensitive response of ice accumulation and spread to climatic conditions is the rapid pulse of ice advance and recession during the 1000 years or so of the Loch Lomond Stadial (Fig. 7.3). Cooling was sufficient for an ice cap to develop over the uplands of western Scotland, and for small ice tongues to flow from the deeper corrie hollows of the Lake District and Snowdonia.

Table 7.2: Late Devensian chronology in the British Isles

Condition	Name	Approximate dates (years BP)
Ice recession and disappearance	Flandrian Interglacial*	Since 10 000
Ice advance	Loch Lomond Stadial	11 000–10 000
Ice recession	Windermere Interstadial	13 000–11 000
Ice advance	Dimlington Stadial	26 000–13 000

*This is the post-glacial or Holocene period, which is still proceeding

1 Explain why depositional landforms are likely to be the result of the most recent ice advance and recession episode.

2 Use Figure 7.3 to suggest what environmental conditions would have been like in your home area during each of the three stadials shown.

3 From Figure 7.4 name the principal ice accumulation areas which fed the Dimlington Stadial ice sheet and describe the patterns of ice movement.

4 Project: Identify and describe the main glacial landforms in your home locality. (NB You may wish to limit your study area to that covered by all or part of a 1:25 000 or 1:10 000 OS sheet or, if available, the 1:50 000 Drift Geology map. You may also wish to include periglacial features in your study. If so, you need to refer to Chapter 8.)

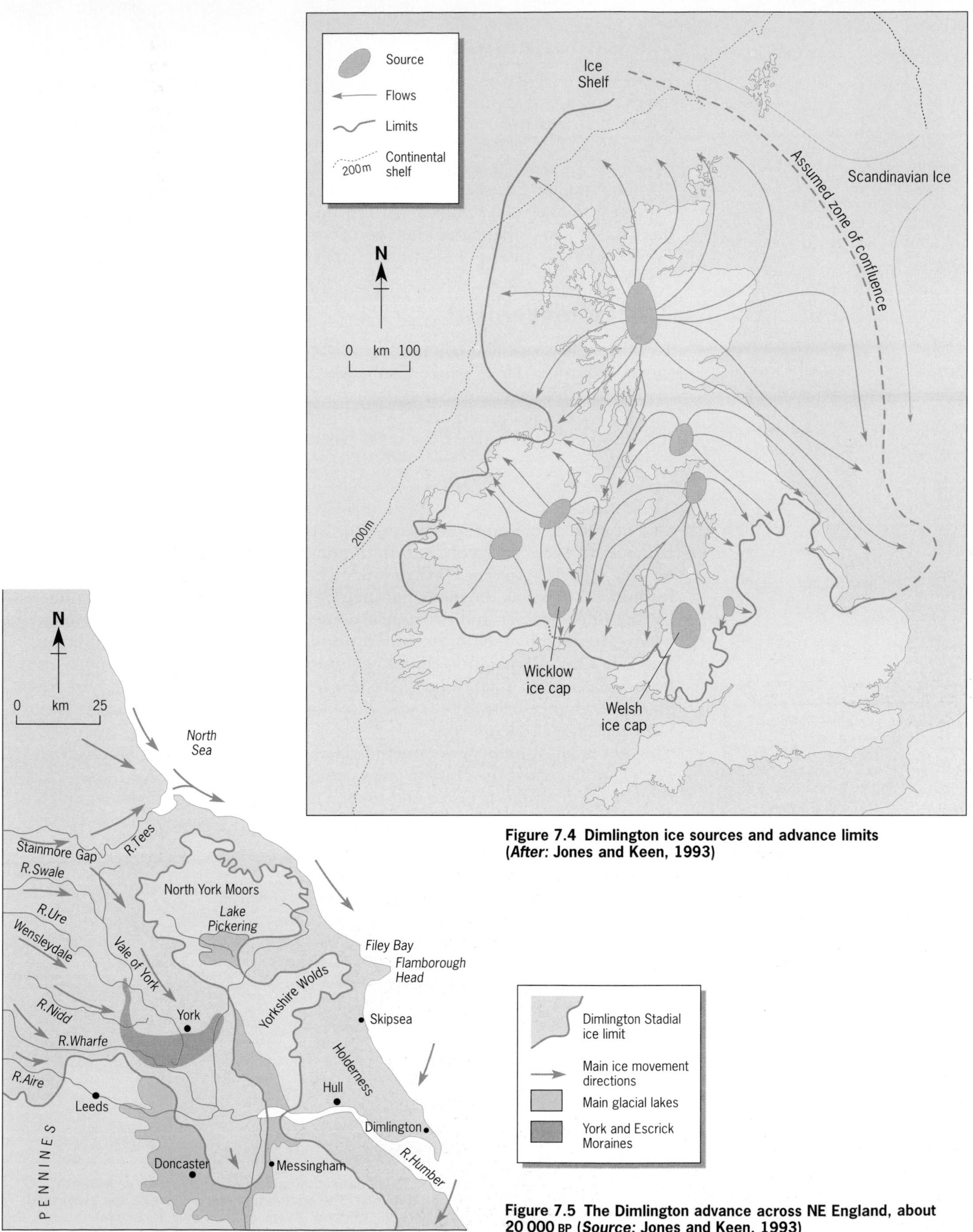

Figure 7.4 Dimlington ice sources and advance limits (*After*: Jones and Keen, 1993)

Figure 7.5 The Dimlington advance across NE England, about 20 000 BP (*Source:* Jones and Keen, 1993)

7.3 Ice formation

Fresh snow is fluffy or powdery, with air spaces between the crystals and flakes making up about 90 per cent of the total volume. That remaining after a year of summer warming followed by winter cooling is more granular, and is known as **firn** or *névé*. The altitude of the *firn line*, i.e. the isoline above which permanent snow and ice persist, varies with latitude.

Each winter, fresh snow covers the older layers which are progressively compressed and recrystallised. Air is gradually excluded and density increases – just as happens when you compress snow to make a snowball. Fresh snow has a density of around 0.06 g/cm^3 and glacier ice may reach 0.9.

7.4 Ice movement

To achieve its work of erosion, transportation and deposition, ice must move. Indeed, a glacier is like a conveyor belt: 'a sedimentary system that is involved with the accumulation, transfer and deposition of mass (snow, ice, water, rock debris) in response to additions and losses of mass and energy' (Chorley et al., 1984). Each ice mass has its unique history and rhythms of movement.

Mass balance

One important piece of information we need about a glacier system is the balance between ice accumulation (input) and ice loss (output) by **ablation**. So – is a glacier growing or shrinking? The year-to-year change in this **ice budget** is known as the **mass balance** of the ice. We calculate this budget by dividing a glacier into two zones, an *accumulation zone* and an *ablation zone*, separated by the *firn* or *equilibrium line* (Fig. 7.6). At the equilibrium line (see Fig. 7.7), ablation equals accumulation. Above this line, more ice accumulates than ablates, i.e. there is a net budget surplus. Ice moves down-glacier, and once below the equilibrium line, ablation exceeds accumulation, i.e. a net budget deficit.

This net balance influences how the glacier works and creates landforms. For example, when the climate grows colder (Fig. 7.8a), the ice mass thickens, movement is faster and the ice front advances. The glacier can erode and transport material more vigorously. In contrast, during warmer

The position of the equilibrium line is crucial in determining the balance between accumulation and ablation. When the equilibrium line is high on a glacier, the ablation zone is large, and the glacier is likely to shrink. Conversely, when the equilibrium line is low, the zone of net accumulation is larger and the ice mass will grow.

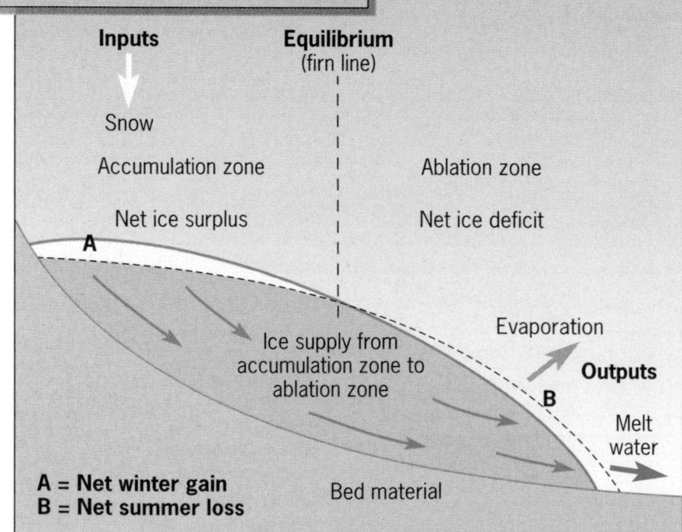

Figure 7.6 The mass balance budget of a glacier

Figure 7.7 Accumulation and ablation zones on an Alaskan glacier in August. The equilibrium line lies across the crevassed area. Note the brighter colour of the upper accumulation zone, the debris mantle of the ablation zone, and the summer meltwater emerging at the 'snout' of the glacier.

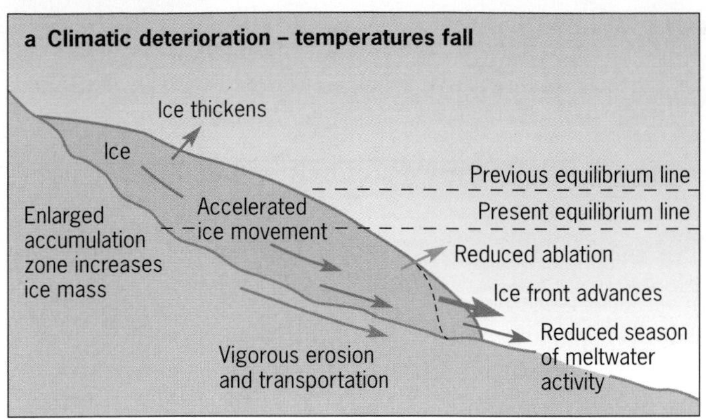

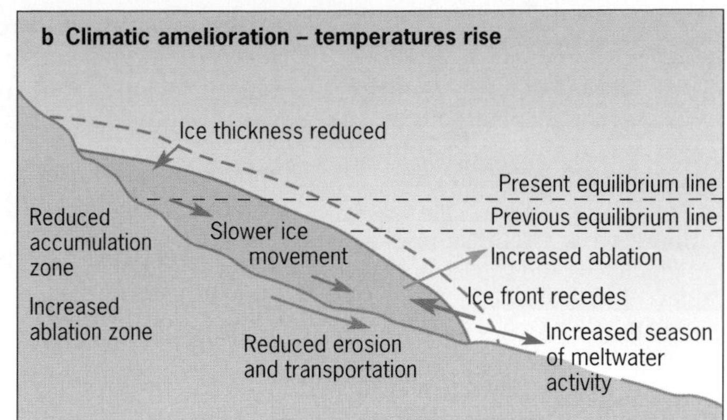

Figure 7.8 The impacts of climatic fluctuations on glacier budgets

5 Use labelled diagrams to summarise the role of the equilibrium (firn) line in the way a glacier behaves.

6 Use Figure 7.8 to describe and explain the relationship between climatic fluctuations and the geomorphological work done by the ice.

periods, the glacier shrinks (Fig. 7.8b), the ice becomes thinner, movement slows down and the ice front recedes. The glacier achieves less erosion, transfers less debris and deposits more of this load. When the rise in mean temperatures is particularly abrupt, ice movement may cease, leaving stagnant ice masses which melt in situ, and dump their debris as hummocky mounds. This occurred at the end of the Loch Lomond Stadial around 10 000 BP (see p.149).

Response times to changing budgets can be rapid, and ice fronts of many North American and European glaciers have receded considerably during recent decades. Portage Glacier in Alaska, for example, has retreated about 5 km during the past 100 years.

The mechanics of flow

As ice moves from an *accumulation source* it behaves with the characteristics of a plastic and a solid. As Figure 7.9 shows, there are three components to this movement: *basal sliding*; *internal deformation* and *bed deformation*. The balance between these components varies over time and space.

As with river flow, ice velocity increases as distance from the rock bed increases, because of diminishing frictional drag (Fig. 7.10). In a valley glacier, maximum velocities are achieved in the central, upper layers.

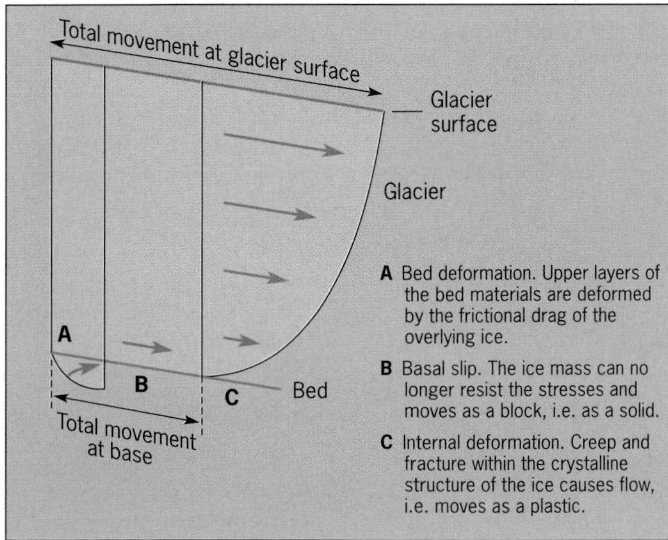

Total movement at glacier surface
Glacier surface
Glacier
A Bed deformation. Upper layers of the bed materials are deformed by the frictional drag of the overlying ice.
B Basal slip. The ice mass can no longer resist the stresses and moves as a block, i.e. as a solid.
C Internal deformation. Creep and fracture within the crystalline structure of the ice causes flow, i.e. moves as a plastic.
A
B
C
Bed
Total movement at base

Figure 7.9 Components of ice movement

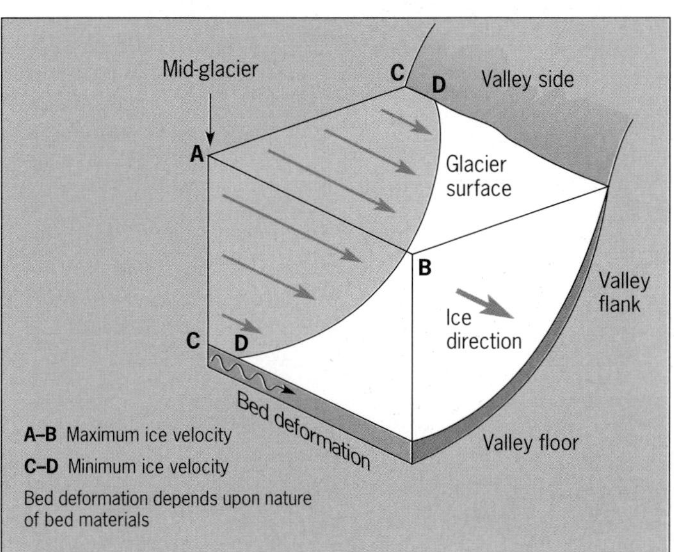

Figure 7.10 Ice velocity within a glacier

7 Essay: Use your understandings of ice movement mechanisms to explain
a why warm glaciers tend to move more quickly than cold glaciers;
b under what circumstances a cold glacier would move more quickly than a warm glacier;
c why flow velocities vary within a glacier.

Internal temperatures of glaciers

Ice movement is also influenced by the *thermal regime*, or temperature characteristics within a glacier. In temperature terms, we can group glaciers into two classes:

- *Warm glaciers*, in which the ice is at the pressure melting point throughout for much of the year.
- *Cold glaciers*, where all or most of the ice remains below the pressure melting point.

Pressure melting point is the temperature at which melting is caused by applying pressure to the ice. As pressure is increased, e.g. by increased ice thickness, the temperature at which water freezes falls. Thus, as ice at 0°C deforms 100 times faster than ice at –20°C, warm ice deforms more readily than cold ice. This influences the speed of flow.

The crucial factor is the temperature of the basal ice in contact with the glacier bed. Basal sliding is made easier where basal warming is sufficient to form a thin film of water between the ice and the bed materials. Conversely, in cold glaciers, ice may be frozen to the underlying bed and basal sliding is much reduced or may cease. Thus, in Antarctica the ice sheet is very thick, yet little erosion may take place where the ice is frozen to the rock bed.

The influence of gradient

A steepening of the slope may cause a glacier to accelerate (Fig. 7.12). However, as we have learned, ice thickness, mass, the mass balance budget and the temperature of the basal ice, all influence ice movement. In consequence, gradient alone is rarely sufficient to explain movement. Remember, too, that because an ice mass is driven by forces from the accumulation zone, ice is capable of flowing upslope (see section 7.8).

7.5 Erosion by ice

The photograph of the Yosemite Valley (Fig. 7.1) is awesome proof of the power of moving ice as an agent of erosion. A glacial trough is only one of a wide range of glacial erosional landforms which occur at all scales (Table 7.3). To explain these we need to understand the interactions between the behaviour of moving ice, the character of the bedrock, the processes at work and the timescale involved.

Table 7.3 Classes of glacial erosional landforms (*Adapted from:* Hambrey, 1994)

Process	Form	Landform type
Abrasion by glacier ice dominant	Streamlined	Areal scouring Glaciated valley, trough or fiord Hanging valley Whaleback Striations Polished surfaces
Combination of *abrasion and rock fracturing* by glacier ice	Part streamlined	Trough head Rock step Riegel/rockbar Cirque Col Roche moutonnée
Rock crushing by glacier ice	Non-streamlined	Crescentic surface fractures
Erosion by glacier ice and *frost shattering*	Residual	Arête Horn Nunatak

Figure 7.11 The 'milky' colour of this lake in the Canadian Rockies is caused by meltwater streams carrying rock flour from glaciers

Figure 7.12 Crevassed icefall cascading from the Kenai icefield, Alaska, USA

Erosion processes

Two groups of erosion processes are at work: at the glacier base or sole, where the ice is in contact with the bedrock, and across rock surfaces exposed above the ice.

Basal erosion

Erosion at the bed of a glacier involves the processes of *abrasion*, *rock fracturing* and *crushing*. Abrasion may be defined as 'the wearing down of rock surfaces by rubbing and the impact of debris-rich ice' (Hambrey, 1994). This grinding process depends mainly on the amount, distribution and character of debris in the glacier base; the speed of the basal sliding; the character of the bed materials; and the thickness of the ice, which affects the forces applied to the bed materials. Additional factors include whether there is a film of water at the ice–rock interface, and how quickly the eroded debris is removed.

The abrasion is done by debris particles embedded in the base of the moving ice. The repeated impacts of the particles in this narrow *basal zone of traction*, with the underlying surface, cause erosion of both the bed materials and the embedded particles. For this reason, abrasion produces a high proportion of *fines* (small fragments). The end product of this grinding process is *rock flour*, the tiny particles which give the milky appearance to glacial meltwaters (Fig. 7.11). (Think of the fine wood dust produced when you rub sandpaper over a block of wood.)

Fracturing occurs especially in well-jointed and bedded rocks, where pressures can wedge ice and debris into cracks, opening them up, and loosening a surface layer. Where rock joints lie roughly parallel to the bed surface, removal of overlying material may allow the new surface layer to *dilate*, or spring open slightly, and so become more easily loosened. Under thick ice, forces may be sufficient to *crush* the surface of even massive, poorly jointed bedrock.

Peripheral or marginal weathering and erosion

Rock surfaces adjacent to glaciers experience extreme environmental conditions, and are likely to have little or no vegetation cover. The dominant process is *frost-shattering*, produced by frequent repetition of the freeze–thaw cycle. How much debris is produced depends upon the character of the jointing and bedding of the rocks, the amount of water present, and the frequency and range of the temperature fluctuations. In mountainous areas, this process provides an ample debris supply to the glacier surface by rockfalls and debris flows (Fig. 7.12).

7.6 The collection and transport of load

Entrainment

Figures 7.7 and 7.12 show that frost-shattered rock debris falls on to the surface and margins of a glacier. The collection of basal load is more complex. Small fragments can be entrained by *basal ice freezing* around the particles, then applying sufficient drag to pull them along. In a similar way, ice may envelope a large boulder by *deformation flow*, and drag it into the glacier base.

Where the underlying bed is irregular, the key processes are pressure melting and **regelation** (Fig. 7.13) as the ice moves over the irregularity. The result of these processes is that the up-ice rock surface is smoothed, while on the down-ice side, ice-gripped blocks are *plucked* from the bedrock and entrained within the glacier base.

Figure 7.13 **Processes at the ice–rock interface: pressure melting and regelation**

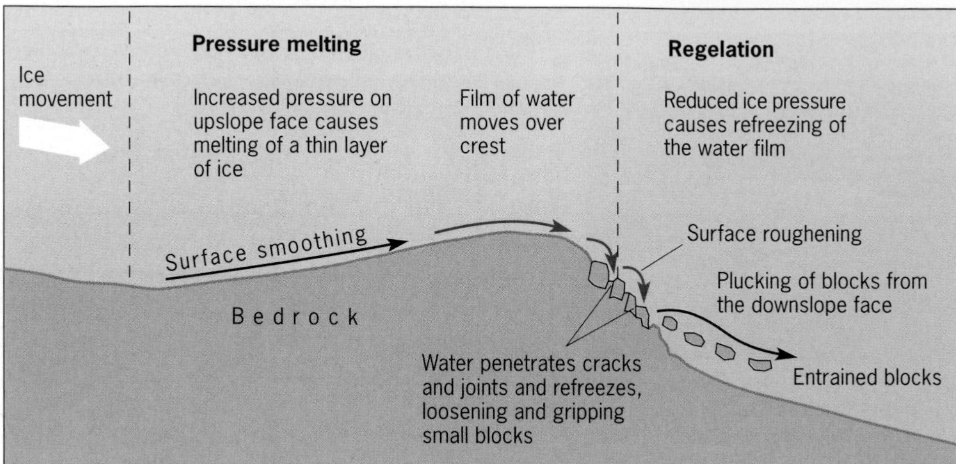

The combined work of these processes means that in the basal traction zone of the glacier, perhaps 1 m thick, debris may make up 50 per cent of the volume. The huge amounts of debris involved help us to understand the great erosive power of a glacier.

8 From Figure 7.14 and your understanding of the erosional processes,
a explain how the glacier obtains its debris load, and
b describe the journey of the debris along the glacier 'conveyor belt'. (Use the formal terms of the diagram, e.g. supraglacial, moraine, etc.)

The transport of load

The glacier 'conveyor belt' moves its debris load in three ways: on the surface of the ice (*supraglacial load*), e.g. lateral and medial moraines; within the ice mass (*englacial load*); and at the base of the ice (*subglacial load*), e.g. till or ground moraine (Fig. 7.14). Because of the complex structure and flow of the ice, debris may change location during its journey, e.g. supraglacial debris may fall through crevasse networks to become englacial and subglacial load. Near the 'snout' of a glacier, where ablation is rapidly reducing the ice mass, these three types of load converge, to give very 'dirty' ice. It is on this part of a glacier that we can appreciate the huge volumes of debris eroded and conveyed by the ice (Fig. 7.15).

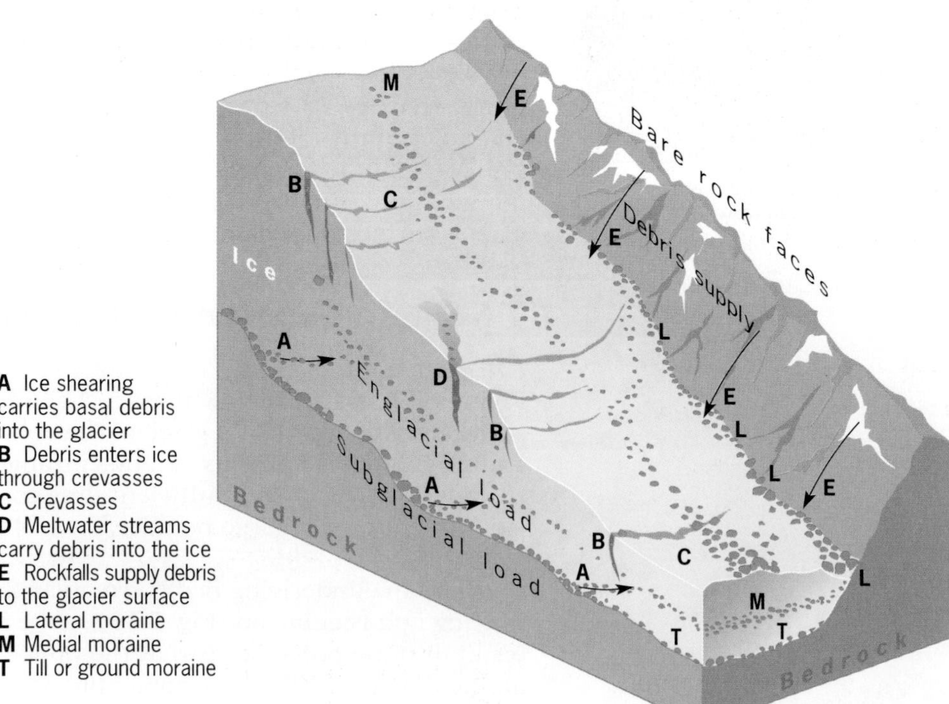

Figure 7.14 **The glacier 'conveyor belt': the transportation of debris**

Figure 7.15 Sheridan Glacier, Alaska, USA. Progressive ablation produces a dirty 'snout' – a surface mantle of moraine debris. Note the tunnel through which meltwater flows into a proglacial lake.

7.7 Landforms of erosion

Ice masses range from extensive ice sheets to small hanging glaciers. However, we can group them into two types in terms of the landforms they produce: ice sheets/caps, and alpine or valley glaciers. Some regions, such as the Canadian Shield, have been dominated by ice sheets. In other, more temperate regions, such as the British Isles, there have been lengthy periods of alpine glacier conditions. This history influences the range of erosional landforms we see today in any specific locality.

Ice sheet erosional landscapes

Erosion beneath an ice sheet varies widely. Key variables affecting this erosion are ice thickness, speed of movement, temperature regime of the ice, the underlying relief and geology, and the timespan involved. The erosional processes produce four main landscape types (Table 7.4).

Areal scouring of extensive surfaces produces landscapes usually of low relief, with smoothed and occasionally knobbly hills separated by depressions. These landforms are well illustrated in north-west Scotland, where scouring of ancient crystalline rocks has produced a 'knock-and-lochan' topography, of rocky knolls ('knocks') interspersed by depressions with small lakes ('lochans') (Fig. 7.16).

Figure 7.16 'Knock-and-lochan' landscape produced by glacial scouring across ancient gneiss in the Outer Hebrides, Scotland

?

9 Explain why areal scouring occurs most commonly beneath warm glaciers (refer back to section 7.4).

Table 7.4 Landscape types of ice sheet erosion (*Adapted from:* Chorley et al., 1984)

Landscape type	Ice character	Typical topography	Typical geology	Areal example
Areal scouring	Warm-based	Low-lying, flattish, with smoothed hills	Crystalline, well-jointed, impermeable rocks	Central Canadian (Laurentide) Shield; western highlands of Scotland; ice-sheet margins of Labrador
Selective linear erosion	Cold-based over plateaus; warm-based in troughs	Upland plateaux	Of little significance	Uplands in intermediate areas of ice sheets. Baffin Island; west Greenland; Cairngorms, Scotland
Few signs of glacial erosion	Cold-based with no basal load	Uplands and peninsulas	Most likely over permeable rocks, but common in 'dry' polar regions	Dry areas of Arctic Canada and Greenland; eastern Scottish lowlands of Buchan
Combination of sheet and alpine glaciation	Glacier types fluctuate over time	Mid-latitude uplands	Hard rocks	Fiord region of Norway

10 Which of the four landscape types of ice-sheet erosion result mainly from • warm glacier and • cold glacier conditions? Explain your choices for both.

Selective linear erosion increases the relief of the pre-glacial landscape. It occurs where the ice moves across uplands. Thicker, active ice above valleys causes further deepening, while thinner, slow-moving ice over the upland summits may produce little erosion. For example, the tors of the Cairngorm summits in Scotland are pre-glacial in origin, yet survived, while surrounding valleys were deepened. Landscapes with few signs of glacial erosion are likely to have lain beneath cold-based ice, towards the dry polar or continental limits. Here, ice is frozen to the bed, and movement is slow. Today, the landscape tends to have a weathered debris mantle with gentle slopes, scattered **erratic** blocks and occasional rocky **tors**, e.g. northern Arctic Canada.

The combination of ice sheets and alpine glaciation produces landscapes which are an extreme form of selective linear erosion, with smoothed upland surfaces broken by overdeepened troughs. Thick, relatively quickly moving ice streams following pre-existing valleys exert powerful erosional forces, capable of carving the spectacular fiords of Norway, Patagonia and South Island, New Zealand.

Do not, however, always think on the large scale, because the work of ice sheets may vary over short distances to produce localised or micro features.

How much work?

One of the most intriguing questions about ice sheets concerns how much they lower the landscape. In the case of troughs carved by vigorous ice streams, the answer may be measured in hundreds of metres. Across the expanses of scoured topography, however, 'there is formidable evidence to suggest that the depth of material removed is of the order of ten to a hundred metres' (Chorley et al., 1984). In north-east Scotland, Hall (1983) has estimated that the Pleistocene ice sheets lowered the landscape by less than 50 m.

a Ice at work

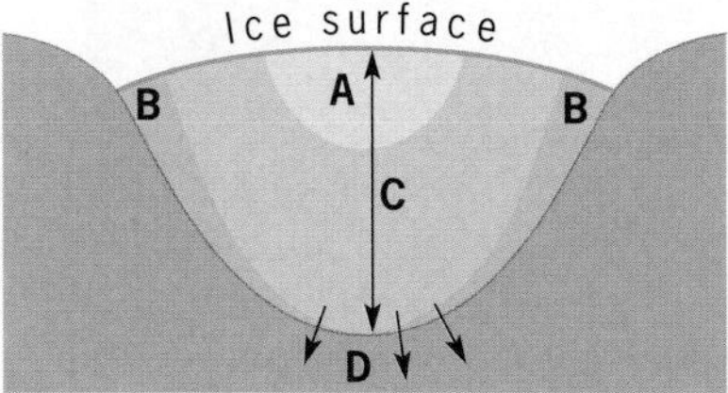

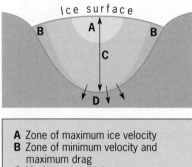

A Zone of maximum ice velocity
B Zone of minimum velocity and maximum drag
C Maximum ice thickness
D Maximum forces applied

b Impact of valley glacier erosion

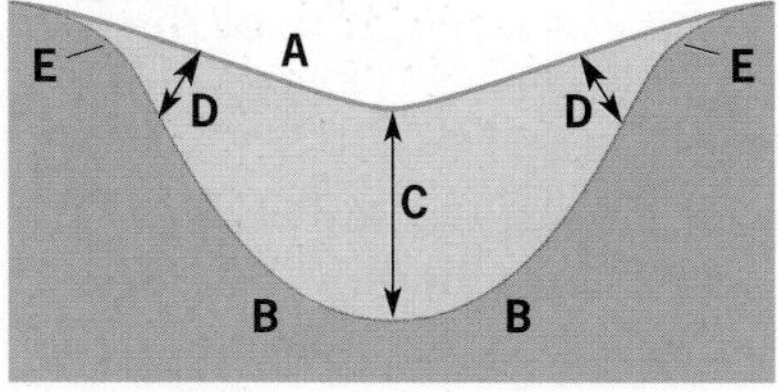

A Pre-glacial fluvial valley cross-section profile
B Post-glacial valley cross-section profile
C Greatest overdeepening
D Minimal deepening and slope retreat
E Shoulder of overdeepening (break of slope)

Figure 7.17 The carving of a glacial trough

7.8 Erosional landforms of alpine (valley) glaciation

In a landscape experiencing alpine glaciation, moving ice masses and hence the erosional forces are concentrated in valleys. These valley glaciers are either outlet glaciers, fed from ice fields (Fig. 7.12), or originate in one or more ice accumulation sources at the valley head.

Overdeepened troughs

The best-known landform is the U-shaped valley or glacial trough. Although some glacial troughs do have a strongly developed 'U' shape (see Fig. 7.1), the majority have a broader parabolic cross-section profile (Fig. 7.18). As Figure 7.17 explains, where the ice mass follows a pre-existing river valley, ice thickness and velocity are greatest over the central part of the valley floor.

Figure 7.18 The broad parabolic profile of a glaciated valley: Loch Doine and Voil, Balquhidder, Scotland

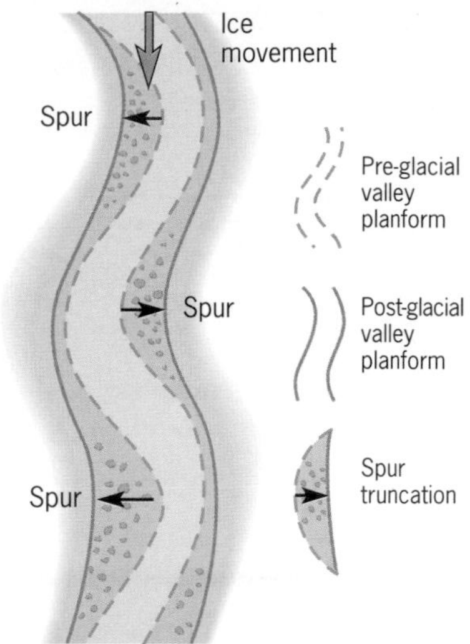
Ice movement
Spur
Pre-glacial valley planform
Spur
Post-glacial valley planform
Spur
Spur truncation
As ice moves down-valley, forces are applied most intensely across the 'nose' of a spur. As a result, abrasion and plucking are vigorous in this zone, and the spur is truncated, or shortened.

Figure 7.19 Truncated spurs: the modification of a valley planform by ice

In contrast, along the valley flanks, ice is thinner, drag is greater and velocity is lower. Thus, the processes of abrasion, plucking and rock crushing are most vigorous across the valley floor. The contrast between floor and flank conditions increases over time.

None the less, valley flanks are eroded by the ice. The valley planform tends to be straightened to a gently sinuous path (Fig. 7.19). In addition to the truncation of spurs, valley sides show frequent evidence of smoothing and plucking.

Factors influencing valley form

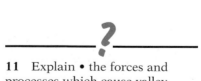

How much erosional work a glacier can achieve depends upon the mass and thickness of the ice, the velocity, the bedrock and nature of the basal load, and the time over which the forces are applied. For example, the asymmetrical shape of the Buttermere–Crummock trough in the English Lake District can be explained by the differential resistance to erosion of the rocks making up each flank (Fig. 7.20).

Because ice thickness and speed of flow are important, the mass balance budget of the glacier controls the erosive power (see section 7.4). For valley glaciers which are independent of ice caps, the main ice accumulation sources are sheltered hollows around the valley head and along the valley sides. These are called *corries*, *cirques* or *cwms*, e.g. Burtness Coombe and Bleaberry Tarn above Buttermere (Fig. 7.20).

Figure 7.20 The asymmetrical Buttermere–Crummock trough, Cumbria, looking north-west. The south-west flank on the left is made up of Borrowdale Volcanic Series and igneous intrusive rocks which are more resistant than the Skiddaw Group sedimentaries of the north-eastern (right-hand) flank. Thus, the valley cross-section is steeper and more rugged on the south-west side. Note the series of corries set into this south-western face. These would have supplied ice to the main valley glacier as it moved north-west.

The development of corries

When fully developed, corries are large, armchair-shaped depressions cut into a mountain face, with a steep back or *headwall* and sharp ridge flanks called *arêtes*. They are created by a combination of vigorous basal erosion caused by rotational movement of the ice, and frost-shattering of the rock faces above the ice. In glaciated uplands such as the English Lake District, we can see these ice-eroded depressions in all stages of development, from shallow nivation hollows on hillsides to large, overdeepened depressions containing tarns (Figs 7.21 and 7.22).

When mean annual temperatures decline the equilibrium line of a glacier falls in altitude (see section 7.4). The ice accumulation area expands, the ice mass grows and, in turn, ice moves more rapidly out of the corries. It is during such periods that corrie formation is particularly active.

The importance of corries in the formation of glaciated landscapes is well illustrated in Glacier National Park, Montana, USA (Figs 7.22, 7.23). Here, corries have supplied the ice which has carved a series of troughs. Today, in the dying stages of glaciation, ice survives only in corrie hollows (see Case Study, p.139).

?

11 Explain • the forces and processes which cause valley overdeepening, and • why the difference in erosional power between central and marginal parts of a glacier increases over time.

a Nivation hollow

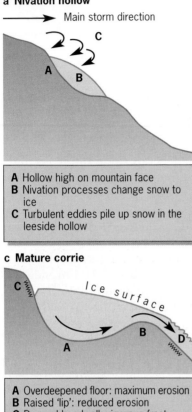

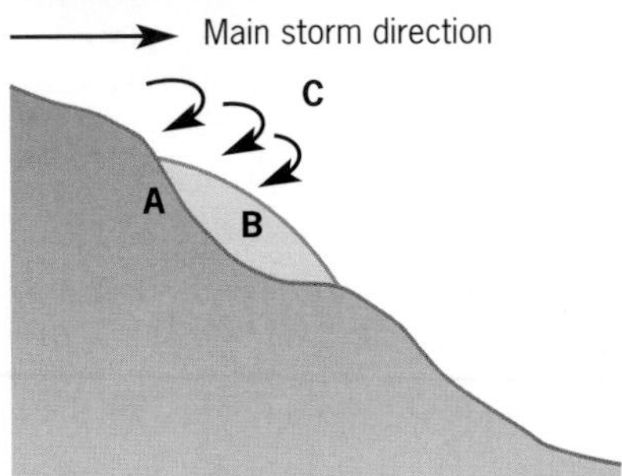

A Hollow high on mountain face
B Nivation processes change snow to ice
C Turbulent eddies pile up snow in the leeside hollow

b Intermediate corrie

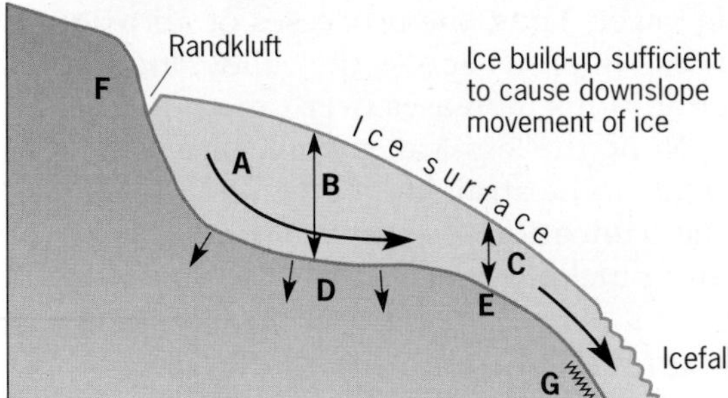

A Rotational slip
B Greatest ice thickness and velocity
C Ice thins as slope steepens
D Maximum erosion across the floor of the depression
E Reduced erosion beneath thinner ice
F Frost-shattering causes headwall to steepen and retreat
G Crevassed icefall produces plucking

Randkluft is the gap between rock and ice at glacier head

c Mature corrie

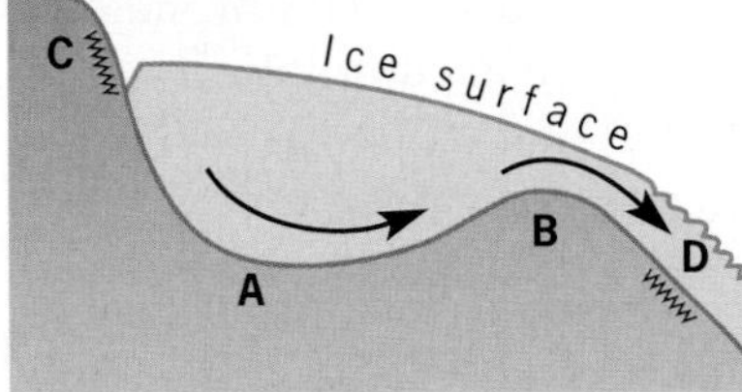

A Overdeepened floor: maximum erosion
B Raised 'lip': reduced erosion
C Rugged headwall: vigorous frost-shattering
D Ice-plucked slope

Figure 7.21 Corrie evolution

Figure 7.22 A massive corrie and hanging valley in Glacier National Park, Montana, USA. It faces ENE.

Figure 7.23 Corries in the Glacier National Park, Montana, USA (*Source:* Klimaszewski, 1993)

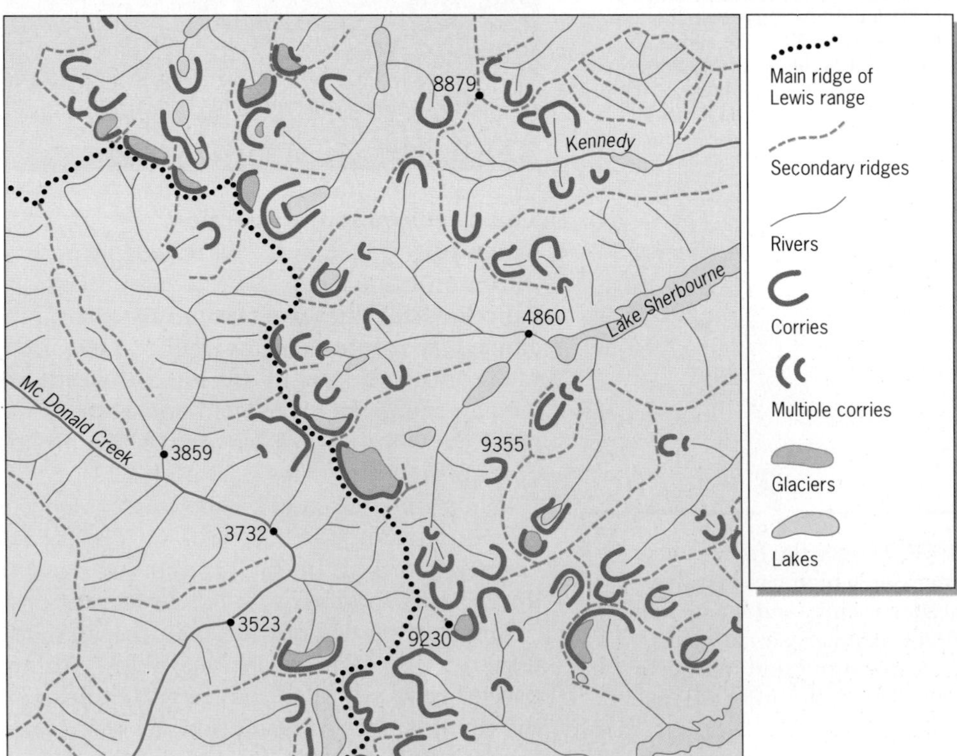

12 Place tracing paper over Figure 7.23.

a Plot • all corries; • ridge lines.

b Draw arrows from the corries and along the glacial troughs to show the patterns of ice movement during full alpine glaciation of this region.

c Describe the pattern of ice accumulation and movement, and how they explain the present-day landscape.

d Use the corries of Glacier National Park to test the following hypothesis: 'Corries have a preferred orientation, facing east and north-east.' (Use the Chi-squared test to assist your answer.)

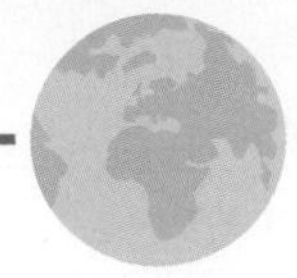

The Coire na Creiche corrie glacier and its landforms, Skye

The Loch Lomond Stadial, or readvance, was the last time mean temperatures fell low enough for long enough for permanent ice to develop in the British Isles. Temperatures were about 6.5 °C lower than today. On the Isle of Skye, an ice cap built up across the Cuillin Hills, covering at its maximum over 150 km^2 (Fig. 7.24). The ice cap fed several outlet glaciers which reached sea level, e.g. at present-day Loch Scavaig and Loch Slapin. In addition, the hill mass was fringed by nine small corrie-fed glaciers. The largest of these, with an area of 4.6 km^2, extended for as much as 4.5 km down Coire na Creiche towards the major glacial trough of Glen Brittle.

The relationships between the glacial processes and present-day landforms are summarised in Figures 7.25, 7.26 and 7.27. The sketch of Figure 7.25 reconstructs the landscape of Coire na Creiche as it looked in the later stages of the Loch Lomond Stadial. At this time, perhaps 10 400 BP, melting was greater than accumulation, the ice mass was becoming thinner, and the ice front receding. The maximum extent of ice advance is indicated by terminal (end) moraines, lines of boulders and fluvioglacial stream terraces (A–D on Figs 7.25, 7.26).

The ice front at a halt stage of the recession is marked by the lines of moraines and boulders, E and

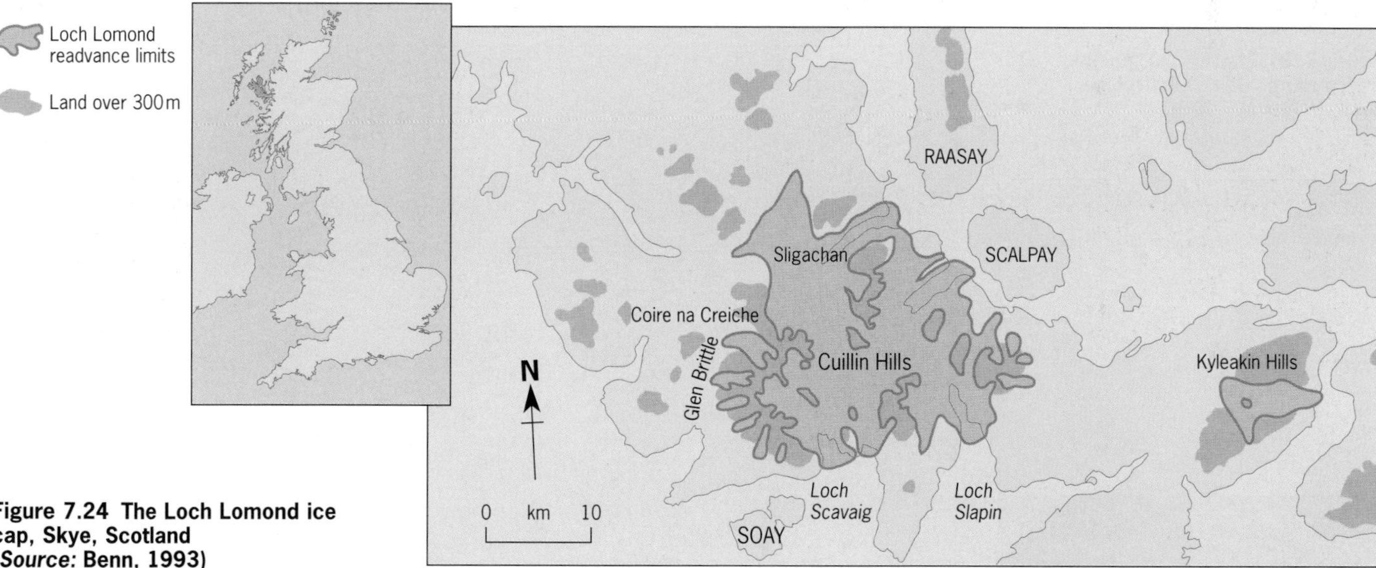

Figure 7.24 The Loch Lomond ice cap, Skye, Scotland (*Source:* Benn, 1993)

A,D	Lines of boulders	Features indicating the maximum advance of the glacier
B	Small moraine ridges	
C	Stream-laid terraces	
E,F	Recessional moraines marking the ice margin as shown on Figure 7.26	
G	Lateral moraine, supplied by scree H	
H,J	Screes from frost-shattering	
M	Medial moraine supplied by the Sgarr an Fleadain nunatak	
S	(On Fig. 7.26 only) rock surfaces eroded by load scouring	

Figure 7.25 Sketch of the glacial landforms at Coire na Creiche (*Source:* Benn, 1993)

Coire na Creiche glacier

F. The debris for these end and stage moraines was supplied mainly by basal erosion, except for the material of the medial moraine (M on Fig. 7.25). This was produced by frost-shattering on the isolated exposed peak or *nunatak* (isolated peak protruding from an ice cap or glacier) of Sgurr an Fheadain. Rock shattering on the corrie headwall and the flanking ridges created the screes (H) which supplied the lateral moraines (G). The moraines and boulder lines which lie within the ice limit of the recessional stage of Figure 7.25 represent even later stages in deglaciation. As the ice front receded up-valley, so the deposits indicate where the ice front halted at a particular location for a few years.

The floor of the corrie and upper valley consists of scoured bedrock (S on Fig. 7.26). This indicates the vigorous basal erosion caused by active ice movement across the accumulation zone.

Yet remember, this small ice mass had a lifespan of not more than 1000 years. It was not capable of carving the two well-developed corries and valley we see today (Fig. 7.27). This degree of overdeepening has required the work of repeated ice advances, during ice cap and valley glacier episodes. In contrast, each of the ice readvances would have removed the depositional evidence of earlier episodes, and the deposits mapped on Figure 7.26 are all from the final Loch Lomond Stadial.

We can see, therefore, that Coire na Creiche is an excellent example of the generalisation that landforms of glacial erosion are often the products of several glacial advances, while depositional features are likely to be the work of the most recent ice advance and recession episode.

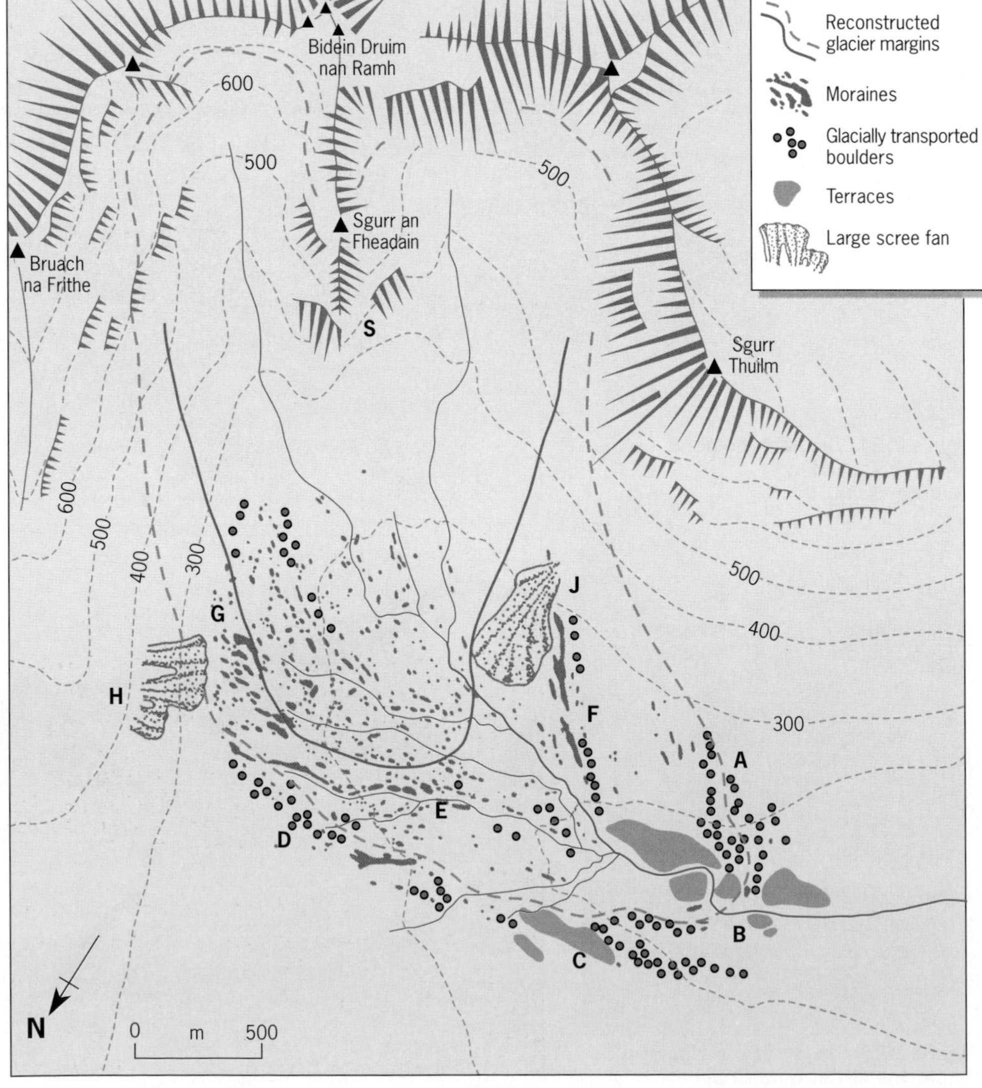

Figure 7.26 Plan of the glacial landforms at Coire na Creiche (*Source:* Benn, 1993)

Figure 7.27 Coire na Creiche (© Crown copyright (433772))

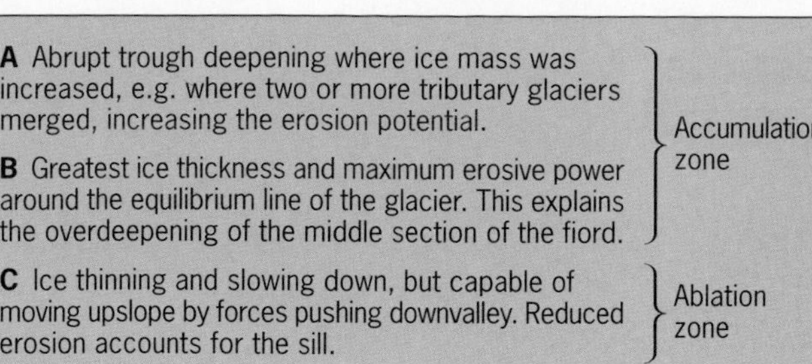

A Abrupt trough deepening where ice mass was increased, e.g. where two or more tributary glaciers merged, increasing the erosion potential.
B Greatest ice thickness and maximum erosive power around the equilibrium line of the glacier. This explains the overdeepening of the middle section of the fiord.
Accumulation zone
C Ice thinning and slowing down, but capable of moving upslope by forces pushing downvalley. Reduced erosion accounts for the sill.
Ablation zone

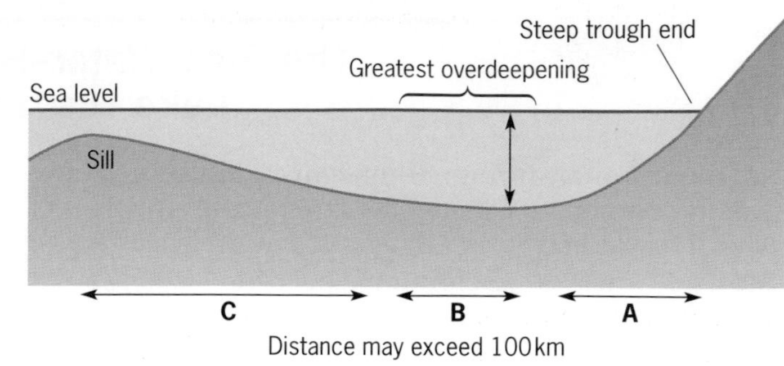

Figure 7.28 Typical long profile of a Norwegian fiord

The long profile of a glaciated valley

A distinctive feature of many glaciated valleys is the irregularity of the long profile. We can explain this by the way ice moves and does its work. For example, Figure 7.28 shows the long profile of a typical Norwegian fiord. Fiords are glacial troughs eroded below sea level and drowned after the disappearance of the ice. In Norway, the Sognefjord has a maximum depth of 1308 m, but at the entrance narrows to only 3 km and with a sill depth of only 200 m.

Another common feature is a *stepped* long profile, where the valley descends in a series of steeper sections separating more gentle, or even overdeepened, stretches. Two main factors influence the development of this stepped profile. First, where a tributary glacier provides additional supply, the enlarged glacier can achieve greater downcutting. Second, where unusually strong rock types cross the valley floor they may resist erosion, and form a step. Where the rock bar is especially pronounced, it is called a *riegel* ('A rock barrier that extends across a glaciated valley, usually comprising harder rock, and often having a smooth slope facing up-valley and a rough slope facing down-valley' (Hambrey, 1994).

In some situations a series of steps, or 'staircase', may develop. Such features may become more pronounced over time because of compression–extension flow over the surface irregularities (Fig. 7.29). The ice-scoured rock exposures may show small-scale features such as *striations* where rock particles protruding from the glacier base have etched grooves in the bedrock. The Blea Water–Haweswater trough in the English Lake District has a well-developed stepped profile (see Case Study, pp.142–143).

Figure 7.29 Compression–extension flow in basal ice

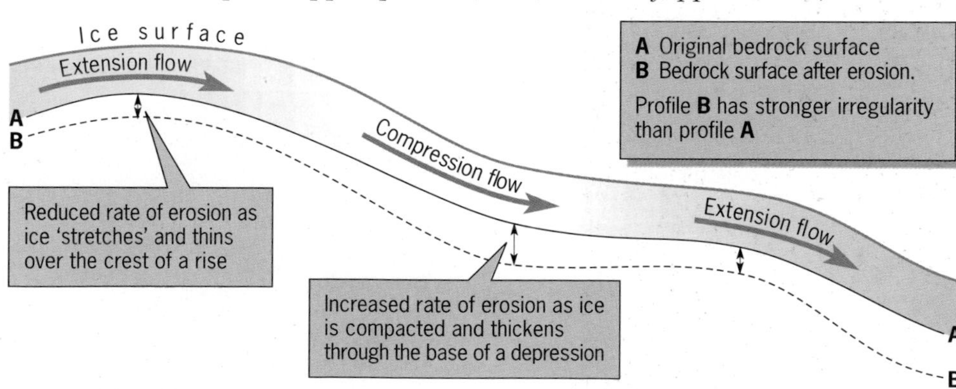

13 From your understanding of processes at the base of a glacier (sections 7.5, 7.6), explain

a how compression–extension flow (Fig. 7.29) may make rock steps more pronounced over time;

b why rock exposures on the valley floor and sides tend to be smoothed on their up-valley faces and roughened on their down-valley faces;

c why tributary valleys 'hang' above a major trough.

Hanging valleys

As a main valley glacier accumulates ice from smaller tributary glaciers, and increases its power to erode, its floor is overdeepened and, when the ice has disappeared, the tributary valleys end abruptly above the main trough floor, i.e. they 'hang' and so are called hanging valleys (Fig. 7.1). In the English Lake District hanging valleys are common, e.g. Lodore Falls cascading from the Watendlath tributary valley into the Borrowdale trough, and Measand Beck dropping into the Haweswater valley (see Case Study, pp.142–143).

The Blea Water–Haweswater trough, Lake District, England

The Haweswater trough is one of a series of glacially deepened valleys radiating from the High Street massif of the south-east Lake District (Fig. 7.30). At times of maximum glaciation, e.g. the Dimlington Stadial, the Lake District supported an ice cap. Across uplands such as High Street, ice movement was slow and erosion slight. However, active ice streams followed pre-existing valleys, including Haweswater, causing further deepening.

During episodes of alpine glaciation, the uplands and ridges were ice-free, and the ice accumulation sources were the trough ends and corries at the heads of the valleys. Ice tongues moving down-valley completed the overdeepening process. The result is that the High Street massif, built of great thicknesses of volcanic lavas and tuffs, 'has been eaten away on all sides like a lump of cheese that has been bitten by numerous mice, leaving only the central boss intact' (Prosser, 1977).

The Haweswater trough is the deepest and widest of the valleys, because it had the largest and most powerful glacier moving down it. Two principal factors influenced this size: first, the glacier was fed

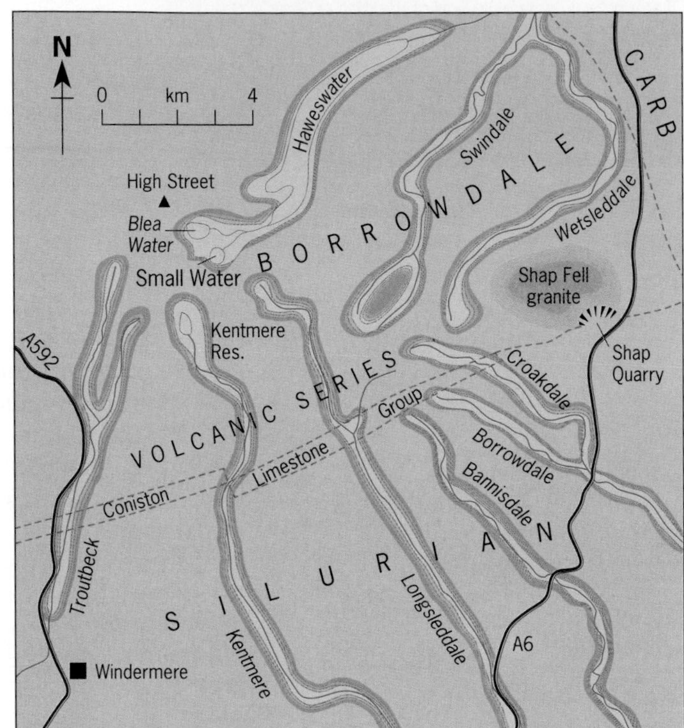

Figure 7.30 Glacial valleys of the south-east Lake District, England

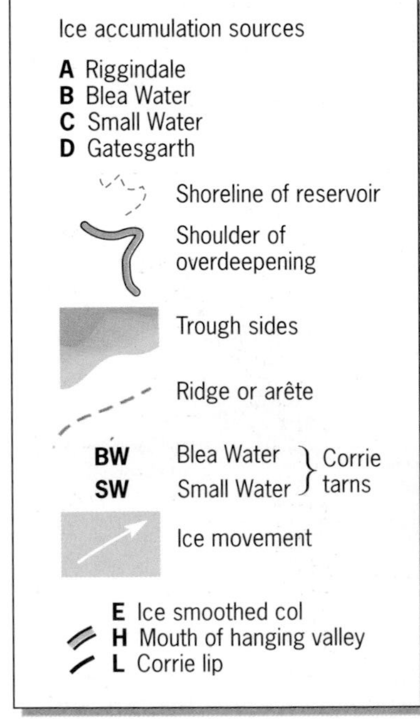

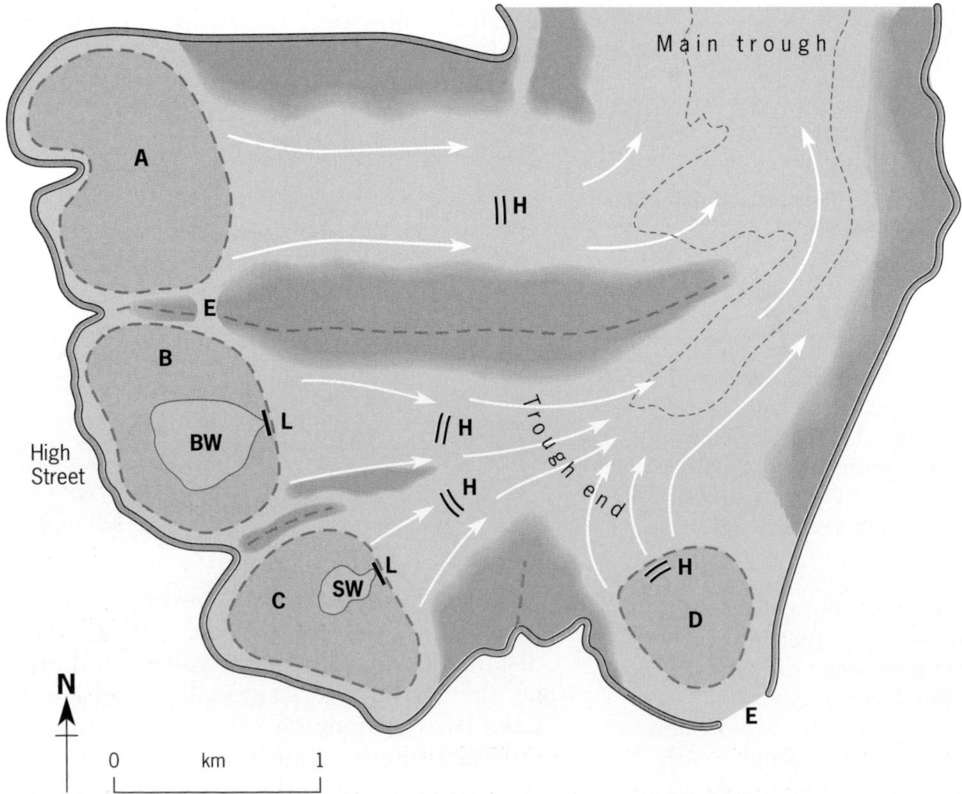

Figure 7.31 Ice accumulationn sources for the Haweswater glacier

from several ice sources (Fig. 7.31); second, the ice accumulation was particularly active on the east and north-east flanks of the uplands. This orientation gave maximum shade and snow delivered by eddies and downdraughts from westerly storms. Throughout the British Isles, corries show a preferred orientation towards the north-east, and develop especially well in well-jointed, strong rocks such as the Borrowdale Volcanic Series of the High Street locality (Table 7.5).

Figure 7.32 Blea Water corrie, the main ice accumulation source for the Haweswater glacier

Table 7.5 Relationship between corries and rock types in the Lake District

Rock type	Igneous intrusive	Skiddaw Slate Group	Borrowdale Volcanic Series
Number of corries	5	28	125
% of surface area	5	46	49

The main ice source was the corrie which today contains Blea Water tarn. This is one of the largest corries in the British Isles, and as much as 500 million tonnes of rock have been removed from the flanks of High Street. It has all the classic features of fully developed corries (Fig. 7.32). The length (lip–headwall foot distance): height (headwall foot–top distance) ratio is 2.8:1, within the typical range for corries of 2.8–3.2:1.

The ice which overdeepened the corrie and moved down-valley has produced an irregular long profile in the upper trough, which consists of a set of steps and basins (Fig. 7.33). Notice that later deposits have modified the eroded form. The main trough eroded by the trunk glacier (today flooded by the reservoir) is strongly overdeepened. Two basins are separated by a rock bar or riegel (Fig. 7.34).

14 From Figure 7.34:

a Plot a long profile section of the main Haweswater trough.

b Select three locations and construct cross-sections of the trough (you may find it easier if you alter the horizontal scale).

c Use your sections and the plan to describe the form of the trough.

d What features typical of glacial troughs are illustrated?

e From your understandings of the way a glacier behaves, explain the form of this trough.

f What are the key differences between the form of this trough and that which you would expect of a valley eroded by streams? (Look again at Chapters 2 and 3.)

g There was a natural lake in this trough before the building of the dam. It was much smaller than the present reservoir. In which part of the trough was this natural lake? Give your reasons.

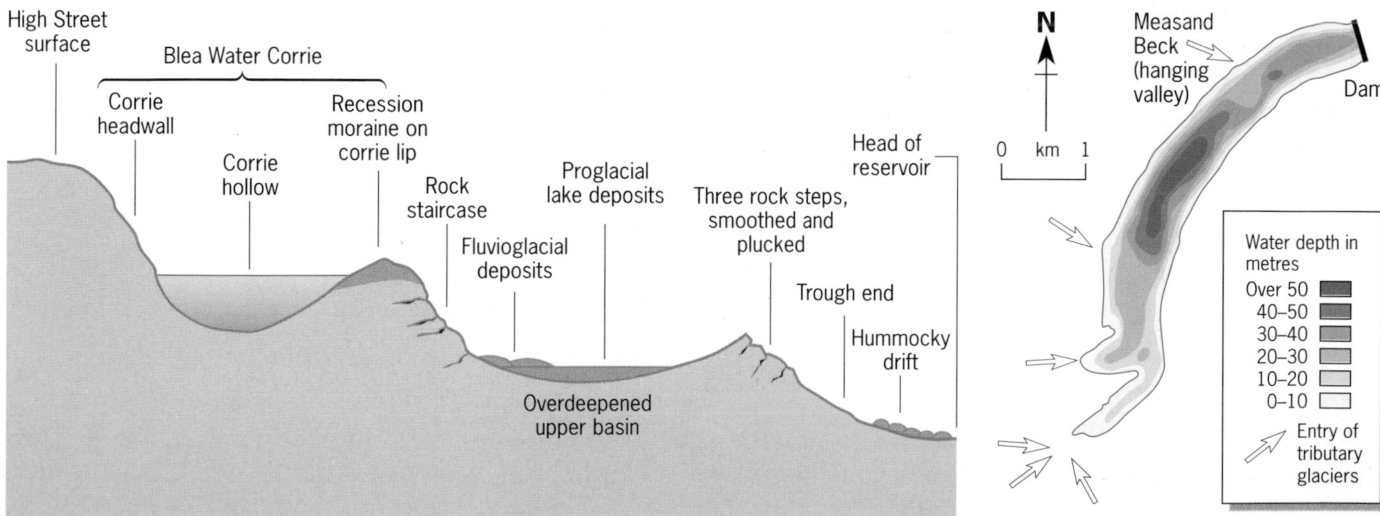

Figure 7.33 Long profile of the Upper Haweswater trough

Figure 7.34 The Haweswater trough: evidence of overdeepening

7.9 Deposition by ice

The landform shown in Figure 7.35 consists of *glacial drift*, i.e. material deposited directly by ice. On the left of the photograph are basic questions we need to keep in mind when studying such features. Drift landforms vary greatly, but fall into three groups according to where they were deposited, e.g. under the ice or marginal to the ice, and what has happened to them (Table 7.6).

Content

- What is it made of?
- Were the materials deposited *under* or *marginal* to the ice?
- Where did the material come from?
- How did it arrive in this location?
- What was happening to the ice when the materials were deposited?

Form

- Where is it located?
- What shape and size is it?
- Is it oriented in a particular direction?
- How does it relate to other landforms around it?

Figure 7.35 Morainic debris, Balquhidder, Scotland

Table 7.6: Main types of till

Type	Landforms	Material and appearance	Processes
Lodgement till	**a** Till sheets **b** Streamlined features, e.g. drumlins	Compact, unsorted, with stones set into a finer matrix. May show successive layers. Long axes of stones often aligned parallel to direction of ice flow	Subglacial *lodgement* from the base of the ice from melting and drag
Melt-out till	Often occurs as hummocks of moraine, without streamlining. May occur on top of lodgement till. Shape may be related to location of deposition, e.g. at 'snout' of the glacier, or along the side	Largely unsorted, and poorly compacted. Stones angular, with some long-axis alignment. Internal structure may show signs of slumping after deposition. Finer materials may be washed out, leaving a coarse fabric	*Dumping* by melting of stagnant or slow-moving ice. May be ice-marginal, e.g. end moraine, or simply dumped from above as the ice melts, e.g. hummocky drift
Modified tills	**a** Low ridges, often transverse to ice direction. Earlier tills, etc., pushed and overridden by later ice readvance	Usually 'reworked' older deposits, with a variety of internal folds and faults, indicating deformation	Deformation by pressure from advancing ice on pre-existing sediments
	b Thin sheets or lobes, with flattish surfaces	Till which, when saturated, has flowed from the ice surface on to other ice-marginal sediments. Frequently compact clays with layered structures	*Flow* as saturated till slurry during the summer melt season

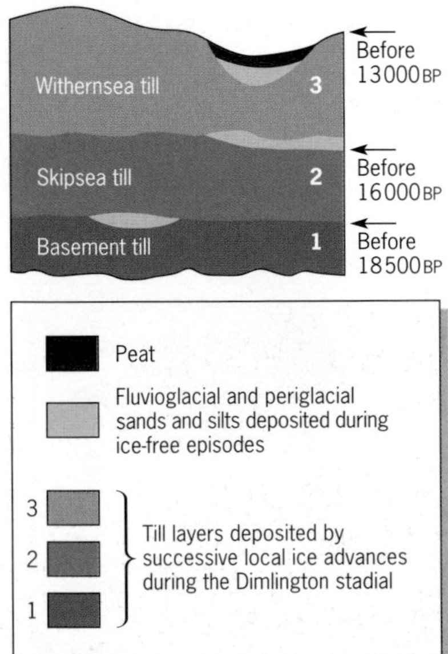

Figure 7.36 Till plain aggradation, Holderness, Yorkshire (*After:* Jones and Keen, 1993)

Sub-ice deposits and landforms

In the zone of traction at the base of a moving ice mass, up to 50 per cent of the volume may be debris load (see section 7.6). Some of this debris is released on to the underlying surface by basal melting or drag, especially beneath the ablation zone of an ice sheet or glacier. Such material is given the general name of **till**.

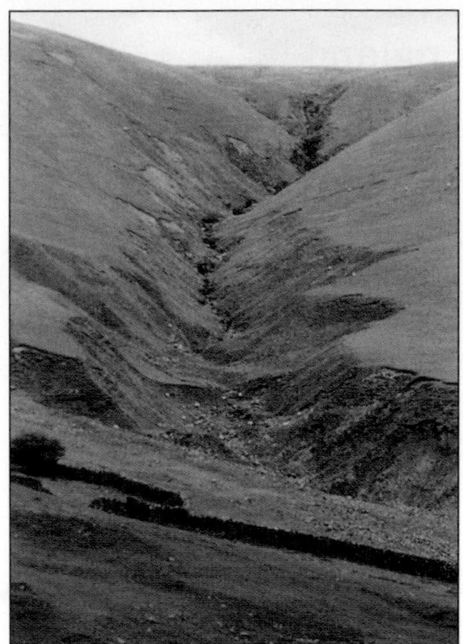

Figure 7.37 Lodgement till mantle, Carling Gill, Howgill Fells, Cumbria. Ice moved south (left to right) across the flank of the River Lune gorge, depositing a lodgement till sheet up to 10 m thick. The head of the post-glacial gully is cut into bedrock, but the lower sections have exposed the till plastered across the irregular underlying rock. Notice the smooth surface of the till slope on either side of the V-shaped incision.

Lodgement till

Over time a till layer may be built up by the process of *lodgement*, as the material becomes plastered on to the underlying surface. After the ice has disappeared, surfaces may be covered with a mantle of lodgement till (Fig. 7.37). Uneven bedrock surfaces are frequently obscured to form extensive spreads of *ground moraine* ('relatively smooth surface topography consisting of gently undulating knolls and shallow closed depressions' (Skinner and Porter, 1987) or *till plains* (Fig. 7.37). Thicker till sheets may have been built up by successive ice advances (Fig. 7.36).

Lodgement till sediments generally consist of angular fragments set into a sandy or silty matrix. There are few signs of sorting. Some of the larger fragments may show some down-ice alignment, caused by the drag of the overlying moving ice. Geomorphologists can use this alignment to find out the direction of ice movement, and so build up a history of the glaciation. This technique is called *till fabric analysis*.

Drumlins

Lodgement till thickness often varies over short distances. This unevenness is the result of bedrock irregularities, variations in debris supply from the ice, and in the character of this supply, e.g. from fine clays to boulders. Where frictional drag of the overlying active ice is sufficient to mould the materials but insufficient to entrain and remove them, thicker deposits may achieve streamlined shapes called *drumlins*. A drumlin may be defined as 'a streamlined hillock, commonly elongated parallel to the former ice-flow direction, composed of glacial debris, and sometimes having a bedrock core; formed beneath an actively flowing glacier' (Hambrey, 1994).

They generally have an asymmetrical long profile, with a steeper, blunter end facing up-ice, and an elongated 'tail' (Fig. 7.38). Elongation varies, with length–width ratios commonly between 2.5:1 and 4.0:1. Greater elongation is believed to be related to faster ice movement. By plotting the orientation of their long axes, we can work out the local direction of ice movement.

Drumlins commonly occur in clusters, called *drumlin fields*, to give a hummocky landscape known as 'basket-of-eggs' topography. Individual drumlins vary from small mounds (2 m high; 10 m long) to several kilometres in length and 50m high. The largest drumlin field, in New York State, USA, has 10 000 drumlins spread across 12 500 km^2. In the British Isles, well-developed drumlin fields are found in southern Scotland, north-west England (see Vale of Eden Case Study, p.146), Wales and north and west Ireland. All were formed during the Dimlington Stadial, i.e. the most recent ice advance, and may well involve the reworking and shaping of earlier deposits.

?

15 Suggest reasons why lodgement till is formed beneath the ablation zone of a glacier rather than the accumulation zone. (An annotated diagram may assist your answer.)

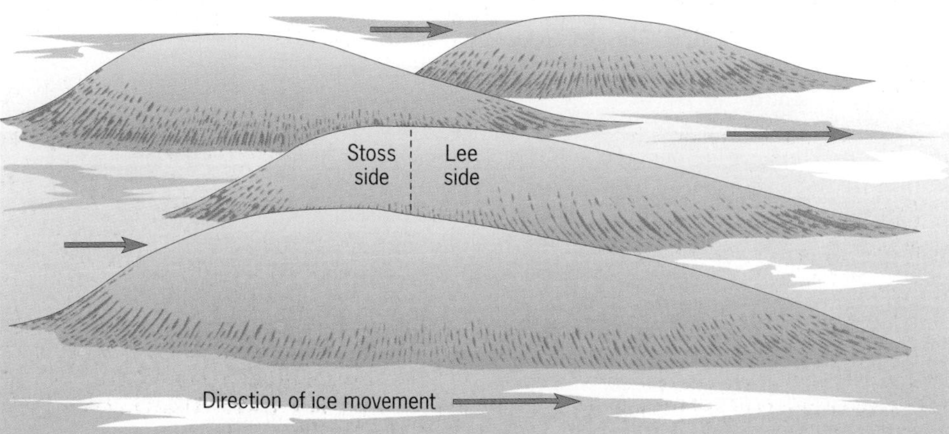

Figure 7.38 Drumlin shape and the direction of basal ice movement

The Vale of Eden drumlin field, Cumbria, England

One of the best-known drumlin fields covers much of the Vale of Eden, between the Lake District and the Cross Fell Pennines in Cumbria (Fig. 7.39). During the Dimlington Stadial (23 000–11 000 BP), an ice sheet spread across the Vale of Eden, fed from several ice accumulation sources. In the southern sector of the vale, south of Appleby, debris-rich ice moved eastwards, to be squeezed through the Stainmore **col**. The drumlin field left behind indicates this movement, and the large volume of material eroded, transported then deposited.

In the 36 km^2 area south and west of Brough, shown on Figure 7.40, 308 drumlins have been mapped. They vary in size between 150 and 1000 m long, and 75 and 500 m wide. A few are rock-cored, but most are moulded entirely from till. The till content and the orientation of their long axes indicate ice merging from three accumulation sources. All the drumlins contain materials derived from the local underlying red sandstones, but examples in the north-west corner contain Cross Fell **clasts**; drumlins of the central sector contain Lake District clasts; in drumlins in the southern sector, debris from the Howgill Fells occurs (Fig. 7.41).

16 Place tracing paper over Figure 7.40. Use an appropriate sampling method to select 30 drumlins (about a 10 per cent sample) and draw their outlines carefully. Add the line of the long axis and number each drumlin.

a Construct and complete a table of 30 rows, with the following columns: • Drumlin number; • 6-figure grid reference (to enter location of central point of the drumlin); • Length; • Width; • Elongation ratio (length–width ratio); • Orientation of long axis.

b Using different colours/symbols, mark each drumlin according to whether you would expect it to contain clasts from (i) Cross Fell; (ii) Lake District; (iii) Howgill Fells.

c Calculate the mean elongation ratio for your sample and record the extreme values. Compare your sample with the 'normal' range for drumlins (p.145). What factors might influence this range and any differences between your drumlins and the 'norm'?

d Elongation ratios are influenced by speed of ice movement. Thus, as ice became squeezed into the Stainmore col east of Brough, we would expect it to speed up. On Figure 7.40, ice velocities would have increased from west to east. So, we can propose the following hypothesis: 'Elongation ratios increase from west to east across the mapped area.' Use your data to test this hypothesis. (Think carefully about the possibilities of statistical analysis.)

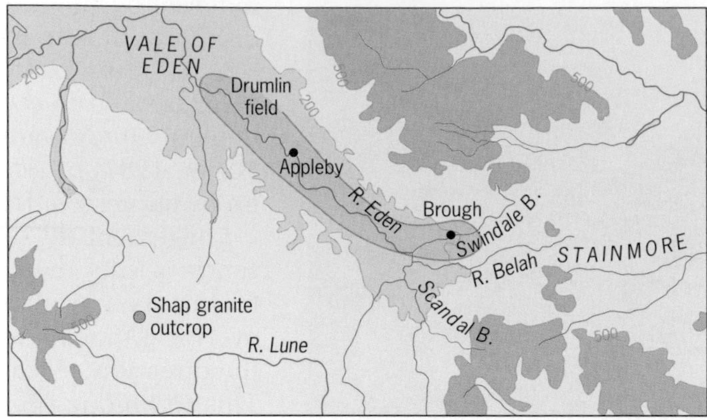

Figure 7.39 The Vale of Eden drumlin field, Cumbria

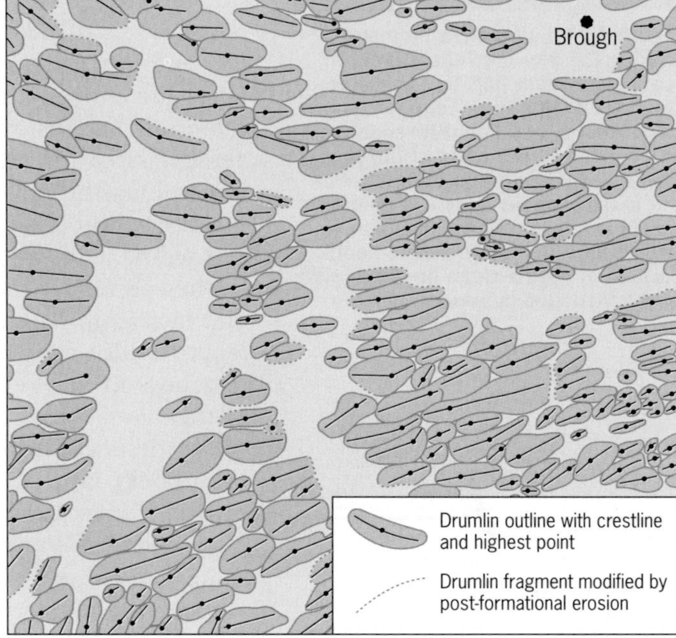

Figure 7.40 Drumlin field near Brough, Cumbria. Each drumlin is shown by its planform, along with the crestline and highest point where these can be identified.

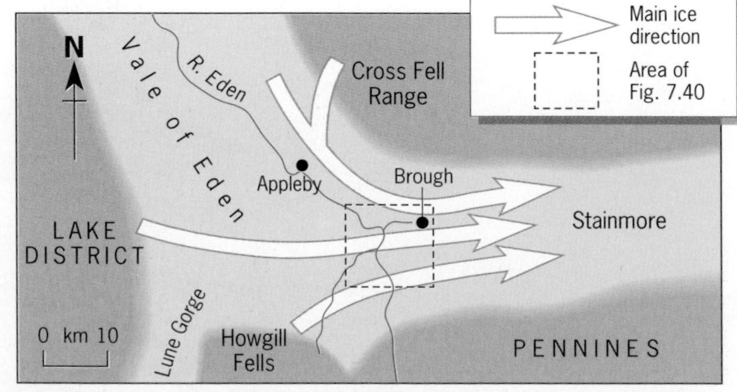

Figure 7.41 Main ice sources and movements in the southern Vale of Eden during the Devensian Stadial

Erratics

The relationship between where glacial drift is deposited and where the materials came from gives us important clues about ice movements. Individual rocks and boulders transported from elsewhere, and found on till deposits or on bare rock surfaces, are known as *erratics*. Rock types known to outcrop as bedrock in a single location are particularly useful. These are called *indicator rocks*. For example, the distinctive Shap Granite outcrops only in the south-east corner of the Lake District (Fig. 7.39). Erratics of this granite are scattered across the southern Vale of Eden area examined in the case study (p.146), signalling the direction of ice movement from the Lake District. Indeed, what happened to the ice sheet after it flowed east from the Stainmore col is indicated by Shap Granite boulders found on the beach at Robin Hood's Bay, along the North Sea coast.

Landforms of ice-marginal deposition

At the ice front, or 'snout', of a glacier, ablation causes the dumping of the supraglacial and englacial load as mounds of *melt-out till*. There may be some finer material near the base, as saturated ground moraine is squeezed from beneath the ice margin, but most is coarse, angular, unsorted material. Melt-out till ridges also develop along the flanks of the glacier, and may be enlarged by the addition of frost-shattered debris from rock faces above.

Terminal and recessional moraines

Ice front melt-out deposition creates *terminal* (*end*) *moraines* and *recessional moraines* (Fig. 7.42). They occur as irregular ridges lying transverse to the direction of ice movement, e.g. the York and Escrick moraines shown on Figure 7.5. End moraines indicate the maximum ice advance where the ice front was stationary for a number of years, e.g. the Cromer Ridge in north Norfolk, 90 m high in places, is an end moraine of the Dimlington advance. Even small, short-lived hanging glaciers mark their presence by curved end moraines.

A recessional or stage moraine marks a temporary halt in the retreat of an ice front. As with end moraines, their size is controlled by how long the ice front was stationary, how much ice melted and how much debris was being supplied. Thus, large Alaskan glaciers such as the Sheriden Glacier (Figs 7.15, 7.43) can create ridges 50 m high and several kilometres long in a few decades. In contrast, small corrie glaciers such as Coire na Creiche, Skye (Figs 7.25, 7.26), leave only low, non-continuous ridges.

Figure 7.42 Terminal moraine, Exit Glacier, Alaska, USA

Figure 7.43 Stage moraine, Sheridan Glacier, Alaska, USA. The debris in the foreground covers a base of 'dead' ice. The ice front lay against the higher moraine less than 80 years ago. The moraine forms a dam for the proglacial lake.

Figure 7.44 The final stage moraine on the lip of the Blea Water corrie, Cumbria. It increases the overdeepening of the corrie and raises the level of the tarn.

Figure 7.45 Lateral moraine ridge of the Athabasca Glacier, Jasper National Park, Alberta, Canada

During the final stages of ice recession, permanent ice may be confined to the corrie accumulation sources. This stage is often marked by moraine piled upon the corrie lip, as in the Blea Water corrie (Fig. 7.44).

Lateral moraines

Debris deposited on the flanks of a glacier in the accumulation zone is transported into the ablation zone as lateral moraine. Melting then dumps this load as a ridge at the rock–ice junction. There may be further additions of frost-shattered debris from rock faces (Fig. 7.45). As the ice level declines, and after its disappearance, slumping, solifluction and other slope processes progressively modify the moraines. Thus, in Britain today, we see lateral moraines as mounds and sheets of drift lying along the eroded flanks of glaciated troughs (Fig. 7.46).

Modified tills

Till may be modified by later ice movement. For example, ice dragging over till sheets can contort the sediments, and may even leave low ridges known

Figure 7.46 Drift landforms along the side of the Haweswater trough. The glacier spilling from the Blea Water corrie has deposited an extensive sheet of what seems to be lateral moraine along the trough flank, below the rock face of the Riggindale Crag arête. Periglacial and post-glacial slumping, solifluction and gullying have modified the drift. (An alternative interpretation is suggested in section 7.11, which warns us of how careful we need to be in explaining landforms.) Photo taken from Blea Water corrie lip.

Figure 7.47 Hummocky moraine, upper Haweswater trough, Cumbria. The rounded mounds lying at the trough end are melt-out till from a patch of stagnant ice left stranded as the glacier from the Small Water and Blea Water corries receded. They are the youngest tills in this trough, dating to the last ice advance–recession episode, 11 000–10 000 BP.

as *rogen* moraines running like ripples across the till sheet. Melt-out tills may be 'bulldozed' by local ice readvance to form *push moraines*, identifiable by their steepened up-ice faces. However, the best evidence for modification is found in internal structures. In the field, therefore, we need to find exposed till faces and check for folding and changes in till fabric alignment.

Melt-out till from stagnant ice
As an ice front recedes, a debris-laden mass of ice near the 'snout' may become separated from the trunk of a glacier. This *stagnant* or *dead ice* downwastes in situ, dumping its supraglacial and englacial load on top of the subglacial as *hummocky moraine*: 'groups of steep-sided hillocks, comprising glacial drift, formed by dead ice-wastage processes' (Hambrey, 1994). Patches of this hummocky moraine are found on a number of Scottish and Lake District valley floors, and are believed to indicate the maximum ice advance during the last, brief Loch Lomond Stadial (11 000–10 000 BP) (Fig. 7.47).

7.10 Fluvioglacial processes and landforms

During the short summers of glacial environments, copious meltwaters flow on, within, beneath, alongside and from the front of a glacier. The landforms this meltwater creates are the result of the fluvial processes covered in Chapters 2 and 3. However, meltwater streams have four distinctive features which control the processes and resulting landforms (Fig. 7.48)

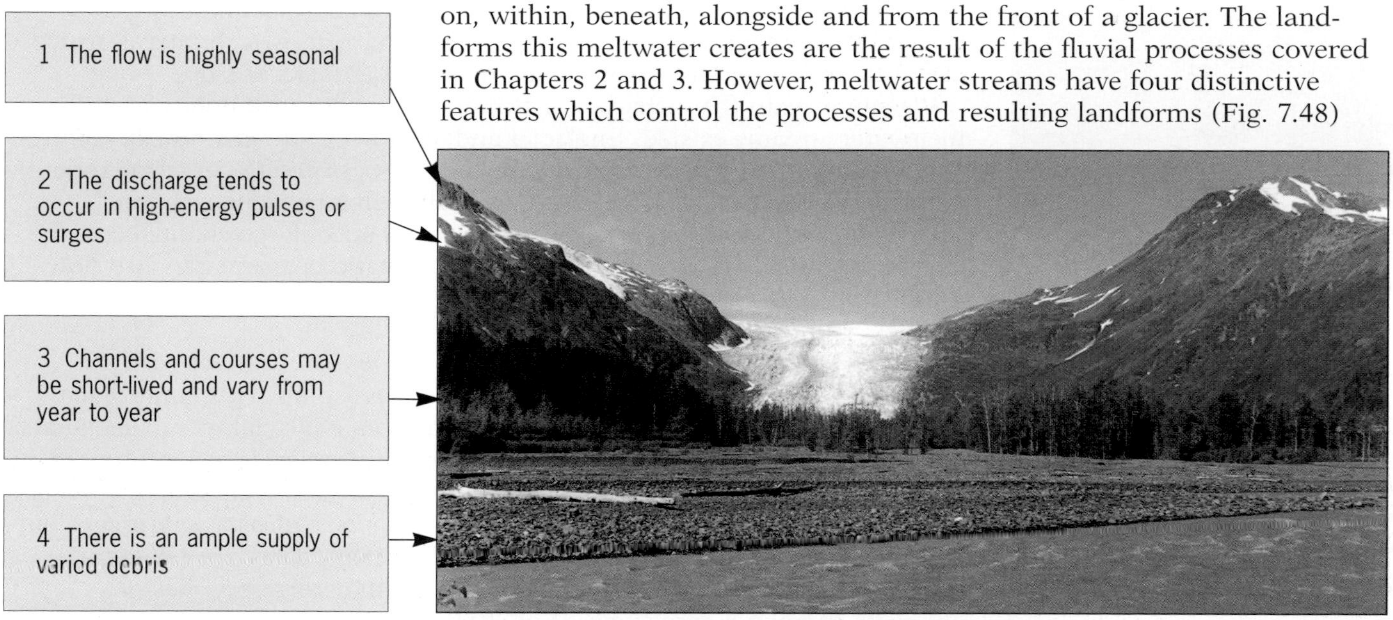

1 The flow is highly seasonal

2 The discharge tends to occur in high-energy pulses or surges

3 Channels and courses may be short-lived and vary from year to year

4 There is an ample supply of varied debris

Figure 7.48 Meltwater stream from Exit Glacier, Kenai Peninsula, Alaska, USA

Fluvioglacial erosional landforms

Meltwater channels are the main landforms of fluvioglacial erosion. They are cut into bedrock and glacial deposits. Channels can originate beneath the ice (*subglacial channels*), at the ice–rock junction (*ice-marginal channels*), or beyond the ice front (*proglacial channels*). All three types can be observed in currently glaciated environments, but it is often difficult to work out their origin long after the ice has disappeared, e.g. to decide whether a channel has a subglacial or ice-marginal origin.

Four key characteristics help us to distinguish meltwater channels from 'normal' stream channels. First, they may be short and discontinuous, e.g. the isolated set of 'in-and-out' channels incised sharply into the Skiddaw Slates of Murton Pike, Appleby, Cumbria (Fig. 7.49). Second, meltwater channels may not be related to present-day drainage patterns. A common type is the *cross-spur channel*, where a dry valley cuts across a spur or ridge and so lies abandoned and totally disconnected from today's drainage. The

?

17 The photograph of Figure 7.48 was taken in late August. State briefly which of the four statements this meltwater river illustrates at this time, and how it may change over the seasons to fit the other statements.

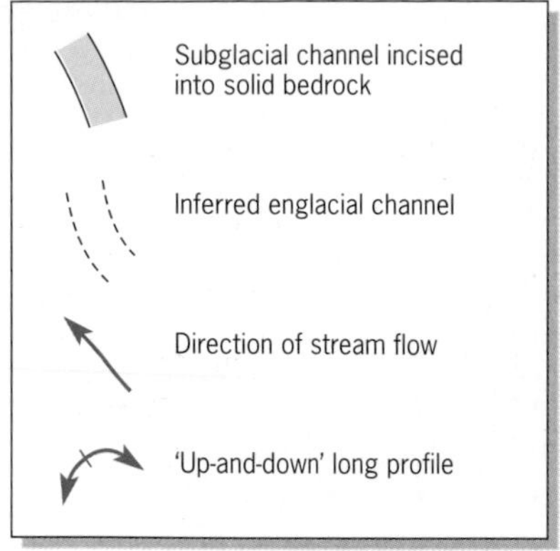

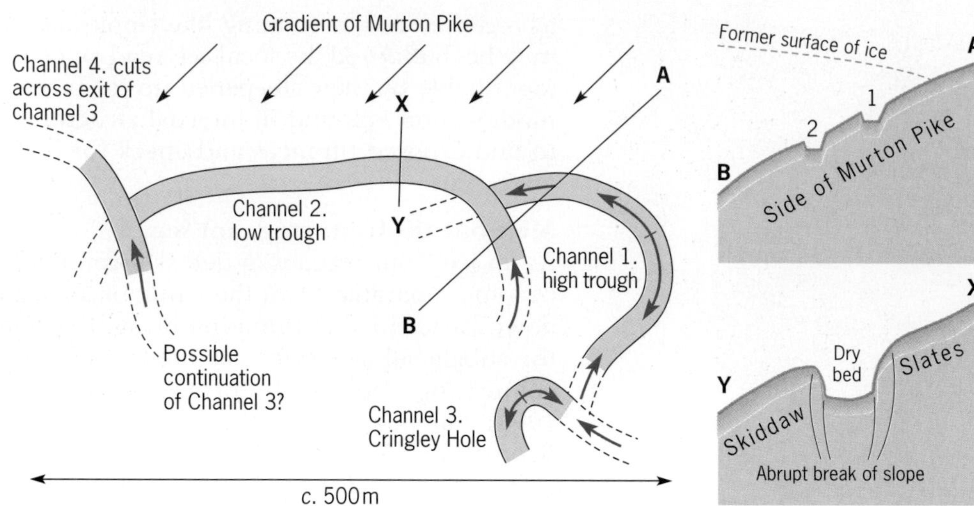

Figure 7.49 The subglacial 'in-and-out' channels of Murton Pike, Cumbria

Figure 7.50 Cocklock Scar meltwater channel, Cumbria

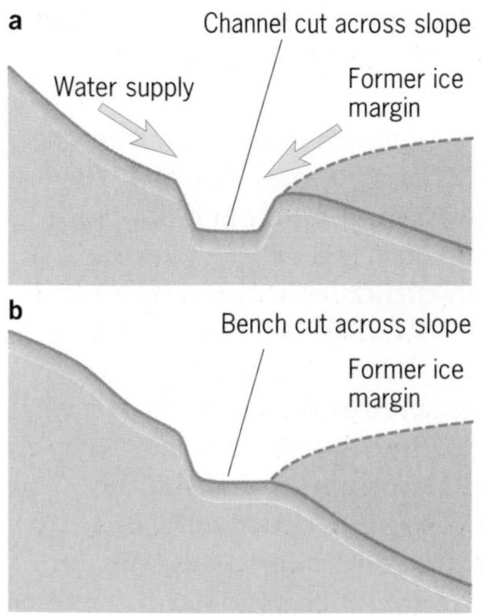

Figure 7.51 Ice-marginal channels

third characteristic is that they may run across, not down slopes. A complex series of channels runs NNW along the flanks of the Cross Fell Pennines, Cumbria. They range in size from short minor features a metre wide and deep, to broad channels such as Cocklock Scar (Fig. 7.50), which winds for more than a kilometre and cuts up to 15 m into bedrock and till. The Cocklock Scar channel illustrates the fourth characteristic, a lateral profile of flat floor and regular sides defined by sudden breaks of slope.

We can explain these characteristics by the conditions under which the meltwater streams existed. Englacial and subglacial streams flow in tunnels, usually under hydrostatic pressure, so they possess great energy to erode and transport sediment. As a result, even though a particular channel course may not last long, because of the moving ice, large channels can be cut in only a few years. Streams under hydrostatic pressure can also flow uphill, and so some subglacial meltwater channels have 'up-and-down' long profiles.

A channel will be cut wherever a stream comes into contact with the underlying surface, e.g. along valley sides, across spurs. This contact may not be continuous as a stream course shifts from englacial to subglacial and back again. This explains the discontinuity and peculiar location of many channels, e.g. the Murton Pike 'in-and-out' channels and cross-spur channels. Ice-marginal channels do not flow under hydrostatic pressure but may have been cut by vigorous, short-lived streams. They are related to stationary positions of the glacier margin and may form benches or channels running across slopes (Fig. 7.51).

Proglacial meltwater channels commonly show a braided form. This is the result of irregular discharge and the volume of debris supply. (Braiding is discussed in the depositional section, p.152.) However, there is another quite different, and often spectacular landform – the *glacial overflow channel*, or *spillway*.

When meltwaters are obstructed by an ice or hill mass, or by a debris fall, a proglacial lake develops. If this lake bursts through the obstruction, e.g. by ice downwastage, outflow across a col, or failure of the debris dam, a powerful flood surge may incise a gorge. In the north-western USA, the canyon incision and scoured landscapes across the Columbia and Snake river basins is an excellent example (Figs 7.52, 7.53). Such violent episodes may cause *watershed breaching* and *drainage diversion*, e.g. the cutting of the Severn Gorge at Ironbridge, Shropshire. Below the gorge, increased meltwater discharges first built and then eroded the well-developed Main, Worcester and Power House terraces along the lower Severn.

Figure 7.52 The Columbia River canyon, Washington, USA

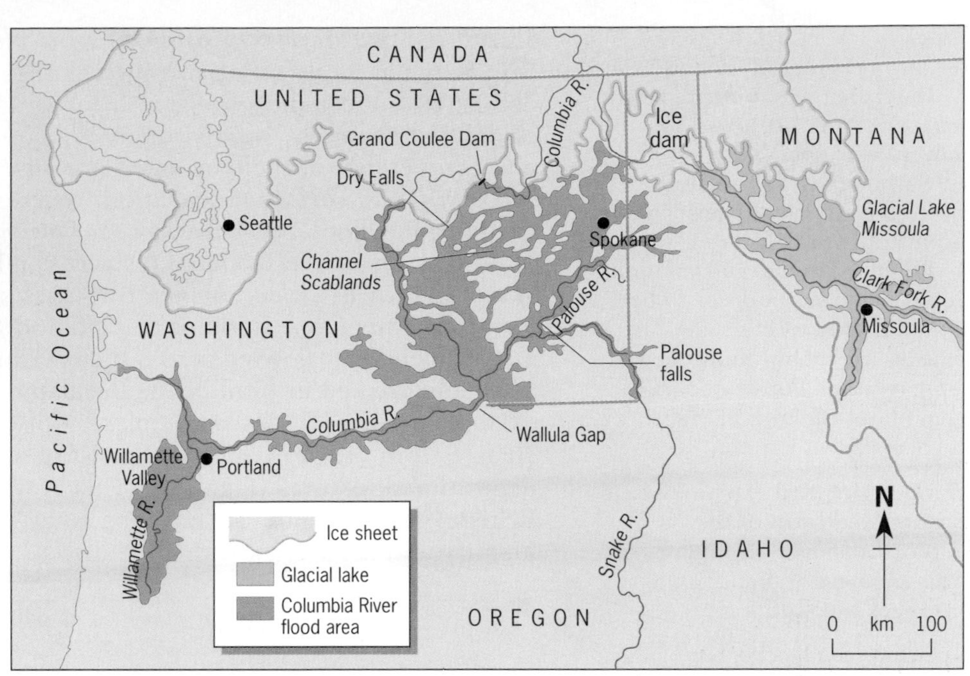

Figure 7.53 The largest known meltwater floods surged across north-west USA, 20 000–15 000 BP. A glacier advanced to block the Clark Ford River, creating glacial Lake Missoula which eventually contained 500 cubic miles of water (about as much as present-day Lake Erie and Lake Ontario combined). The combination of a powerful subglacial river and the enormous pressure exerted by the lake waters eventually caused the ice dam to collapse. A wall of water, with a volume 60 times that of the River Amazon, roared downstream at up to 90 km an hour. It eroded canyons and scoured surfaces across thousands of square kilometres of Washington State, creating what are known as the Channel Scablands. The flood surge lasted only a few weeks, yet cut canyons up to 500 m deep, and scoured rock surfaces across areas 100 km wide. This sequence of ice dam → glacial lake → dam burst → flood surge was repeated many times over several thousand years.

7.11 Fluvioglacial deposits

As with erosional processes, fluvioglacial deposition takes place in three locations: beneath, marginal to and in front of an ice mass. Deposition occurs where debris load exceeds the energy available for transportation.

Sub-ice depositional landforms

Subglacial streams flowing in short-lived tunnels at the ice–rock junction fluctuate rapidly in velocity, hydrostatic pressure and gradient. As a result, depositional landforms left behind tend to be discontinuous ridges and mounds, sitting on valley floors or glaciated surfaces. The best-known fluvioglacial depositional landform is the *esker*, 'a long, commonly sinuous ridge of sand and gravel, deposited by a stream in a subglacial tunnel' (Hambrey, 1994). Eskers vary from a few metres in length to long, undulating ridges snaking across the landscape, e.g. the Munro esker winds for 400 km across the Canadian Shield. Such longer examples may be explained by the lowering of a supraglacial stream on to the land surface as the ice melted. At a much smaller scale, rounded ridges sitting on the floor of the upper basin of the Blea Water–Haweswater trough, in Cumbria, appear to have esker characteristics (Fig. 7.34).

Ice-marginal depositional landforms

The area alongside the ice is a complex, constantly changing environment. It is often difficult to look at the landscape today and distinguish lateral moraines or sub-ice fluvioglacial features from genuine water-lain marginal deposits. The most common such landform is the *kame*, defined simply as 'a steep-sided hill of sand and gravel deposited by glacial streams adjacent to a glacier margin' (Hambrey, 1994). Kames vary widely in form and location. They may be mounds on a valley floor and deposited by streams in front of the glacier snout, and many kames seem to be old deltas of subglacial streams. In contrast, extensive deposits along the flanks of a trough are known as *kame terraces*. For example, the deposits along the side of the upper Haweswater trough (Fig. 7.46) are probably lateral moraine. An alternative suggestion is that they are a kame terrace (Fig. 7.54).

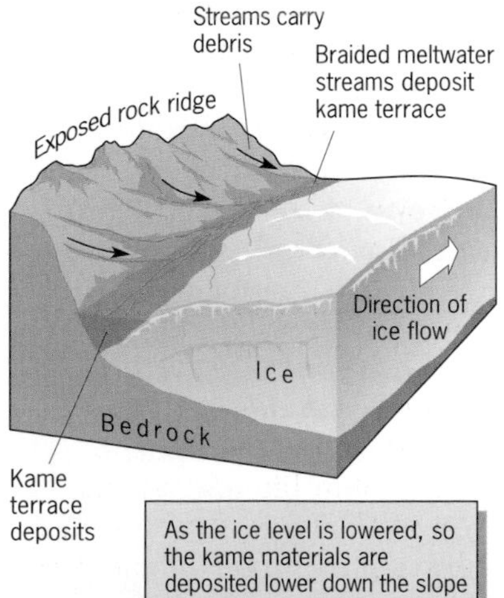

Figure 7.54 The upper Haweswater deposits as a kame terrace

?

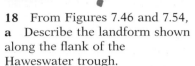

18 From Figures 7.46 and 7.54,
a Describe the landform shown along the flank of the Haweswater trough.
b Summarise the processes involved for each of the suggested origins of the deposits.
c What additional information would you need in order to be able to decide whether the features are lateral moraine or kame terrace? For example – what about the internal fabric of the deposits?

19 Describe and explain the location and form of the landforms shown on Figure 7.55. (You will need to include the processes involved.)

20 Once the glacier has disappeared from the landscape of Figure 7.55, suggest *two* landforms we might expect to find in the area covered by the ice in the diagram. Explain why you would expect to find these landforms.

Fluvioglacial deposits in front of a glacier

During the summer months, large volumes of meltwater, heavily laden with debris, criss-cross the surface in front of an ice-front in a complex pattern of braided channels. Debris supply is frequently greater than the capacity and competence of the meltwater streams, resulting in extensive *outwash deposition* and surface aggradation. Where this sedimentation infills a glacial valley floor, it is called a *valley train* (Fig. 7.42). Where the streams build a series of overlapping fans across a land surface in front of an ice sheet, an *outwash plain* or *sandur* (plural, *sandar*) develops (sandar: 'Laterally extensive flat plains of sand and gravel with braided streams of glacial meltwater flowing across them' (Hambrey, 1994)). Extensive sandar are being created in front of the Breidamerkurjökull ice sheet in Iceland, for example and illustrate the complex nature of proglacial landscapes: they change from year to year, and consist of an assemblage of erosional and depositional features (Fig. 7.55). This is why, in environments such as the British Isles where the ice has long since disappeared, it is often difficult to identify and explain individual landforms.

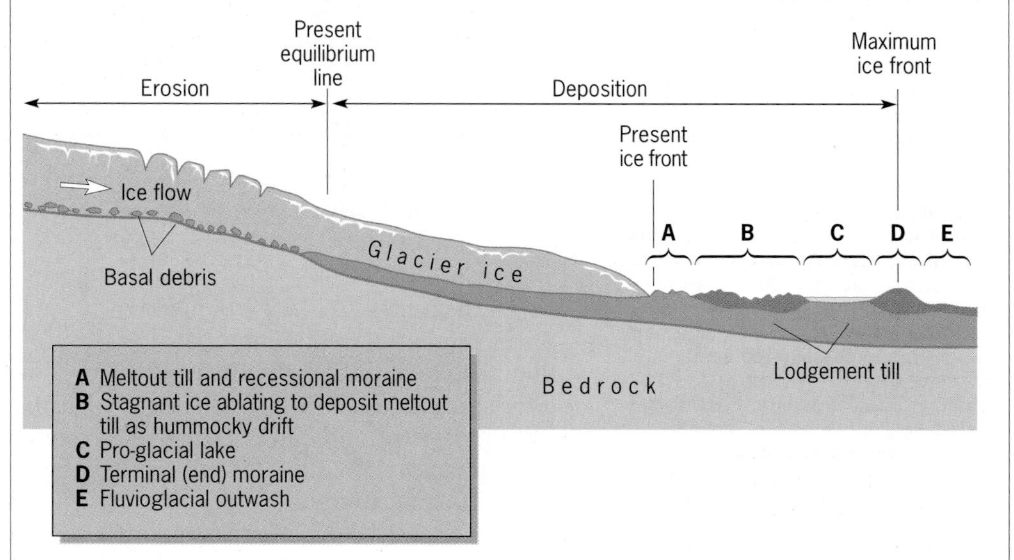

Figure 7.55 Typical glacier deposits during a recession phase of a temperate glacier (*After:* Hambrey, 1994)

7.12 Human interactions with glacial environments

Humans influence the process of glaciation through their role in global warming, e.g. CO_2 emissions to the atmosphere. Higher global temperatures will cause shrinkage of ice masses. We influence the ice itself through our use of technology and our changing lifestyles. Ice sheets and glaciers are becoming increasingly accessible, and are attractive to tourists (Fig. 7.56). For instance, year-round glacier skiing is booming. Environmentalists are concerned about pollution from increased tourist numbers arriving by boat and plane on the Antarctic ice sheet. As water demands grow, we hear of schemes for towing icebergs from the polar ice caps to populated regions. Conversely, icefields and glaciers are seen as hazards to human activities, e.g. icefalls and glacier advance can threaten settlement and transport routes; the failure of ice dams can cause flood surges.

Figure 7.56 Athabasca Glacier, Jasper National Park, Canada. Several thousand tourists visit this glacier each day during the summer.

Landform impacts

Glacial landforms are valuable tourist resources. The glacial trough is the main attraction of California's Yosemite National Park (Fig. 7.1). Some of the most popular ocean cruises are to the Norwegian and Alaska fiords.

Figure 7.57 Erosion by walkers along an arête in the Trossachs, Scotland

Surveys of visitors to the Lake District National Park tell us that the most attractive features are the lakes and valleys, which are the products of glaciation, while walkers love the challenge of arêtes such as Striding Edge. In the Swiss Alps, pyramidal peaks such as the Matterhorn lure climbers as well as armchair viewers. Glacial valleys provide transport corridors through mountains but may present major engineering problems and environmental hazards, e.g. landslides, rockfalls, etc. The overdeepened troughs have also become attractive sites for dams, reservoirs and HEP generation schemes, e.g. the Haweswater dam and reservoir (see Case Study, pp.142–143). There is much controversy about such developments in environments of high scenic value, e.g. the building of the Hetch Hetchy Dam in Yosemite National Park, across a valley regarded by many as being as beautiful as the main Yosemite valley with which we began this chapter.

Impacts upon large-scale erosional features are usually localised, e.g. erosion along the most heavily used paths, climbs, and 'honeypot' sites (Fig. 7.57). Depositional features, however, are much more vulnerable. In mountainous regions, till and fluvioglacial deposits may provide the only soils and relatively flat surfaces for agriculture, and so are much modified, e.g. terraces, drainage. In lowland areas, till plains and drumlin fields may be intensively farmed, even though they need much stone clearance and drainage, e.g. the Vale of Eden, Cumbria (see Case Study, p.146). Proglacial rivers create outwash plains, sandar and valley infill deposits of sands and gravels. These are previous sources of raw materials for the construction industry and are extensively quarried, with serious environmental impact, e.g. Thames and Trent valleys.

?

21 As a review exercise, make a list of the glacial and fluvioglacial landforms mentioned in this chapter. Beside each, suggest the nature of the interactions between people and the landform, and the possible effects this interaction may have on the landform.

Summary

- The Pleistocene glaciation lasted over 2 million years, ended in the British Isles by 10 000 BP and contained a number of ice-front advances and retreats.
- The ice budget or mass balance of an ice mass is determined by the relationship between accumulation and ablation, and controls the growth or shrinkage of a glacier or icesheet.
- Moving ice erodes bedrock by abrasion, rock fracturing and crushing, to provide subglacial and englacial load through entrainment.
- Frost-shattering and debris fall supply supraglacial load carried as lateral and medial moraines, or via crevasses as englacial load.
- The glacial erosional landforms range from extensive scoured surfaces to large overdeepened hollows, troughs and fiords.
- Valleys eroded by ice have distinctive U-shaped cross-profiles and irregular long profiles.
- Material deposited by melting ice is known as glacial drift, whose principal landforms are till plains, morainic ridges and drumlins.
- The work of meltwater produces erosional and depositional fluvioglacial landforms.
- The principal fluvioglacial erosional landforms are meltwater channels which may have developed under or marginal to the ice, and may be disconnected from present-day drainage systems.
- The principal fluvioglacial depositional landforms are kames, eskers and outwash plains (sandar), and may have either subglacial or proglacial origins.
- Human activities are having increasing impacts upon ice masses and glacial and fluvioglacial landforms, which are perceived as both attractive resources and dangerous hazards.

8 Periglacial processes and landforms

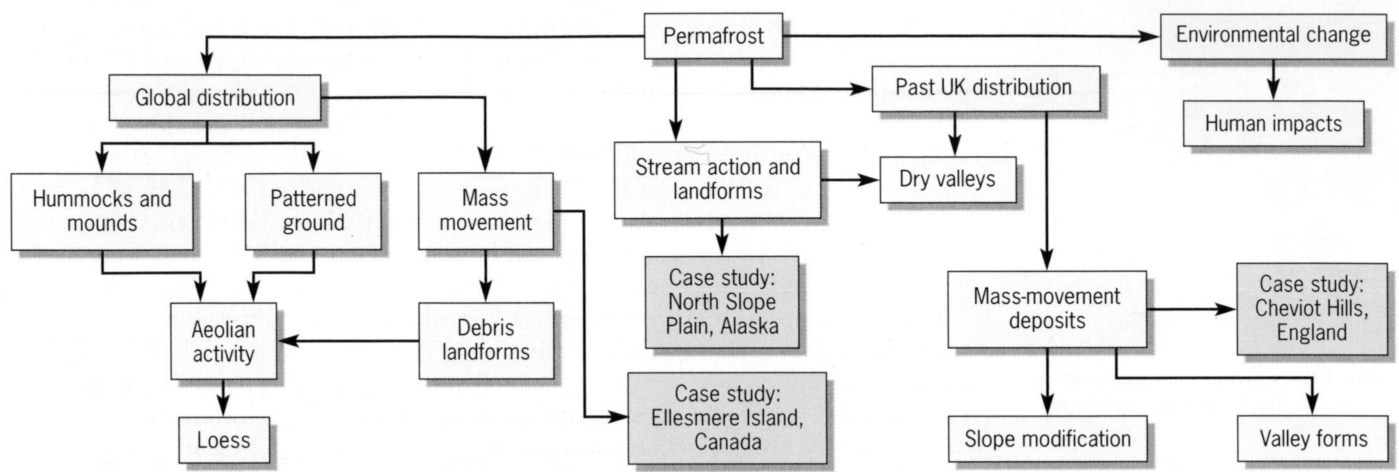

8.1 Introduction

The landforms shown in Figures 8.1 and 8.2 are both in England. However, they cannot be explained by processes that are at work in these localities today. They are **relict** or 'fossil' features, formed when environmental conditions were quite different from the present. Like glacial troughs, moraines, etc., they were formed during the Pleistocene glaciation (Chapter 7). They are *periglacial landforms*, which developed in the zones beyond the edges of the ice sheets, where mean annual temperatures would have been perhaps –8°C to –10°C.

In this chapter, we shall examine regions where the processes which formed these and other periglacial features are still at work. We can then return to the British landform evidence with a better understanding of how they evolved.

Figure 8.1 The Stiperstones tors in Ordovician-age quartzite rock, Shropshire, England. Frost-shattering of the jointed rock during the Pleistocene has removed large amounts of material. The more resistant areas of rock remain as upstanding tors.

Figure 8.2 A ramparted depression at Walton Common, East Anglia, England

8.2 Periglacial regions

The term 'periglacial' refers to the areas around ice sheets and glaciers. The landforms produced by periglacial processes occur in a range of cold, non-glacial environments, not always adjacent to present-day ice masses (Table 8.1). A simple definition which takes into account both present and past conditions is: 'Non-glacial processes and features of cold climates regardless of age and any proximity to ice sheets' (Whittow, 1984). Frost action and the resulting weathering and mass movement are found not only in periglacial regions, but in these regions they are especially intense and dominant. Deciding on what is a 'periglacial environment' is not easy. The most common environmental characteristic used to delimit present-day periglacial conditions is the presence of *permafrost*. Permafrost is perennially frozen ground, i.e. ground frozen continuously for two or more years.

Table 8.1 Classification of periglacial climates (*Adapted from:* Summerfield, 1991)

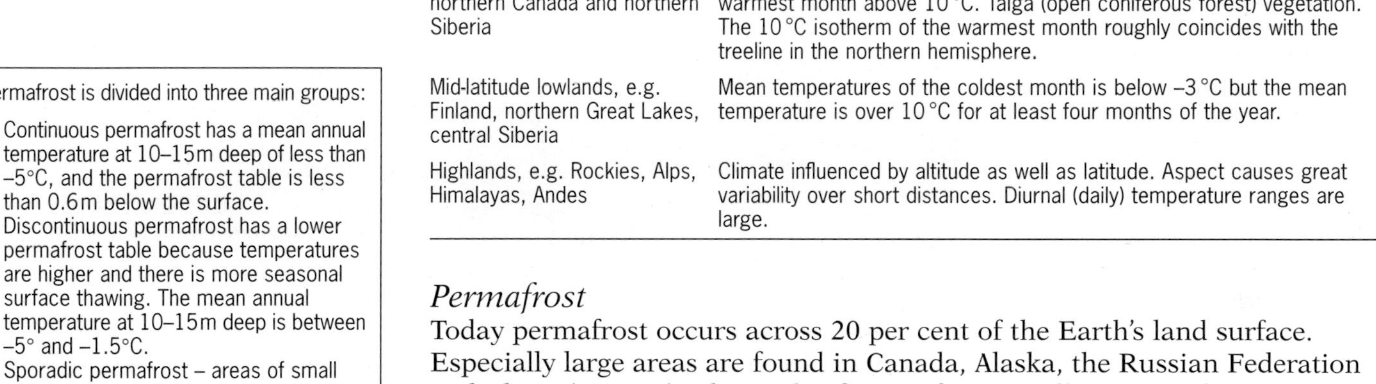

Periglacial region	Climatic characteristics
Polar lowlands, e.g. southern Greenland and Labrador	Mean temperature of the coldest month is below −3 °C. The area is characterised by ice caps, bare rock, and tundra vegetation.
Subpolar lowlands, e.g. northern Canada and northern Siberia	Mean temperature of the coldest month is below −3 °C and of the warmest month above 10 °C. Taiga (open coniferous forest) vegetation. The 10 °C isotherm of the warmest month roughly coincides with the treeline in the northern hemisphere.
Mid-latitude lowlands, e.g. Finland, northern Great Lakes, central Siberia	Mean temperatures of the coldest month is below −3 °C but the mean temperature is over 10 °C for at least four months of the year.
Highlands, e.g. Rockies, Alps, Himalayas, Andes	Climate influenced by altitude as well as latitude. Aspect causes great variability over short distances. Diurnal (daily) temperature ranges are large.

Permafrost is divided into three main groups:

1. Continuous permafrost has a mean annual temperature at 10–15m deep of less than −5°C, and the permafrost table is less than 0.6m below the surface.
2. Discontinuous permafrost has a lower permafrost table because temperatures are higher and there is more seasonal surface thawing. The mean annual temperature at 10–15m deep is between −5° and −1.5°C.
3. Sporadic permafrost – areas of small islands of permafrost in a generally unfrozen area. These are mountain regions or may represent relics of former colder climates.

Permafrost

Today permafrost occurs across 20 per cent of the Earth's land surface. Especially large areas are found in Canada, Alaska, the Russian Federation and China (Fig. 8.3). The study of permafrost is called *geocryology*.

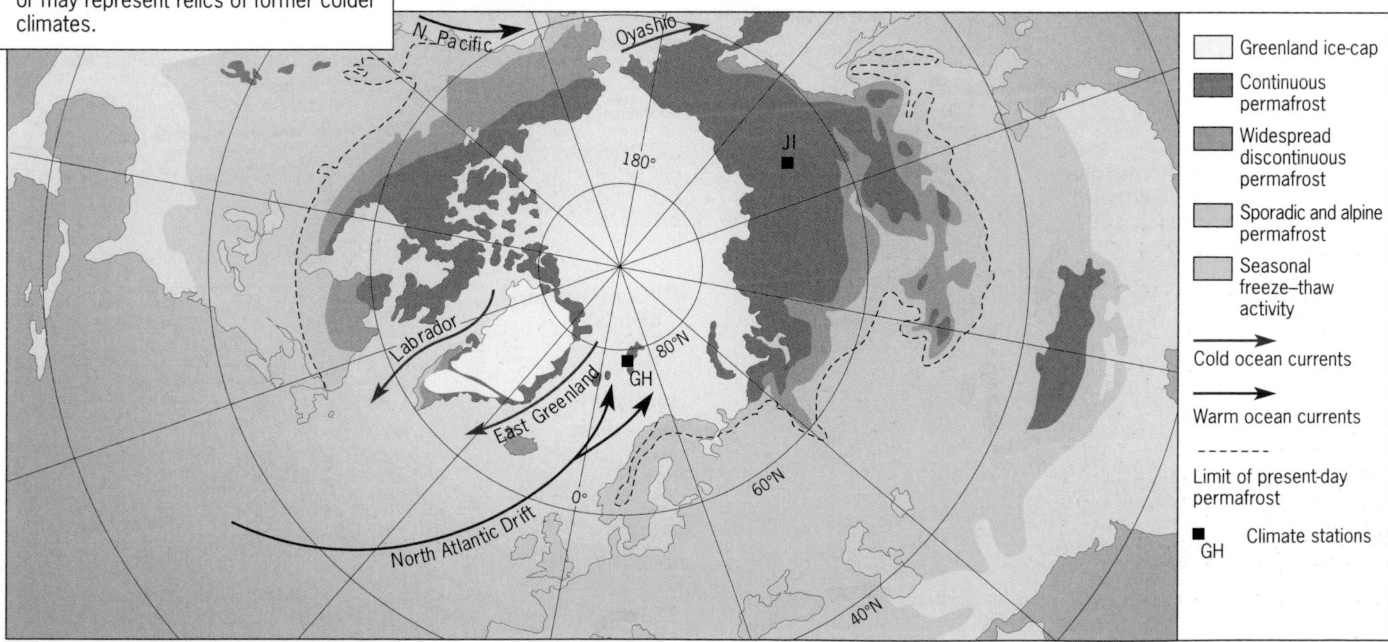

Figure 8.3 The distribution of permafrost and related phenomena in the northern hemisphere (*Source:* Cooke and Doornkamp, 1990)

1 Describe the distribution of continuous, discontinuous and sporadic permafrost in the northern hemisphere.

2a Compare the climate for the two areas of continuous permafrost shown in Figure 8.5.
b Use Figure 8.3 to suggest reasons for the differences in the climates.
c Suggest how the two areas may differ in the depth of the permafrost and active layer.

3 Use the information provided to suggest reasons for the variations in the southern limits of the permafrost. Consider the influence of latitude, ocean currents and continentality in your answer.

4 Suggest how the factors listed in Figure 8.4 will affect the depth of the active layer.

Throughout this chapter, keep in mind the key variables which influence the nature of the permafrost zone and the associated landforms: climate; subsurface materials (soil, weathered regolith, rock type); amount, distribution and state of water (liquid or solid, i.e. frozen as ice). The crucial role of the presence and behaviour of water and ground ice is illustrated in Figure 8.4.

Fully developed permafrost zones need air temperatures below 0 °C for at least nine months of the year, and below –10 °C for six months or more (Fig. 8.5). Even in extreme conditions, the permafrost zone thickens downwards from the surface at only a few centimetres a year. As permafrost depths can reach 1500 m, they have taken thousands of years to form, and are related to past as well as present climatic conditions. Permafrost thickness tends to decrease with decreasing latitude.

Two further important understandings are illustrated in Figure 8.3. First, water in its liquid and solid states is an important component of permafrost, yet the largest expanses are found in continental interiors with severe winters but low mean annual precipitation totals. Second, regular seasonal frost–thaw processes occur well beyond the permafrost regions, and hence outside present-day periglacial environments: for instance, the British Isles experience regular seasonal freeze–thaw conditions but are not within the periglacial boundary.

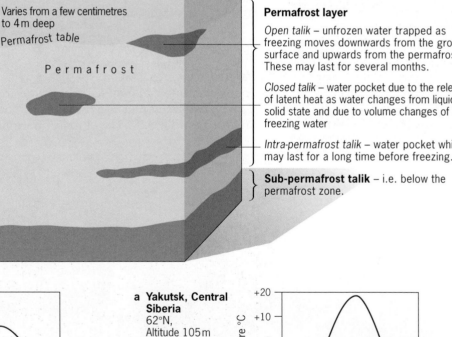

Permafrost depth – maximum thickness of 1500 m in Siberia. Most areas are less than 600 m. In areas of discontinuous permafrost, depths are less than 60 m. Locally depth of permafrost varies due to:

- present climatic conditions
- past climate (may take thousands of years to develop)
- soil type (if any)
- vegetation cover
- water bodies, i.e. rivers, lakes, oceans
- snow cover
- aspect
- wind.

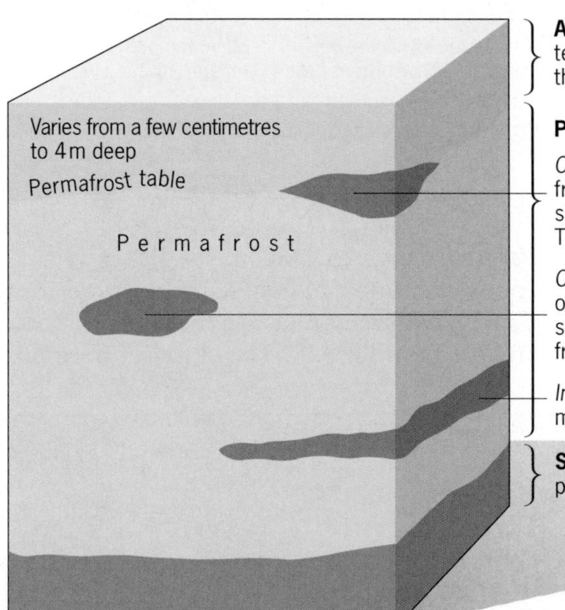

Varies from a few centimetres to 4 m deep
Permafrost table
Permafrost
Active layer – the zone of greatest temperature fluctuation which freezes and thaws seasonally.
Permafrost layer
Open talik – unfrozen water trapped as freezing moves downwards from the ground surface and upwards from the permafrost. These may last for several months.
Closed talik – water pocket due to the release of latent heat as water changes from liquid to solid state and due to volume changes of freezing water
Intra-permafrost talik – water pocket which may last for a long time before freezing.
Sub-permafrost talik – i.e. below the permafrost zone.

Figure 8.4 Permafrost terminology, and factors influencing permafrost depth (*After:* Summerfield, 1991)

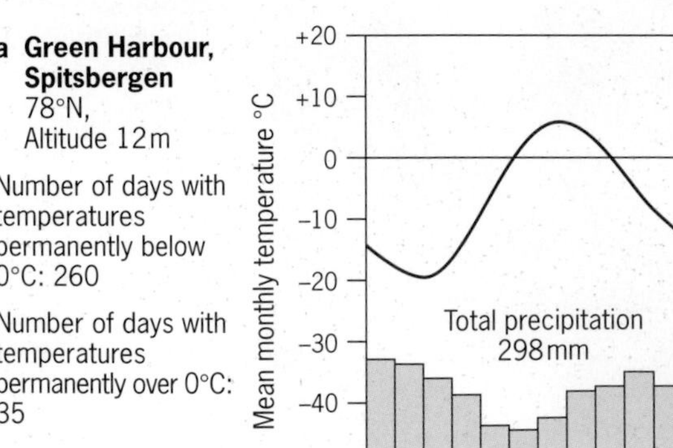

a Green Harbour, Spitsbergen
78°N,
Altitude 12 m
Number of days with temperatures permanently below 0°C: 260
Number of days with temperatures permanently over 0°C: 35
Mean monthly temperature °C
+20
+10
0
-10
-20
-30
-40
-50
Total precipitation 298 mm
50
40
30
20
10
0
Mean monthly precipitation mm
J F M A M J J A S O N D

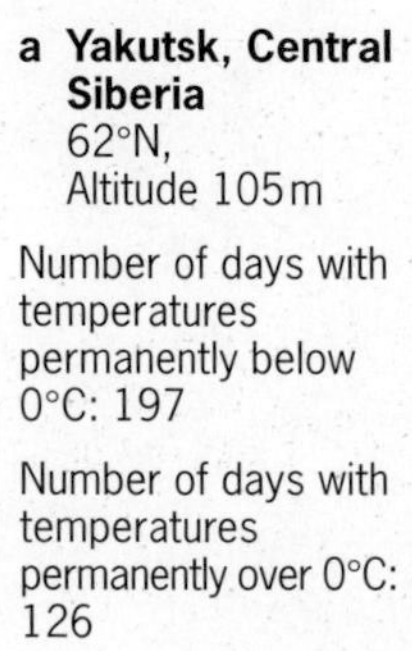

a Yakutsk, Central Siberia
62°N,
Altitude 105 m
Number of days with temperatures permanently below 0°C: 197
Number of days with temperatures permanently over 0°C: 126

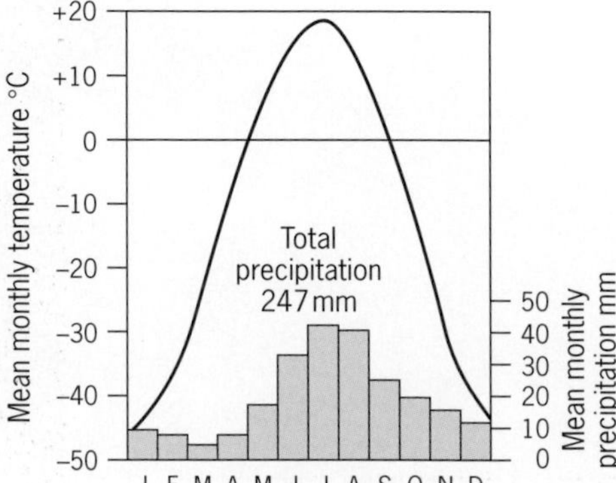

Mean monthly temperature °C
+20
+10
0
-10
-20
-30
-40
-50
Total precipitation 247 mm
50
40
30
20
10
0
Mean monthly precipitation mm
J F M A M J J A S O N D

Figure 8.5 Climate graphs for two periglacial areas

Local conditions are equally important. For instance, microclimate influences the frequency and intensity of the freeze–thaw cycle. This will affect the active layer most significantly (Fig. 8.4), and it is the character and behaviour of this layer which is so crucial in landform development. Freeze–thaw processes may also occur locally at temperatures below 0 °C because of salts in solution, soil type and pressure conditions.

Ground ice and the structure of frozen ground

During the summer, the active (i.e. seasonally thawed) layer deepens and is often waterlogged. As temperatures fall in the autumn, some of this water moves by capillary suction towards the two freezing fronts, i.e. from the ground surface downwards and from the permafrost upwards. There it freezes, as lenses, thin needle shapes, or layers within the pore spaces of the materials. This is called *segregated ice*.

Figure 8.6 Ground ice (black) in tunnel wall of silt (buff) with ice crystals on roof, Canada. Large masses of ice, which can be many metres thick, are called 'massive ice'. These form when the water which is greater than the amount that could be held in the soil pores (i.e. excess ice) collects together and freezes as a single mass.

Where too much water has accumulated within the active layer to be held within the pore spaces, it may collect as subsurface ponds or layers, or may be dispersed throughout the materials. On freezing it can be seen in a variety of forms, from thin sheets to masses several metres thick (Fig. 8.6). This is known as *excess ice*. The combination of segregated and excess ice constitutes as much as 80 per cent of the mass of the frozen ground. Ice also contributes as **taliks** (Fig. 8.4). The volume and proportion of ice decreases with increasing depth.

How much ice is held depends upon how much water is present and the texture of the active layer materials. Fine-grained clays hold water well, but temperatures need to drop well below 0 °C before extensive freezing takes place. Coarse-grained sands are poor retainers of water, but freezing does occur at 0 °C. For example, a study in Alaska found that in clays at –5 °C ground temperature, only 50 per cent of the water had frozen. In coarse sands, all the water froze at around 0 °C. We can therefore conclude that the optimal materials for ice formation in the active layer are sediments of intermediate grain size, with moderate permeability, e.g. silts, with relatively high suction potential and where water freezes at near 0 °C.

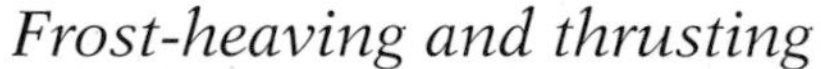

Frost-heaving and thrusting

The freezing water in the active layer is unevenly distributed. Thus, as it freezes, so the expansion associated with this change to the solid state will be uneven. Spatial variations in sediment particle size, and pressure variations in and around trapped pockets of unfrozen soil and water, also help to

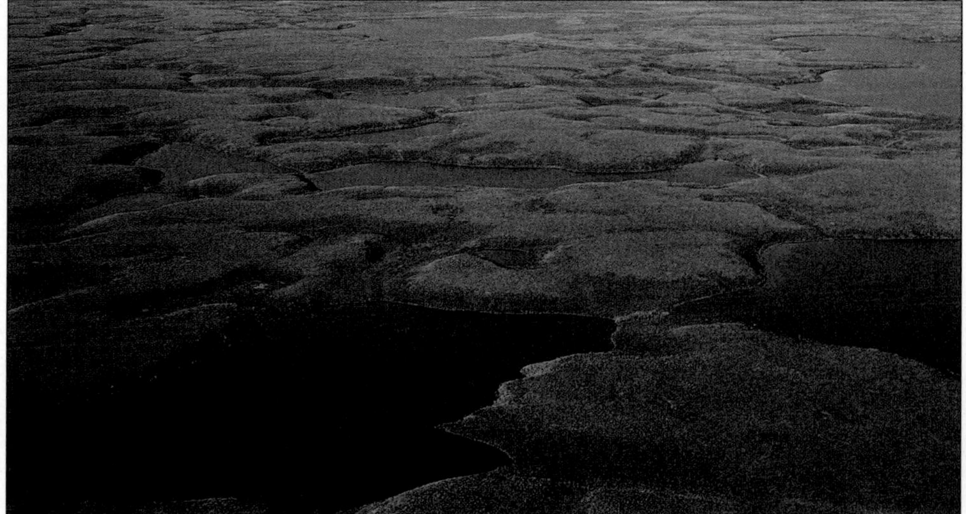

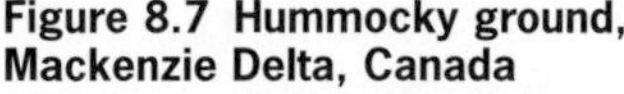

Figure 8.7 Hummocky ground, Mackenzie Delta, Canada

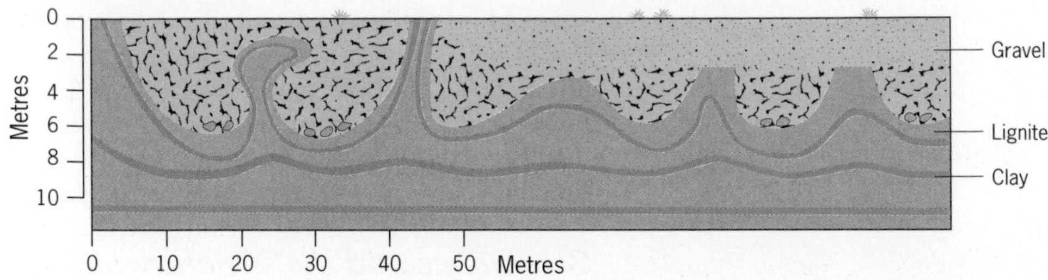

Figure 8.8 Deformation of lignite and clay, Zievel claypit, Rhineland, Germany. These flow-like cryoturbations, called *periglacial involutions*, have been used to delineate the extent and depth of permafrost in central Europe during the Pleistocene. Lignite is quarried as a fuel and is thus exposed in many locations in central Europe. These cryoturbations have been identified as being due to density inversions occurring in water-saturated thawing permafrost (*Source:* Strunt, 1983).

generate differential movement. The differential expansion results in movement of material (particles and water) by the process of horizontal *frost-thrusting* and vertical *frost-heaving*. The vertical movement is the more pronounced, mainly because there is less resistance from the ground surface.

Typical landforms resulting from this uneven pattern of heaving and thrusting are hummocks and lobes, which occur both singly and in clusters (Fig. 8.7). The internal structure of these disturbed active layers shows a complex churning pattern caused by the interaction of the variables of water, sediment and pressure. The churning motions create displacements of the materials, called *cryoturbations* (Fig. 8.8). As the movements recur each thaw season, so the details of the topography steadily change.

8.3 Aggradational landforms

The regular repetition of the freeze–thaw, expansion–contraction processes causes the churning and movement of materials. This movement causes two of the most distinctive groups of landforms found in periglacial regions – *hummocks* and *patterned ground*. A wide variety of hummocky landforms results from the uneven distribution of thrusting and heaving; patterned ground develops because particles of different sizes migrate through the active layer at different speeds.

Hummocks and mounds

Thufur and palsas

Thufur are earth hummocks up to 0.5 m high and 1–2 m in diameter. They occur in clusters, often with a regular spacing, to give a form of patterned ground which may cover extensive areas. Their shape is influenced by slope angle. On gentle slopes they are circular, but become elongated as gradients increase. Thufur develop best in soils with uniform particle size. *Palsas* are larger mounds, most commonly 1–6 m high and 10–30 m wide. They are widely and irregularly spaced, and develop best on unfrozen peat bogs in areas of sporadic permafrost. The broad dome shape of palsas is caused by the formation of segregated ice and differential frost heave.

Figure 8.9 Pingo, Mackenzie Delta, Canada

Pingos

The extreme form of mound is the *pingo*, which is the Inuit word for 'hill'. Pingos are large mounds with permanent (perennial) ice cores. They are roughly circular in plan, 30–600 m in diameter and 3–70 m high (Fig. 8.9). Two distinct types, **closed-system pingos** and **open-system pingos**, have been identified, related to the character of the underlying permafrost.

Although pingos can grow as much as 1 m in a year, large examples may be thousands of years old, e.g. radiocarbon dating of two pingos in the Canadian Arctic gave ages of 4500 and 7000 years. The largest concentration of pingos, about 1450, are found in the Mackenzie Delta region of Canada. Here, 98 per cent have developed in or close to lake basins (*alases*). The presence of abundant water, the uneven distribution of freezing and

Figure 8.10 An ice wedge

expansion, and recurrent freeze–thaw cycles which progressively add to the ice core, are crucial factors.

Frost-cracking and polygons

When ground temperatures continue to fall well below zero, the frozen ground materials may contract. This causes sets of cracks to appear, often polygonal in arrangement, and known as *frost-crack polygons*. They are up to 10 mm wide and 8 m deep, with each polygon 5–30 m in diameter. As the active layer thaws during summer, the cracks may fill with water. In autumn, refreezing causes expansion and widening of the cracks. Over time a crack may be enlarged to become an *ice wedge* (Fig. 8.10).

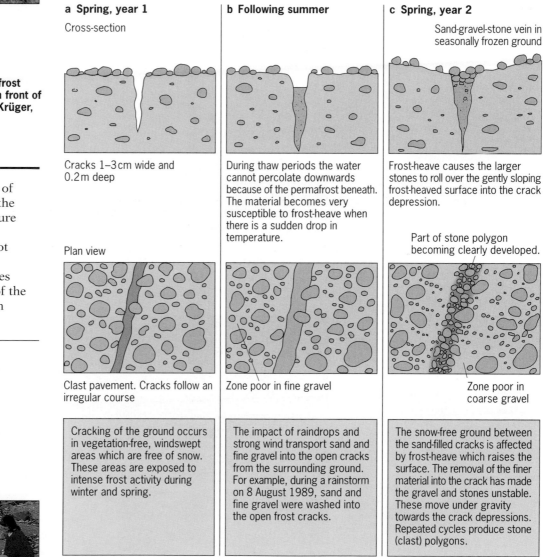

a Spring, year 1

Cross-section

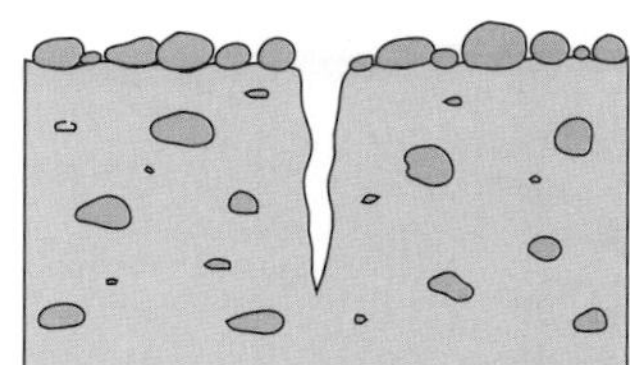

Cracks 1–3 cm wide and 0.2 m deep

Plan view

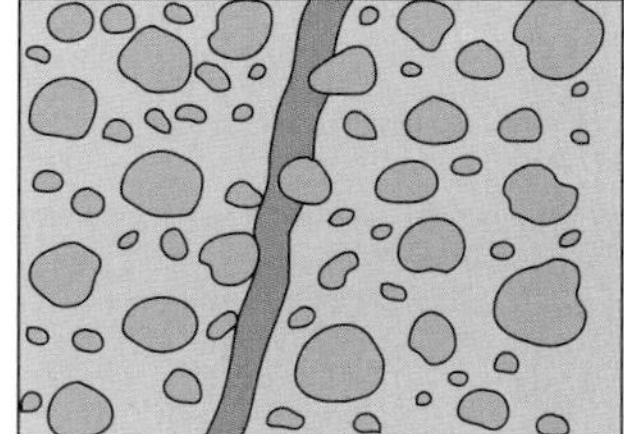

Clast pavement. Cracks follow an irregular course

Cracking of the ground occurs in vegetation-free, windswept areas which are free of snow. These areas are exposed to intense frost activity during winter and spring.

b Following summer

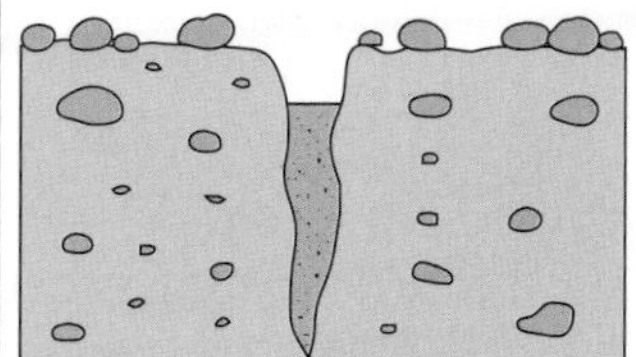

During thaw periods the water cannot percolate downwards because of the permafrost beneath. The material becomes very susceptible to frost-heave when there is a sudden drop in temperature.

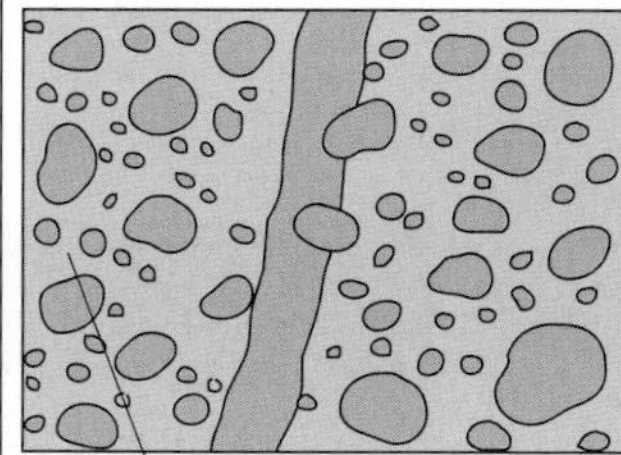

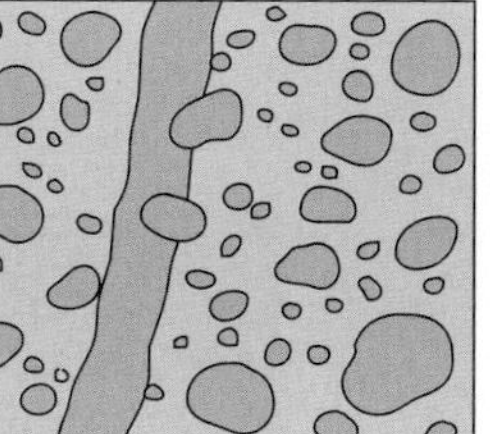

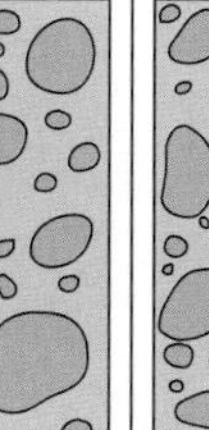

Zone poor in fine gravel

The impact of raindrops and strong wind transport sand and fine gravel into the open cracks from the surrounding ground. For example, during a rainstorm on 8 August 1989, sand and fine gravel were washed into the open frost cracks.

c Spring, year 2

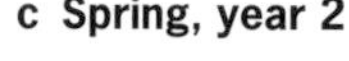

Sand-gravel-stone vein in seasonally frozen ground

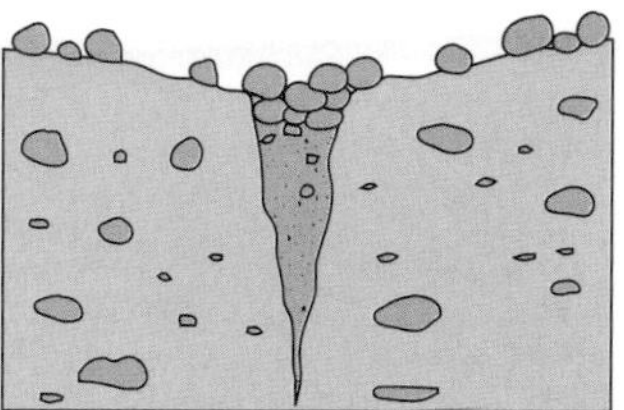

Frost-heave causes the larger stones to roll over the gently sloping frost-heaved surface into the crack depression.

Part of stone polygon becoming clearly developed.

Zone poor in coarse gravel

The snow-free ground between the sand-filled cracks is affected by frost-heave which raises the surface. The removal of the finer material into the crack has made the gravel and stones unstable. These move under gravity towards the crack depressions. Repeated cycles produce stone (clast) polygons.

Figure 8.11 The evolution of frost cracks and sorted polygons in front of Myrdalsjökull, Iceland (*After:* Krüger, 1994)

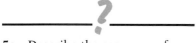

5a Describe the sequence of events in the formation of the frost-crack polygons in Figure 8.11.
b Why had frost cracks not developed prior to 1984?
c Identify the key processes involved in the formation of the sorted polygons in southern Iceland.

An example of the evolution of frost-crack polygons occurred in the highland area of Maelifellssandur, southern Iceland, during the mid-1980s (Fig. 8.11). An expanse of vegetation-free, drumlinised ground moraine has been exposed over the past 40 years by the recession of the Myrdalsjökull ice cap. Until 1982, permafrost did not form, and no cracks appeared. From that year, however, frost action intensified locally, and between 1984 and 1989 a well-developed set of frost-crack polygons evolved. The processes involved are summarised in Figure 8.11.

Figure 8.12 Frost-crack polygons

Figure 8.13 **The two hypotheses for the upward movement of stones in permafrost**

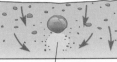

a Frost-pull hypothesis

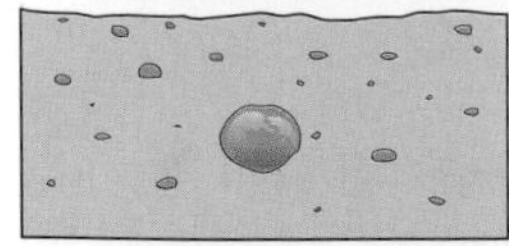

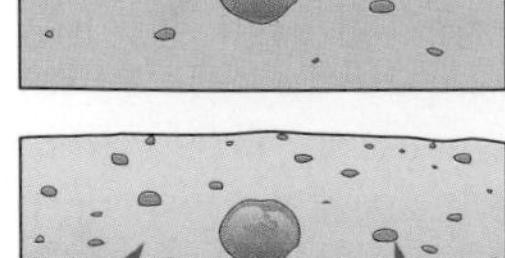

Ground still frozen

b Frost-push hypothesis

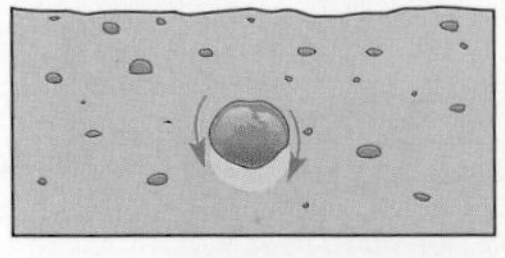

Unfrozen active layer. Soilwater flows round the stone and collects underneath.

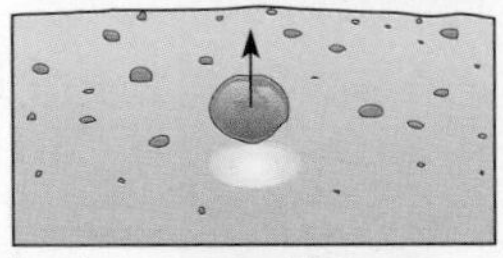

As the active layer freezes the ice lens formed beneath the stone pushes the stone upwards. Uplift most effective with rapid freezing.
Ice lens or needle ice

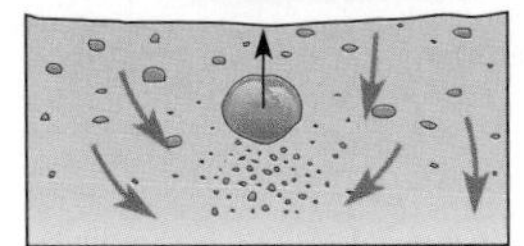

As active layer thaws finer materials fill the gap beneath the stone.

Patterned ground

Repeated freeze–thaw cycles gradually rearrange soil surface sediments into a variety of patterned shapes: stripes, lobes and polygons (Fig. 8.12). For patterned ground to develop, the active layer materials must contain a range of particle sizes. These are then sorted by the freeze–thaw action, which generates cracking, wedging and heaving. The coarser fragments (clasts) are moved slowly towards the surface (Fig. 8.13), where they cluster to form the patterns. The type of patterning is influenced by the thickness of the active layer, the particle dimensions and the slope angle.

Stone stripes (Fig. 8.14) are defined as 'patterned ground with a striped pattern and a sorted appearance due to parallel lines of stones, and

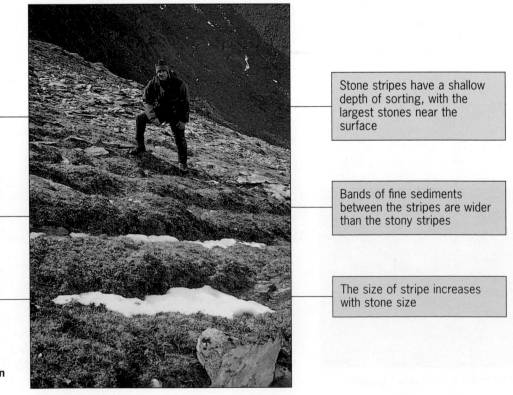

Stripes commonly develop by the downslope extension of sorted polygons or the linear alignment of soil buds occupying level areas
Stripes are found on gradients of more than 3° but rarely on slopes steeper than 30°
Stripes curve around blocks and outcrops
Stone stripes have a shallow depth of sorting, with the largest stones near the surface
Bands of fine sediments between the stripes are wider than the stony stripes
The size of stripe increases with stone size

Figure 8.14 Stone stripes. Sorted stone stripes are common features on sloping terrain in periglacial regions.

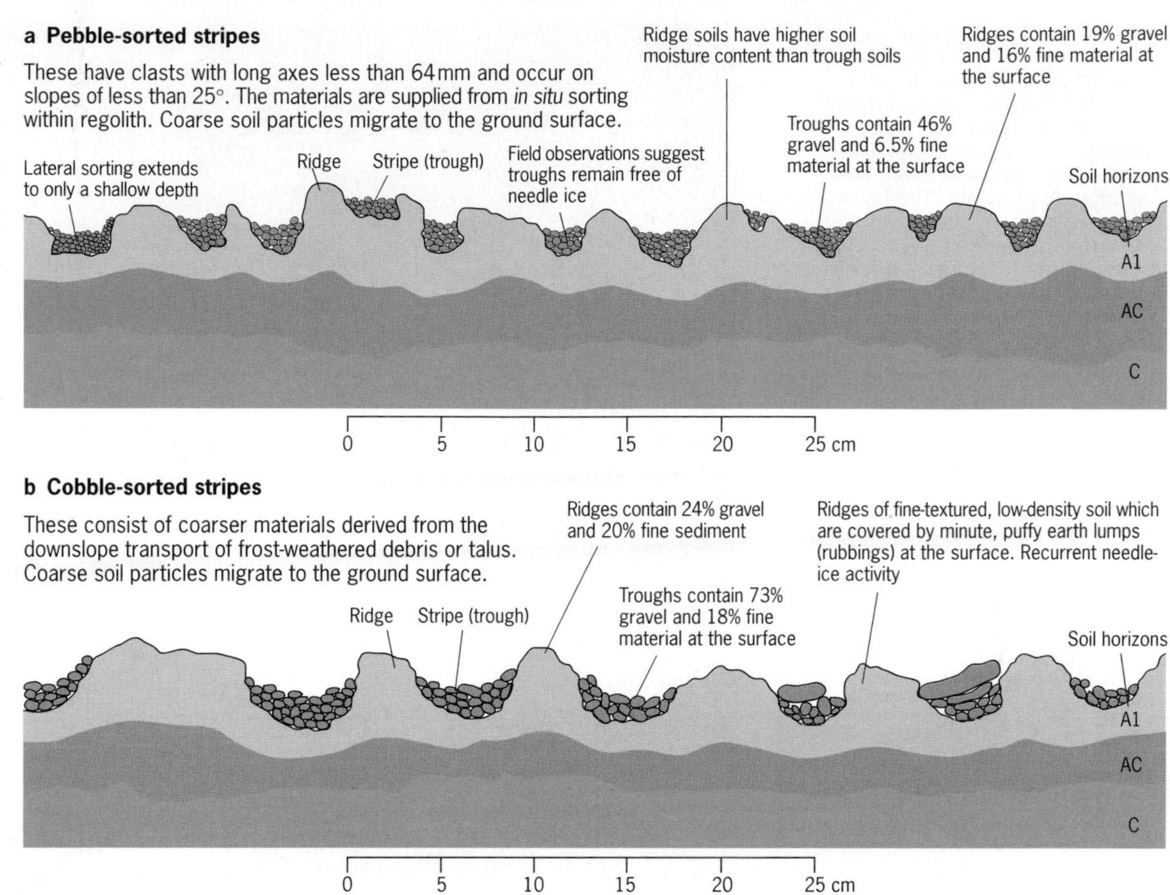
a Pebble-sorted stripes
These have clasts with long axes less than 64mm and occur on slopes of less than 25°. The materials are supplied from in situ sorting within regolith. Coarse soil particles migrate to the ground surface.
Ridge soils have higher soil moisture content than trough soils
Ridges contain 19% gravel and 16% fine material at the surface
Troughs contain 46% gravel and 6.5% fine material at the surface
Lateral sorting extends to only a shallow depth
Ridge
Stripe (trough)
Field observations suggest troughs remain free of needle ice
Soil horizons
A1
AC
C
0
5
10
15
20
25 cm
b Cobble-sorted stripes
These consist of coarser materials derived from the downslope transport of frost-weathered debris or talus. Coarse soil particles migrate to the ground surface.
Ridges contain 24% gravel and 20% fine sediment
Ridges of fine-textured, low-density soil which are covered by minute, puffy earth lumps (rubbings) at the surface. Recurrent needle-ice activity
Troughs contain 73% gravel and 18% fine material at the surface
Ridge
Stripe (trough)
Soil horizons
A1
AC
C
0
5
10
15
20
25 cm

Figure 8.15 Cross-sections of the two types of stone stripes, Venezuelan Andes (*Source:* Pérez, 1992)

?

6 Compare the particle size characteristics of the ridges and troughs (stripes) in Figure 8.15, i.e. the material size and origin for the two types of stripes.

7a Measure the width of the stripes (troughs) and ridges observed in the Andes. Do the pebble-sorted stripes differ in size from the cobble-sorted stripes?
b To what depth does lateral sorting into ridges and stripes occur?

8 Use Table 8.2 to calculate the mean and median of the maximum and minimum rates of movement of the main particle mass and the outliers.

9 Using appropriate graphical and/or statistical techniques, investigate the relationship between slope angle and the rate of particle movement using the data in Table 8.2. Justify your choice of techniques.

Table 8.2 Mean movement rates for painted gravel particles (placed on the soil surface) on plots with pebble-sorted stripes,* Venezuelan Andes. Period of measurement: December 1985 to December 1987. (*Source:* Pérez, 1992)

		Main particle mass			Outliers – larger pebbles which moved faster than the main marker masses
Plot number	Slope angle (degrees)	Maximum rates (cm yr^{-1})	Minimum rates (cm yr^{-1})	Average rates (cm yr^{-1})	Average rates (cm yr^{-1})
1	25.0	—	—	—	27.5
2	20.0	9.7	1.5	3.76	24.4
3	22.0	12.4	2.5	5.53	23.0
4	22.0	20.8	2.9	7.76	33.6
5	22.0	19.7	4.6	9.51	34.3
6	21.0	11.7	2.3	5.14	29.2
7	21.0	11.5	2.9	5.81	21.7
8	18.5	13.8	0.6	2.92	24.8
9	22.0	17.3	5.1	9.42	34.6
10	21.0	13.9	5.0	8.32	36.1

*The gravel particles were painted so that individual stones could be identified and their movements recorded.

intervening strips of fine material oriented down the steepest available slope' (Washburn, 1956). Stone stripes can vary greatly in size. A study between 1985 and 1990 into small-scale stripe formations in the Venezuelan Andes shows how these features develop. The area is at a high altitude (4500 m) with a large number of freeze–thaw cycles per year. Two stripe patterns were observed (Fig. 8.15). The climatic conditions allow only a very shallow layer of ground to be affected by frost action. The sorting and downhill movement of the stones were recorded (Table 8.2).

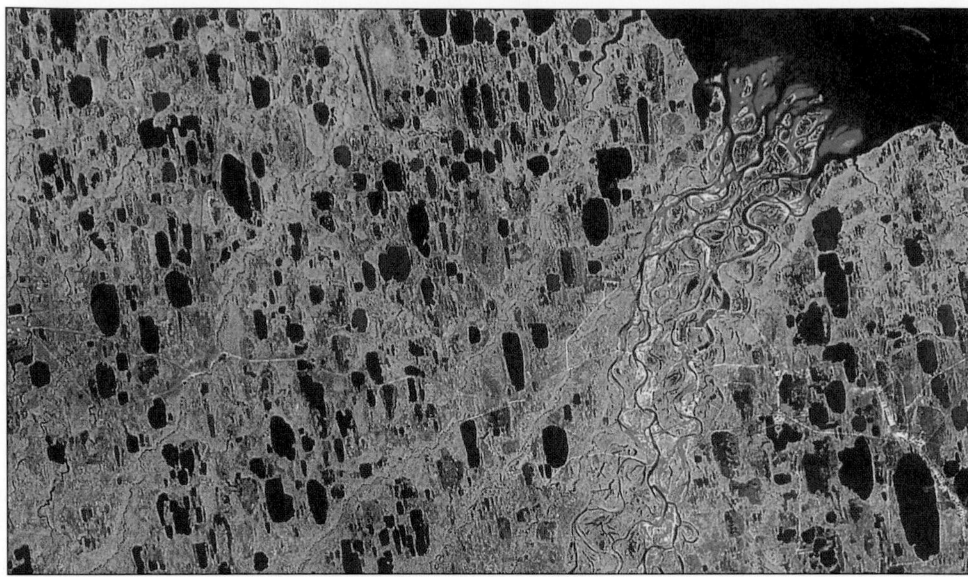

Figure 8.16 Thermokarst depressions, North Slope, Alaska, USA. The lakes may be several kilometres long and tens of metres deep, indicating the scale of subsidence. They are especially common across the Alaskan and Canadian Arctic plain. The lakes are usually elliptical in shape, with the long axis oriented parallel to the prevailing wind. The reason for this apparent relationship is uncertain.

8.4 Degradational landforms

The contraction and shrinkage associated with the longer-term melting of permafrost produces distinctive landforms. The melting can be the result of natural factors, e.g. climatic change, stream erosion, or human activity (see section 8.9), and may occur over varying timescales.

The most widespread result of the thawing of ice-rich permafrost is ground subsidence. Because of the uneven distribution of ground ice, the ground will settle differentially to produce an uneven land surface pitted with depressions. Such surfaces have been given the name *thermokarst* because of the resemblance to certain features of limestone karst landscapes. Thermokarst depressions develop most extensively across level terrain, where the thawing water collects as ponds and lakes (Fig. 8.16). These are called *thaw lakes*, and occur where the active layers contain more ice than sediment.

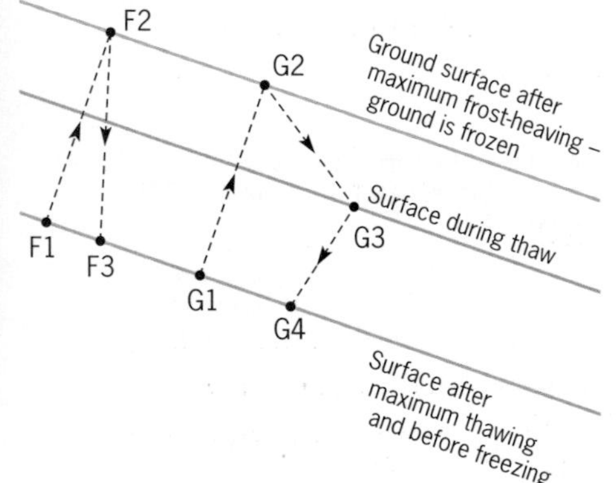

F1–F3
Particle movement during frost-heaving.
F1 to F2 – on freezing particle is heaved at right angles to the slope surface.
F2 to F3 – particle settles vertically as the ground thaws. The particle has moved downslope F1 to F3.

G1–G4
Particle movement during gelifluction.
G1 to G2 – on freezing particle is heaved at right angles to the slope surface.
G2 to G3 – particle moves downslope in the thawing active layer.
G3 to G4 – thawed layer settles back against the slope rather than vertically. The particle has moved downslope G1 to G4.

Figure 8.17 Solifluction and the downslope movement of soil by frost-heaving and gelifluction

8.5 Mass movement and resulting landforms in periglacial environments

The landforms discussed in earlier sections (e.g. hummocks, patterned ground, depressions) develop best on relatively flat terrain. On more strongly sloping surfaces in periglacial regions, a range of mass-movement processes occur, from gradual soil creep through flow surges and slope failures to abrupt rockfalls.

Solifluction

A key mechanism is *solifluction*. This is not restricted to periglacial environments, but does occur in two distinctive and closely related forms in such environments. First, *frost heave* causes the gradual downslope movement of particles through the operation of each freeze–thaw cycle (Fig. 8.17, F1–F3). Second, *gelifluction* is the downslope creep caused as the active layer thaws during the summer (Fig. 8.17, G1–G4). Both occur above permafrost and generate migration of up to 10 cm a year at the surface. The movement decreases rapidly with depth, and generally ceases by 2 m below the surface.

Rates of movement are related to the angle of slope, the amount of surface snowmelt and infiltration, the ice content of the active layer, and the

Table 8.3 Average soil movement rates (1975–9) in the Swiss National Park, eastern Swiss Alps (cm/yr)

Depth (cm)	Vegetation-covered	Partly covered	Vegetation-free
0	0.9	3.2	5.1
5	0.2	2.1	2.8
10	0.1	1.9	2.3
20	0.0	1.2	1.8
30	0.0	0.8	1.3
40	0.0	0.6	1.0
50	0.0	0.0	0.05

10a Use the data in Table 8.3 to plot depth/velocity profiles for solifluction rates in the Swiss Alps. Plot the three types of surface on the same axes, using different colours.
b Compare the depth and rates of movement with the amount of vegetation cover.
c Explain how vegetation cover can influence the rate of mass movement.

vegetation cover (Table 8.3). Most of the clasts settle with their long axes aligned downslope, which increases their ability to slip. Once at the valley bottom or slope base the materials build up as masses of unsorted, angular debris which may be tens of metres thick. These slope-foot deposits are known as *head*. Dependent on local conditions, they may occur as sheets, benches or lobes, and may remain as landforms long after the periglacial conditions have ended (see section 8.8. for UK examples).

Flows and slides

Debris flows are more rapid, intermittent downslope movements of the active layer. Researchers in the Slims River Valley of the St Elias Mountains, Yukon Territory, observed five debris flows during 1984–5. They recorded average velocities of 5 m/sec in surges which occurred every 2–5 minutes, and lasted for up to three days. Slope angle, the depth of the active layer, the amount of ground ice and snow cover were shown to be influential factors.

Where a shallow skin of the active layer across a slope fails, extensive sheets of material may slide over the underlying permafrost surface. These are known as *active layer detachment slides* or, more simply, *skinflows*. Unlike debris flows, they do not involve much deformation of the materials. They can be triggered by sudden rises in temperature, summer rainstorms, rapid snowmelt, or surface disturbances such as earthquakes. The crucial ingredients seem to be the sudden and abundant loading of water into the active layer, and a sharp junction between the permafrost zone and the melting front (see Ellesmere Island Case Study below).

Not all slope failures are restricted to the active layer. For instance, if a permafrost zone is exposed by stream action it melts rapidly, causing collapse which may involve both frozen and unfrozen ground.

Active layer detachment slides, Ellesmere Island, Arctic Canada

Sites in three valleys near Eureka, Ellesmere Island (Fig. 8.18), were studied over a 40-year period, 1950–1990. Data was based on analysis of aerial photos taken between 1950 and 1986, followed by fieldwork, 1988–90. The research found that although active layer detachment slides occurred at an average of 3.5 per year over the 40 years, there were marked variations from year to year. For instance, 1988 was a particularly active year: 75 new slides started in an area of 5.3 km² along Black Top Creek, and there was a 32 per cent increase in slide activity along Hot Weather Creek (Fig. 8.19).

The problem the researchers faced was how to explain both the timing and the location of the slides. The 1988 slides were not triggered by heavy rains or earth tremors, and were not related to the type of bedrock. Most slides were located near to the edge of the plateaus, on the upper or central parts of slopes with angles of more than 5 °, and often with a concave profile. The key variable appeared to be the rapidity of thawing in the active layer. In the 1988 summer of frequent slides, thaw rates were around 1 cm per day near the base of the active layer, which

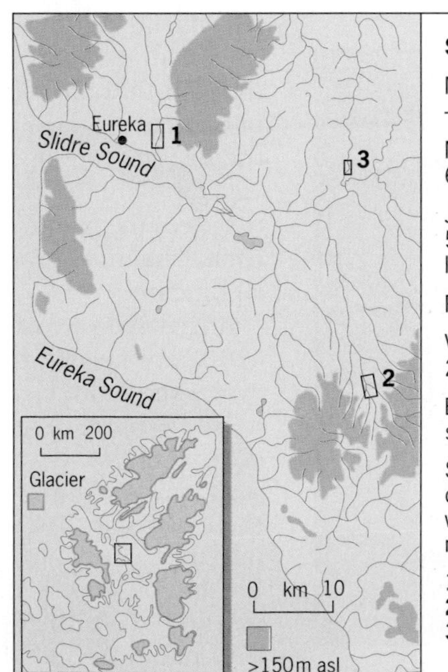

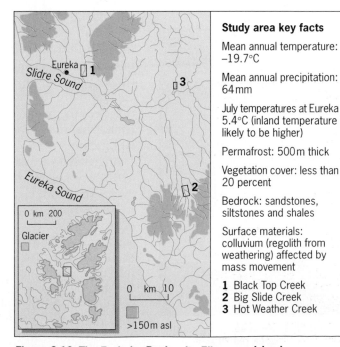

Study area key facts

Mean annual temperature: –19.7°C

Mean annual precipitation: 64 mm

July temperatures at Eureka 5.4°C (inland temperature likely to be higher)

Permafrost: 500 m thick

Vegetation cover: less than 20 percent

Bedrock: sandstones, siltstones and shales

Surface materials: colluvium (regolith from weathering) affected by mass movement

1 Black Top Creek
2 Big Slide Creek
3 Hot Weather Creek

Figure 8.18 The Fosheim Peninsula, Ellesmere Island (*Source:* Lewkowicz, 1992)

Ellesmere Island

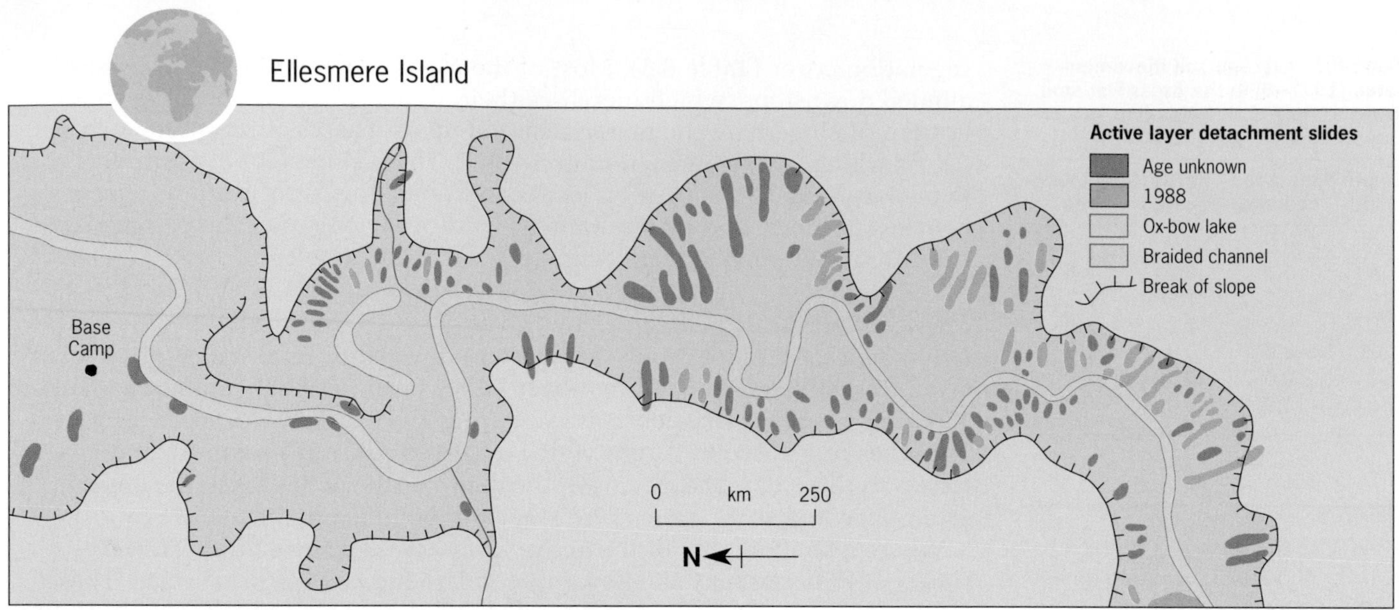

Figure 8.19 Active layer detachment slides in lower Hot Weather Creek (*Source:* Lewkowicz, 1992)

was 56–65 cm deep. In 1989 and 1990, the active layer thickness was similar, but thaw rates were much slower, generally less than 0.5 cm per day.

When the thaw plane advances rapidly through the ground ice, the active layer is suddenly loaded with water. Unless this water can drain away, it has to bear the weight of the material above it. Water pressure increases, causing loss of soil strength, resulting in slope failure.

?

11 Use Figure 8.18 to describe the physical environment of the Fosheim Peninsula study area.

12 Describe the distribution of the active layer detachment slides (Figure 8.19) in Hot Weather Creek. Refer to slope position and aspect in your account. Identify any slides which may have been due to fluvial undercutting.

13 Why was 1988 a year with a large amount of slide activity?

Weathering processes and debris landforms

Expanses and piles of angular rock debris are common landforms in periglacial regions, and occur on a range of slope types (Fig. 8.20). These are the result of vigorous mechanical weathering processes. Until recently it was thought that the repeated expansion–contraction cycle of the freeze–thaw process at work in pore spaces, cracks and joints was sufficient to explain the disintegration. Today, however, researchers know that at least four other variables are important: the rate of freezing; the minimum temperature reached; the frequency of the freeze–thaw cycle; the moisture content of the rock. Chemical processes such as hydration and hydrolysis could also be significant.

A freezing rate of at least 0.1 °C per minute and temperatures of –5 °C to –10 °C are needed for shattering to occur. The more frequent the freeze–thaw cycles the more vigorous will be the shattering, e.g. rock disintegration is more frequent in mountain and coastal arctic environments where there are diurnal cycles, than at inland sites where the cycle is more likely to be seasonal. Moisture content is important because if the rock is not saturated then, as ice forms, some of the pressures created by expansion can be spread into air cavities. In saturated rock, on the other hand, there is no such outlet for the pressure and so shattering occurs more readily. For example, researchers in the Colorado Rockies, USA, recorded less frost-shattering than they had expected. From further fieldwork, they found that the rocks were generally non-saturated, and that the freeze–thaw cycle was infrequent and of low intensity.

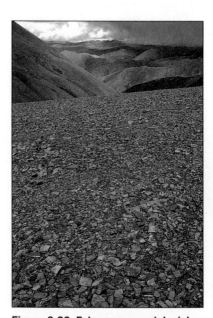

Figure 8.20 Felsenmeer periglacial block field, Pang La (north side of Mount Everest), Tibet

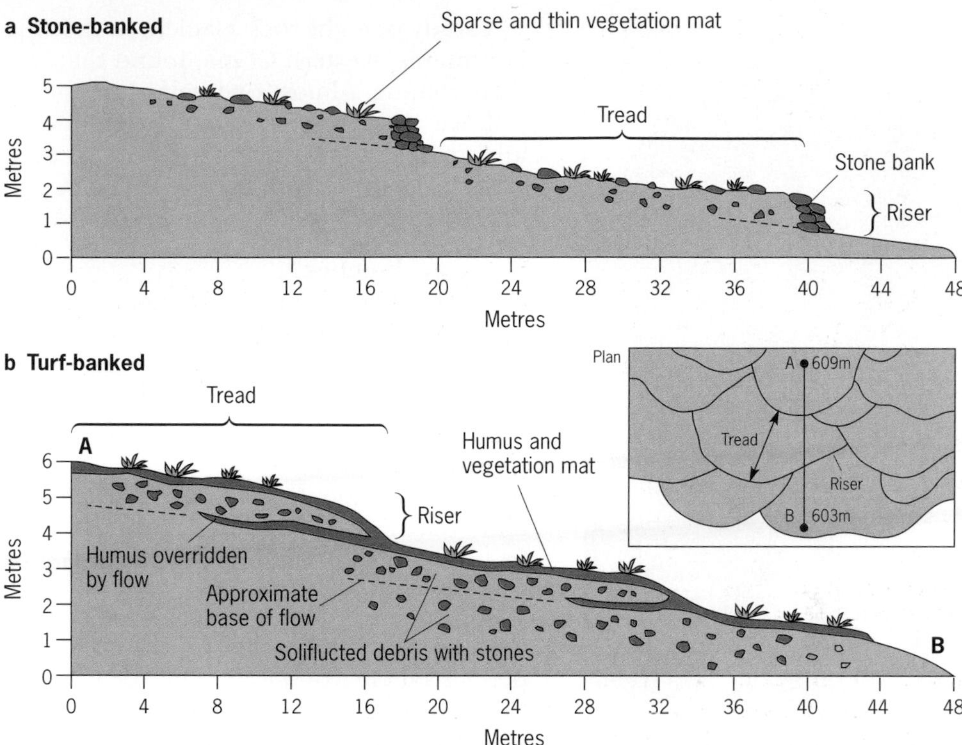

Figure 8.21 Stone-banked and turf-banked solifluction lobes (*Source:* Clowes and Comfort, 1987)

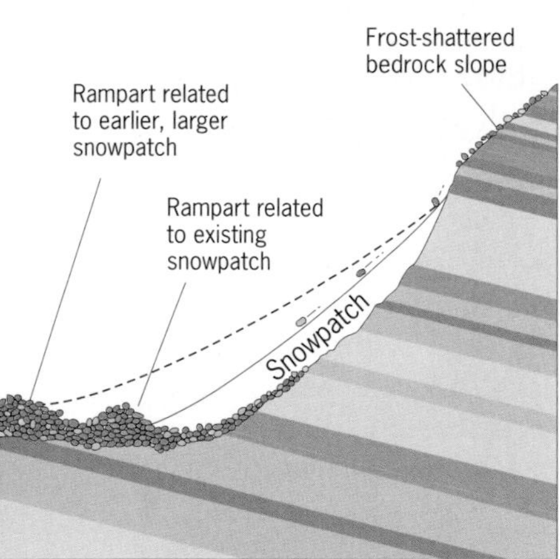

Figure 8.22 The formation of a protalus rampart (*Source:* Goudie, 1990)

Figure 8.23 A rock glacier, Rainbow Mountain, Alaska Range, Alaska, USA

The influence of slope angle

On relatively flat terrain, long-continued mechanical weathering can produce a surface cover of inactive (static) angular boulders. These are called *block fields* (Fig. 8.20). As slope angle increases, so angular blocks up to 3 m in size move spasmodically downslope as *block slopes*. Individual blocks move downslope more rapidly than the finer materials in which they rest. The rolling and tumbling motions are aided by frost-heaving and snowmelt lubricating the ground around the block. Scattered boulders may leave furrows as they plough downslope.

Slope character as well as rock type influence the detail of landforms. For example, where depressions run downslope, boulders may concentrate as *block streams*. If resistant rock strata form cross-slope barriers, blocks may collect to form *stone-banked terraces* (Fig. 8.21a). On relatively gentle slopes (5°–20°), movement may be slow enough for a vegetation mat to develop over the boulders in their soil matrix. Over time, uneven slumping and flow may produce *turf-banked lobes and terraces* which may be 2–15 m high and 50–500 m in length (Fig. 8.21b).

On steeper slopes, and especially below rock bluffs, freeze–thaw weathering produces rockfalls of angular scree known as *talus*. These build up at the slope base to form *talus slopes* with steep angles, up to 38°. Where the sliding and rolling are assisted by the presence of snow patches below the bluffs, the talus may build as a *protalus rampart* below the snow patch (Fig.8.22).

Rock glaciers

Rock glaciers are an extreme and spectacular type of debris landform (Fig. 8.23). These are tongues of rock and ice, up to 2 km long and several hundred metres wide, which move at as much as 100 cm a year. Three conditions are needed for their development: permafrost; vigorous freeze–thaw to produce the abundant boulder supply; a relatively dry climate.

Figure 8.24 A snow avalanche path and course of summer debris flow, Mount Revelstoke National Park, British Columbia, Canada

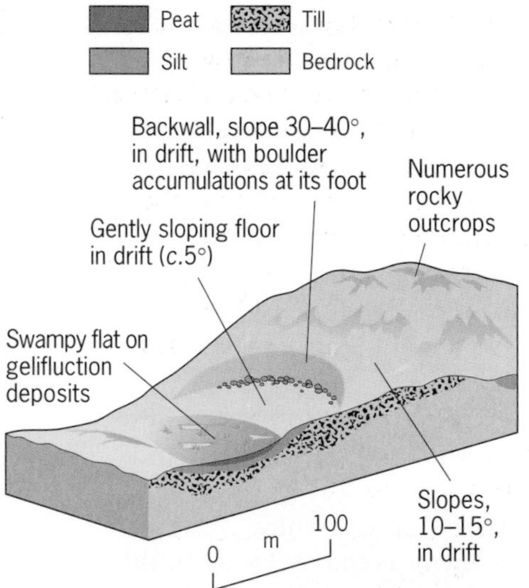

Figure 8.25 A nivation hollow (*Source:* Embleton and King, 1975)

A study of eight rock glaciers in the Jing Xian valley of the Kunlun Shan Mountains, western China, found that ice 'cement' made up 57 per cent of their volume. Almost one-half of the blocks were 0.5–2 m across, and 10 per cent were over 2 m. A combination of creep and surface meltwater lubrication produced flow rates of up to 3 cm per year. These rock glaciers, oriented N to NE, are believed to have originated in nivation hollows when the climate was more severe in the Late Pleistocene period.

Snow avalanches

As well as being a hazard to people, snow avalanches are a significant mass-movement mechanism in mountain periglacial areas along with mudflows and slides. They tend to occur along well-defined *snow avalanche paths*, which become the sites of increased debris flows or debris avalanches (Fig. 8.24). A fan-like structure of debris is commonly found at their base. The velocity of snow avalanches provides great erosion and transportation potential. Although not all snow avalanches transport debris as well as snow, there may be sufficient impact in the valley bottom to produce erosional features called boulder holes, or impact tongues or pools.

Nivation

Permanent snow patches are common features in slope hollows in periglacial environments. They enable increased mechanical weathering through supplying water for the freeze–thaw process (**nivation**). Where there is no underlying permafrost, this weathering can take place beneath and marginal to the snow. On permafrost surfaces, the accelerated weathering is restricted to the snow fringes. This zone expands as the snow patch shrinks. Both conditions result in progressive deepening of the *nivation hollows*, provided that gelifluction and soil creep are capable of removing the weathered material (Fig. 8.25). Some theories see these nivation hollows as the early stage of corrie formation (see Chapter 7).

Asymmetrical valleys

Researchers in periglacial regions have noted that many of the valleys are asymmetrical in shape, i.e. one slope significantly steeper than the other. Although rock structure and lithology may explain this, the large number occurring in periglacial areas has led to the search for an explanation in periglacial processes. Table 8.4 summarises the factors which may be involved.

Table 8.4 Factors which may influence slope processes producing asymmetry

Insolation	North-facing slopes will remain in shadow for longer periods (in the northern hemisphere). Thus they will remain frozen for longer with less gelifluction; this maintains slope steepness. South-facing slopes will experience more thawing and thus more gelifluction, which will reduce the slope angle.
Prevailing wind	Snow will accumulate to greater depths on more sheltered lee slopes, and by providing more meltwater encourage downhill movement of debris. This increases mass movement, and slope angle declines.
Stream action	Streams undercut valley sides, causing collapse and therefore the steepening of slopes. If a stream is flowing more towards one side of a valley due to greater debris supply pushing its course over, then asymmetry could result.
Plant cover	Vegetation cover can be influenced by aspect, since this affects the rates of ground freezing and thawing. In the Ruby Range of Alaska, the south-east-facing slopes had a shallow active layer and less gelifluction, as the thicker vegetation cover provided better insulation than the less vegetated north-west slope which thawed more quickly and deeply.
Animals	Of minor importance, but burrowing animals can be significant locally. In the Ruby Range, Alaska, colonies of arctic ground squirrels favour the south-east-facing slopes. Their burrowing and tunnelling accelerate mass movement.

8.6 Fluvial activity and landforms in periglacial environments

Fluvial activity in periglacial regions is controlled by the generally low precipitation totals and the length of the season during which water is 'stored' as ice and snow. In consequence, river regimes are highly seasonal, with little or no flow for up to eight months a year. This rhythm is reflected in seasonal contrasts in the workings of the hydrological cycle.

During the months when the ground remains frozen, water will not move towards the river channel by the processes of throughflow and groundwater flow. Precipitation inputs to the local hydrological cycle will be stored on the ground surface as snow cover. As temperatures rise the surface snow cover and the active layer will begin to thaw. Water will start to be released and move towards the river channel as saturated overland flow from snowmelt, and as throughflow in the active layer. As the active layer thaws some groundwater flow may occur. Precipitation inputs are highest in the summer, and occur mainly as rainfall. Due to the saturated nature of the ground this will quickly reach the river channel as overland flow. Temperatures begin to cool as winter approaches, and water movement slows down until it becomes locked in a frozen state again for many months of the year. Relatively low temperatures throughout the year and a small seasonal vegetation cover result in low evapotranspiration outputs throughout the year, but with some rise during the brief summer.

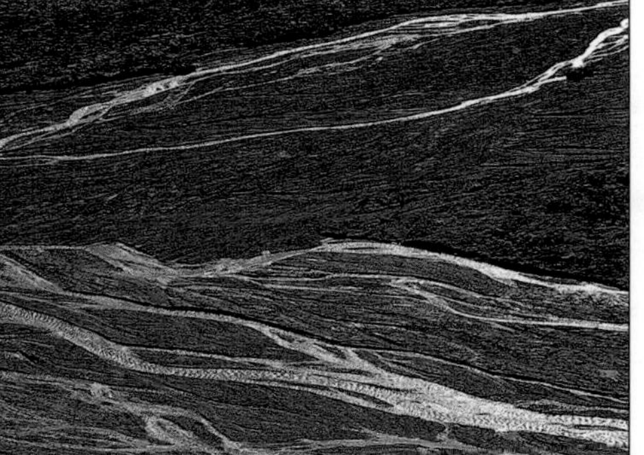

Figure 8.26 A braided river, Denali National Park, Alaska, USA

14 Draw two annotated diagrams to show the hydrological cycle in (**a**) winter and (**b**) summer. Add notes on the key hydrological processes operating and relate these to the periglacial river regime shown in Figure 8.29.

As snow and ice melt in the early summer, stream discharges increase rapidly, i.e. hydrographs have steep-rising limbs. If the upper reaches of a drainage basin thaw before the lower sections, water may be dammed back. It then flows downstream as a series of surges as the ice melts. Whether the spate discharge occurs as a single event or as a series of surges, for a few weeks each year periglacial streams have considerable energy. As a result, they are influential in landscape formation (see the case study below).

Many periglacial rivers flow in braided courses, with broad, poorly defined valleys. The channels are wide and shallow. The weathering processes provide an abundant debris supply for the brief period of spate discharge. The rivers fill their valleys, scour the channels and transport copious load, especially bedload. By late summer, discharge and energy are declining rapidly, i.e. steep-falling limbs on hydrographs. Capacity and competence of the streams decline, causing deposition and infilling (aggradation) of the valley floor with coarse debris. At this stage the braided channel pattern is most pronounced (Fig. 8.26).

In general, therefore, the broad, debris-filled valleys with braided channels reflect the inability of the streams to cope with the amount of debris supplied by the weathering processes active today and in the past.

Stream characteristics on the North Slope Plain, Alaska, USA

The rivers of Alaska's North Slope rise in the foothills of the Brooks Range, and cross the northern plain in poorly defined channels to the Beaufort Sea (Fig. 8.28). The surface features of the plain constantly change with the seasonal fluctuations in ground ice volume and thaw lake development above permafrost (Fig. 8.16). The stream pattern is often beaded, where channels connect thaw lakes and pools. Flow is concentrated in the June–September period, with all discharge ceasing by January.

Mean annual temperatures are –9 °C to –16 °C, and precipitation decreases northwards from 380 mm in the Brooks Range to 125 mm on the plain. Most falls as snow. Continuous daylight from mid-April to the end of August provides high insolation totals. This assists rapid snow- and ice-melt, with the thaw generally occurring within a two-week spell. River discharges increase rapidly, reflected in the steep-

North Slope Plain, Alaska

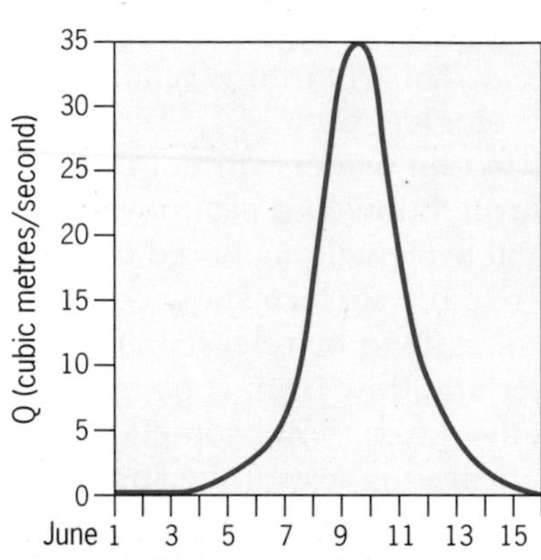

Figure 8.27 Sakonowyak River hydrograph, June 1981 (*Source:* Drage et al., 1983)

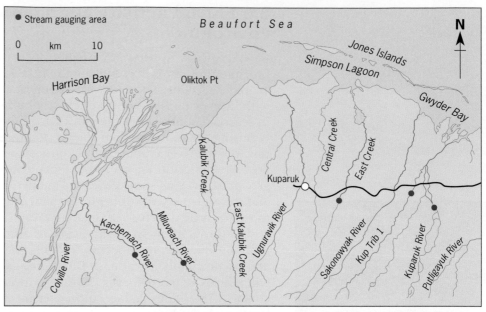

Figure 8.28 Hydrology of the North Slope, Alaska, USA (*Source:* Drage et al., 1983)

rising arms of hydrographs (Fig. 8.27). As the thaw zone moves northwards, the water builds up behind snow-drifts and ice dams. When they melt, pulses of discharge surge downstream, causing extensive flooding across the shallow river valleys. After these initial surges, flows are much lower for the rest of the summer, e.g. 78 per cent of the annual discharge of the Kuperuk River occurs in June, immediately after the thaw.

Because of the irregularity of the streamflows, channel form is unstable. Channel cross-sectional area, bed roughness (Manning's n values, 0.019–0.062), flow velocity, ice load and sediment load all fluctuate widely and rapidly. Spate flows enable bed scouring (degradation), followed by sudden deposition (aggradation) as discharge and available energy fall.

The hydrograph of the Colville River illustrates the North Slope regime vividly (Fig. 8.29). From mid-May, the ice break-up moves northwards at 0.3–1.0 km an hour, causing the flow surges indicated on the rising arm of the hydrograph. Water levels in the delta may rise 5 m in less than 10 days. The snow meltwaters are at first sediment-free. As melting

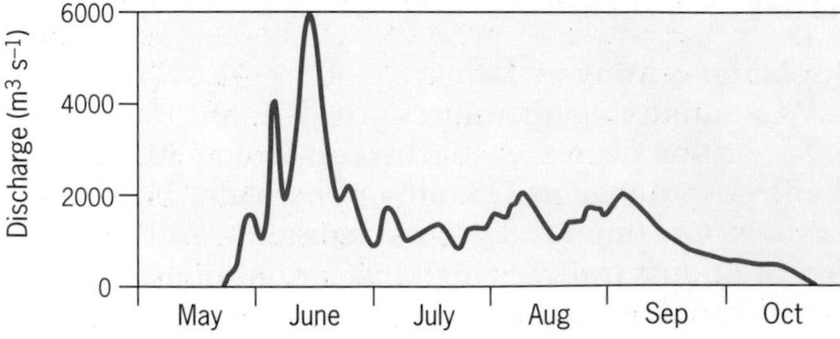

Figure 8.29 Colville River hydrograph, 1962 (*Source:* Summerfield, 1991)

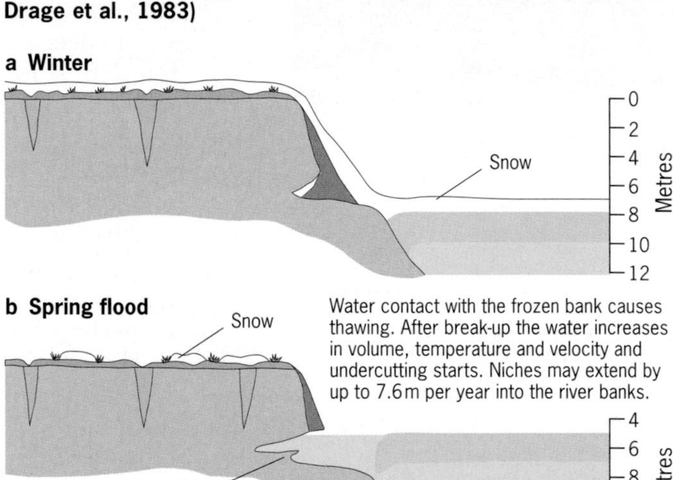

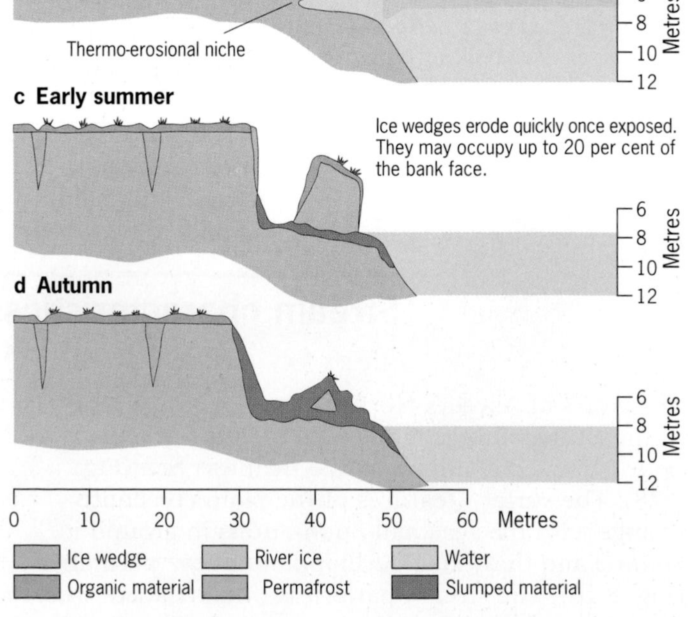

Figure 8.30 Riverbank erosion on the Colville River (*Source:* Richie and Walker)

accelerates and the active layer thaws, sediment from the tundra surface, channel beds and banks are vigorously entrained. As the channel beds are scoured, river bank erosion develops. This can occur for up to four months a year, but the bulk of the erosion, transportation and deposition takes place in June. Most is gradual, but the exposure and melting of ice wedges in the banks can cause sudden collapse. Bank materials influence retreat, which may reach 2 m a year in mineral materials such as old lake beds and pingos. Most bank failures occur just after water levels begin to drop (Fig. 8.30).

15 Use the hydrograph (Fig. 8.29) to explain why the major work of the Colville River takes place within the month of June.

16 Explain why periglacial rivers are effective geomorphological agents even though they may flow for only three or four months of the year.

17 Describe and fully explain the regime of the Colville River (Fig. 8.29).

18 Why is the breakup of ice along the rivers and streams so rapid in the area?

19 Describe and explain how the permafrost in the Colville River banks affects fluvial erosional processes (Fig. 8.30).

8.7 Aeolian activity

?

20 What two sets of factors are important for aeolian activity in an environment? How does the periglacial environment meet these requirements?

21 What is the economic significance of loess?

For wind action to be significant in an environment, there must be suitable climatic conditions and a supply of sand- and silt-sized sediment. Strong winds occur in periglacial environments, especially near to the ice margins. The highest persistent wind speeds recorded on earth were measured in east Antarctica at Port Martin where wind speeds exceeded 30 m/s over a 24-hour period. Gusts of up to 50 m/s were recorded in 1992 at Summit Lake, Canada, during two unrelated storm events.

The availability of dry, loose materials is more problematic. Periglacial areas have low precipitation totals, but also very low evaporation losses, with snow cover for much of the year. However, these conditions also ensure that there is a lack of vegetation to protect the surface. Sediments from periglacial and nearby or previously existing glacial and fluvio-glacial activity provide an ample supply of suitable material.

Evidence of wind erosion is provided by grooved bedrock surfaces, and *deflation surfaces* where the wind removes finer material from the exposed surface, leaving behind the coarser materials as **lag gravels**. Particles carried by the wind abrade exposed stones and boulders, modifying their shape. Such faceted debris is known as *ventifacts*, whose surfaces may be polished or pitted. If the stones have not been disturbed they can be used to provide information about present or past wind directions. Blowing snow can have a similar abrasive effect, especially at very low temperatures.

The wind transports two sizes of particles: sand (0.061–1.00 mm) and silt (0.004 mm–0.06 m). Sand is moved by bouncing movements called *saltation*. Sand deposits of periglacial origin are called *coversands* and may show dune-bedding. Silt-sized particles are carried suspended in the air at heights of up to 3 km and are therefore carried longer distances than sand before deposition. The term *loess* or *limon* is used for the silt-sized deposits which cover large areas of the USA, the European Plain and central Asia. These are in areas which were not covered by ice during some or all of the Pleistocene glacials. The formation of these huge loess deposits up to 80 m thick is regarded as evidence of great aeolian activity during the Pleistocene when periglacial and glacial activity would have provided ample surfaces of unvegetated silt-sized material. Loess is a grey, mineral-rich silt that weathers into a yellowish soil which is excellent for agricultural purposes, and these areas are some of the world's most fertile farmlands, e.g. the palouse soils of Washington and Idaho, USA (Fig. 8.31).

Figure 8.31 The rich soils of the palouse landscape, Idaho, USA, developed on wide expanses of loess

8.8 Periglacial landforms in the UK

As Figures 8.1 and 8.2 illustrated, the UK landscape shows evidence of many periglacial processes and landforms. We can use the understandings gained through this chapter to identify and to explain the location and character of these features.

The Stiperstones ridge has ten tors, e.g. Devil's Chair and Cranberry Rock. These rise abruptly from smooth slopes of 9–12° with slope angles over 50°. The Stiperstones consist of hard, quartzite rock, which is moderately to highly jointed.

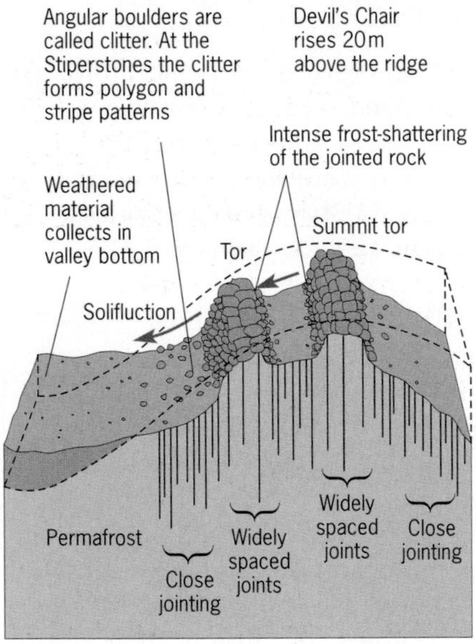

Figure 8.32 Possible mechanisms for tor formation, such as the Stiperstones, Shropshire, England

Periglacial environments of the late Pleistocene

The periglacial landforms evident in the UK landscape today were formed during the later stages of the Pleistocene glaciation, and for a period after the disappearance of permanent ice. This glacial period saw a succession of ice advances and recessions across the UK as climate fluctuated over more than two million years. Each successive readvance removed or modified many features left by the previous advance–recession cycle (see Chapter 7). Thus, most periglacial features visible today are associated with the most recent cycle, known as the Devensian. This reached its maximum advance around 18 000 BP, when mean annual temperatures in southern Britain would have been –8 °C to –10 °C.

It is important to remember that as the Devensian ice sheets receded, so permafrost crept northwards across northern Britain. This resulted in the development of periglacial features in the uplands of the Lake District, the Cheviot Hills (see Case Study, p.172) and the Scottish Highlands. The last permanent snow and ice disappeared less than 10 000 BP, and permafrost conditions survived for perhaps another 2000 years. Most of the periglacial features seen in the UK landscape therefore developed between 18 000 and 8000 BP. Even today, conditions are severe enough on a few exposed summit areas, e.g. the Cairngorms, for patches of patterned ground to develop (pp.175–6).

Features of permafrost ground conditions and frost action

Polygons and stripes are found in mountain areas such as the Lake District, Snowdonia, the Pennines and the Stiperstones of Shropshire. The Stiperstones are a series of projecting rock masses called tors to the west of the Long Mynd. Tors occur in several locations in the UK, and there is much controversy as to how they formed. Several researchers suggest that periglacial processes are the key to their development (Fig. 8.32).

In lowland Britain the most widespread form of patterned ground is former ice wedges which have been preserved by infilling with fine sediment to form **ice-wedge pseudomorphs**. These are characteristically 0.5–3 m deep and up to 1.5 m across, sometimes forming patterned ground such as polygons (Fig. 8.33). Lowland areas also show evidence of pingo and related mound features such as palsas in parts of Britain. The remnants of the ramparts and depressions have been identified (Fig. 8.34). Involutions or cryoturbations of fine silt and clay materials are found in sediments in much of lowland Britain.

Mass-movement deposits

Solifluction (gelifluction) deposits are widespread throughout the UK. In south-west England, the granite slopes of Dartmoor and Bodmin Moor have well-developed block streams, and in southern England block streams are evident in the Marlborough Downs of Wiltshire and the Portesham area of Dorset.

Head deposits formed by gelifluction, frost creep and other periglacial processes are found on lower scarp and valley slopes throughout much of the country. They are particularly evident in eastern England on various bedrocks including chalk, limestones, sands and clays. The *coombe deposits*

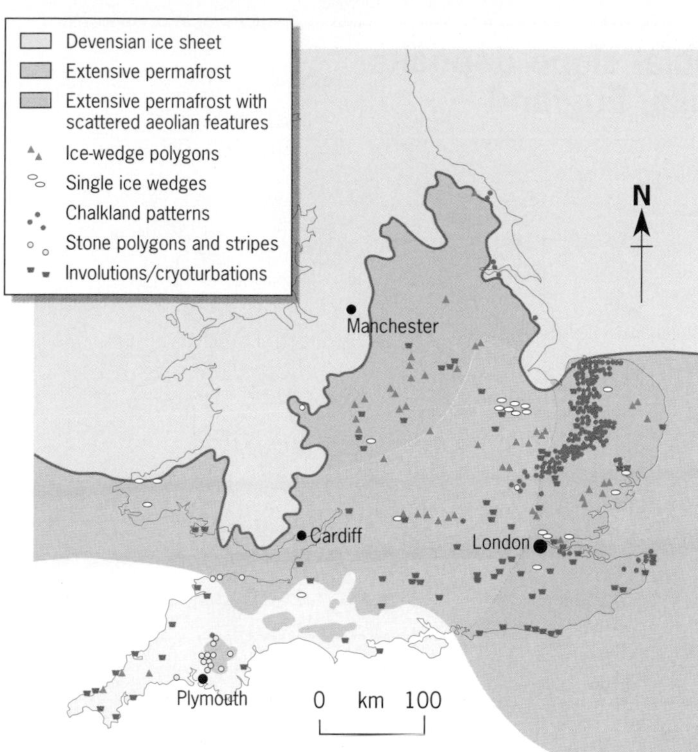

Figure 8.33 Ice sheet, permafrost and patterned ground relationships at the maximum of the Devensian advance (18 000 BP). Ice recession over the following 8000 years or so allowed periglacial conditions to extend northwards (*Source:* Goudie, 1990)

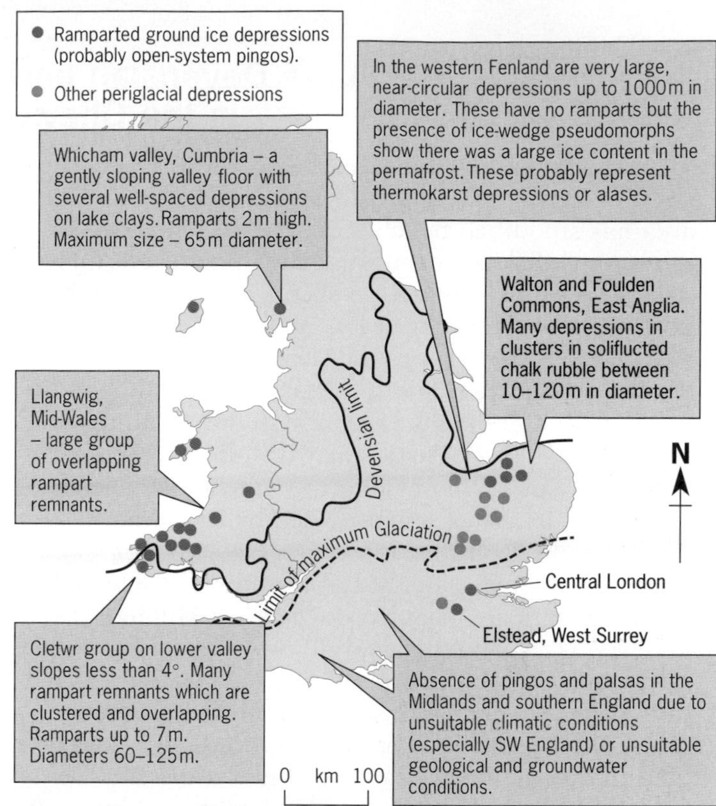

Figure 8.34 Distribution of areas of ramparted ground, ice depressions and other periglacial depressions which provide evidence of former pingos and palsas

found on the floors of chalk dry valleys (see p.174) are made up of frost-shattered chalk formed by rolling, frost creep and mass sliding by the melting of ice lenses or gelifluction. Clay vales, e.g. the Weald Clay of south-east England and the Tertiary Clays of the London Basin, have thin spreads of stony loams. These deposits occur on very gentle slopes, but contain stones derived from rock escarpments several kilometres away. They probably originated from periglacial mudflows on the steeper scarp slopes. The rock debris from the upper scarp mixed with saturated clay in the active layer and was moved long distances.

Upland areas show many examples of screes (talus) from frost weathering of high-angle mountain slopes. Many upland screes are now vegetated and are not being actively formed (Fig. 8.35), although some areas have active screes today (see Fig. 8.41). In the Lake District, protalus ramparts are developed at heights of between 300 m and 600 m with a north-easterly aspect. They vary in height from 1 m to 10 m and up to 550 m in length. Other examples occur in Wales and were probably formed during the Loch Lomond advance.

?

22 Use Figures 8.33 and 8.34 to describe the type and location of the former periglacial features in Britain.

23 What do the presence of pingos, alases and ice wedges tell us about the specific ground conditions at the time of their formation during the Devensian?

Figure 8.35 Vegetated scree in the Aran Ridge, southern basin of the Bala Mountains, North Wales

Late Devensian periglacial slope deposits in the Cheviot Hills, England

The Cheviot Hills in north-east England (Figure 8.36) show evidence of glacial and periglacial activity. The area has smoothed benches or solifluction sheets on the lower hillslopes of most of the upland valleys. In wide valleys the solifluction sheets have developed on both sides of the valley at low angles (3–10 °C). In narrower, incised valleys the sheets are usually only developed on one side of the valley at slope angles up to 31° with a bluff end 3–20 m high due to fluvial erosion (Fig. 8.37). Particularly well-developed sheets are found upstream of Belford. Evidence of glacial deposition in the area is provided by morainic deposits and drumlins.

The soliflucted material has a smooth profile and the sediments are oriented downslope. Much of the material is reworked glacial till or weathered granite (growan). The gelifluction of rock only accounts for a thin surface veneer. The solifluction sheets show no preferred aspect and their distribution appears to be determined by the availability of materials.

The sequence of events can be determined by study of a section of the solifluction sheets (Fig. 8.37).

- Valley glaciers flowing northwards down the Bowmont Valley deposit large amounts of till in the valley bottom (Unit 1) and on the surrounding hillslopes, during the glacial maximum 18 000 BP. This represents in situ till which consists of angular and subangular fragments with the clasts oriented down the valley in the direction of ice flow.
- During deglaciation about 15 000 BP, rapid mass-movement processes reworked the valley-side tills and redeposited them on to the tills at the valley bottom as Unit 2. Temperatures rose and the hillslopes were stabilised by vegetation. This soliflucted till consists of angular, subangular, and sub-rounded material.

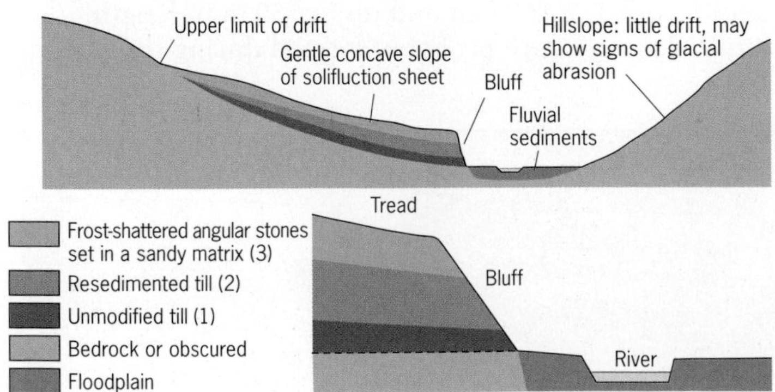

Figure 8.37 Cheviot solifluction sheet, showing characteristic morphology (*Source:* Harrison, 1993)

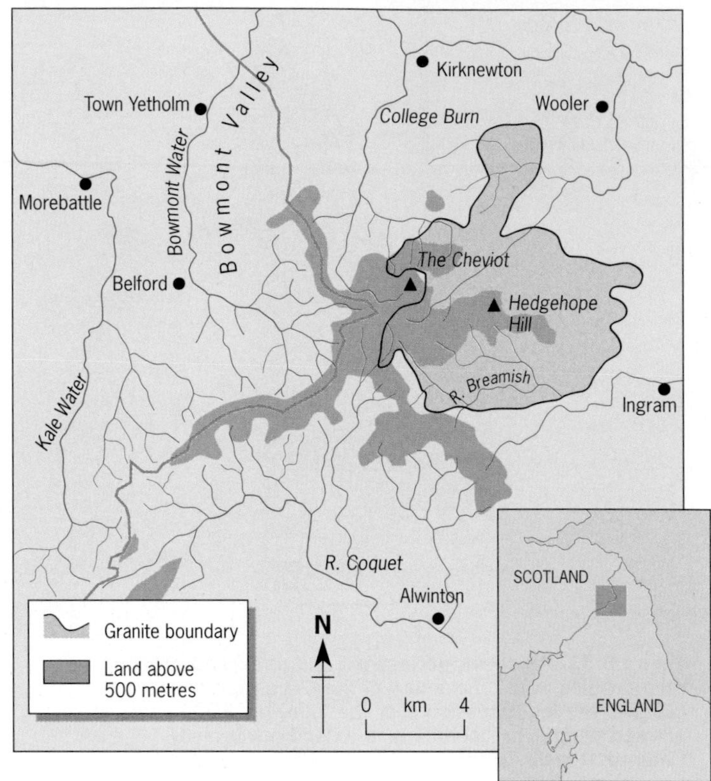

Figure 8.36 Bowmont Valley, Cheviot Hills (*Source:* Harrison, 1993)

- About 11 000 BP, during the Loch Lomond Stadial, periglacial processes began operating on the drifts and a thin layer of frost-shattered material was soliflucted on top of Units 1 and 2 to form Unit 3. The clasts are oriented downslope and consist of very angular and angular till, and soliflucted growan in deposits up to 1 m thick. This phase lasted for about 1000 years.
- The Flandrian period (10 000 BP to present) has seen the deposits stable on the slopes, but fluvial erosion has incised into the sheet fronts to form the bluffs in the narrow valleys.

?

24 Explain how the nature and orientation of the clasts in the deposits of the Bowmont Valley were used to identify the processes involved in their formation.

25 Suggest reasons for the differences in the form of the solifluction sheets in the broad and incised valleys.

26 Suggest why the solifluction deposits have shown little modification during Flandrian times.

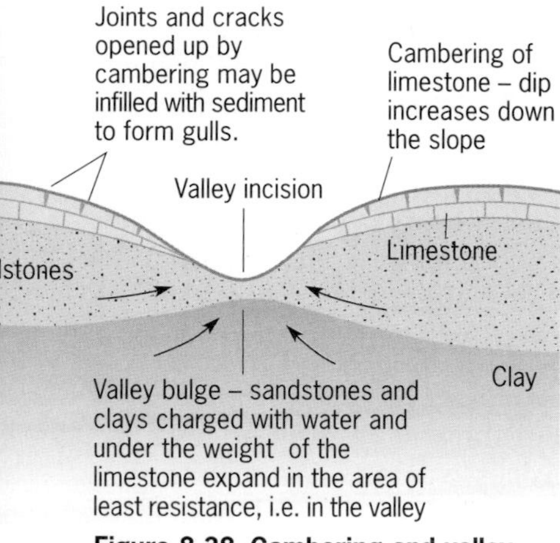

Figure 8.38 Cambering and valley bulging

Periglacial modification of slopes and valleys

Cambering and valley bulging

Many scarp and valley slopes show structures which have been caused by periglacial processes. *Cambering* is where the dip of rocks is increased by movement on rocks susceptible to frost action and saturated flow such as clays. Overlying resistant rocks will be split and moved downslope. The dip of these rocks will be steepened and blocks may break away and move downslope. Cambering is often associated with *valley bulging*, for example in the Jurassic rocks of the Midlands where limestones and clays occur together (Fig. 8.38).

Many British river valleys are asymmetrical with the west- or south-west-facing slopes steeper than other orientations. Valley asymmetry is well developed in the Chilterns, and the North, South and Dorset Downs. The steeper south- and west-facing slopes are often bare of deposits, with the valley floors covered with solifluction deposits known as coombe rocks, a form of head deposit. The more gentle north- and north-east-facing slopes have a solifluction mantle along their whole length. In the north Dorset Downs, near Tolpuddle, the tributary valleys show asymmetrical development and there is also a greater density of tributaries on east-facing slopes (Fig. 8.39), although many of these are now dry.

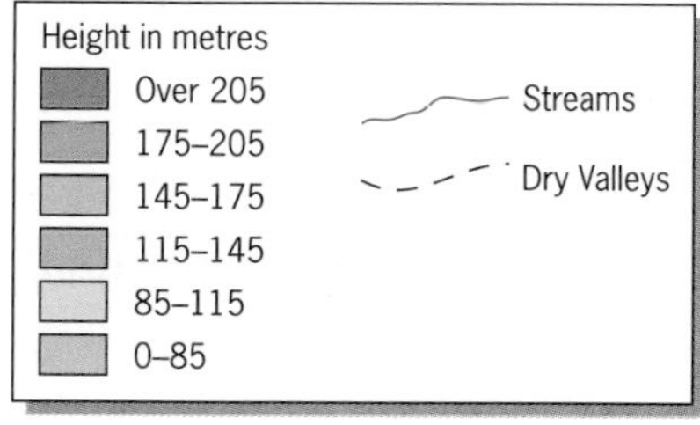

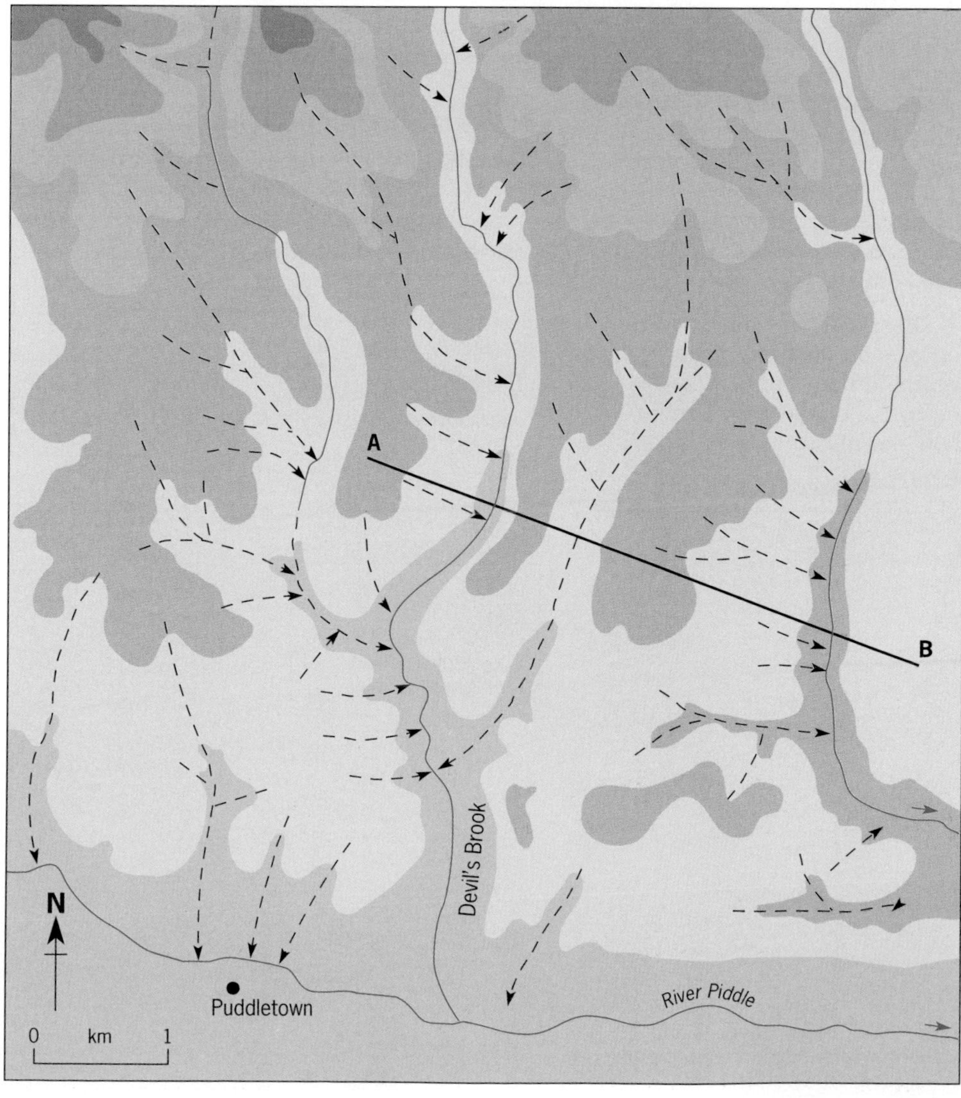

Figure 8.39 Asymmetrical valleys and dry valleys in the north Dorset Downs near Tolpuddle (*Source:* Goudie, 1990)

?

27a Draw a cross-section from A to B on Figure 8.39. On to your section label the following: present-day streams, the dry valley, asymmetrical valley profiles, the slope aspect with the most dry valley development.

b What is the orientation of the steep and gentle slopes in the asymmetrical valleys?

28 Use the information on page 166 to suggest reasons for the asymmetrical valley development in the north Dorset Downs.

29 Study Figure 8.41. Classify the periglacial features shown into: • weathering and mass movement features; • features associated with freezing of the ground; • wind action.

30 Describe and explain the field relationships of five of the features shown. Include the influence of slope angle, wind direction and rock type.

31 Which landforms shown in the diagram are relict features from the Devensian? What evidence is there to support this conclusion?

32 Essay: Describe and explain how the processes operating in permafrost result in landforms specific to periglacial areas.

33 Essay: 'By far the most widespread and important periglacial process is frost action. Actually frost action is a "catch-all" term for a complex of processes involving freezing and thawing' (Washburn, 1980). Discuss this statement with reference to specific examples you have studied.

34 Essay: What evidence would you present to support the argument that many of the landforms of southern Britain show modification by periglacial processes?

Dry valleys

Many areas in the UK, e.g. Cretaceous chalk, Triassic sandstone, Carboniferous limestone, have dry valleys, i.e. typical valley profiles and features but without a stream. There are many explanations for dry valleys and probably each area is unique, and the same reasons cannot be given for all British dry valleys. Theories involve changes of base level, changing groundwater levels, scarp retreat, spring sapping and river capture. However, the most widely held explanation of dry valleys is Pleistocene periglaciation. The colder climatic conditions would have reduced evaporation, and produced a greater proportion of runoff. Snow melt running over the impermeable permafrost could have cut valleys even on relatively permeable rocks such as chalk (Fig. 8.39). Support for the periglacial hypothesis is provided by many dry valleys showing evidence of gelifluction and block streams. Many of the dry valleys are also asymmetrical.

River terrace aggradation

The middle and upper reaches of many UK rivers show aggrading of river terraces which cannot be explained by fluvial processes alone. Work in the upper Thames valley shows that the terrace sediments contain evidence of colder conditions with mollusca and pollen characteristic of periglacial areas, as well as ice-wedge pseudomorphs, involutions, and braided channels. The sediments themselves are often subangular to sub-rounded, rather than the more rounded shape associated with fluvial deposits. This suggests only a short time in transport before they were deposited.

Aeolian deposits

Although the UK does not have periglacial wind deposits to the same spatial extent or depth as the European Plain and the USA, there is ample evidence of aeolian activity (Fig. 8.40). Other evidence for wind action comes from ventifacts found in the Midlands and the Cheshire Plain.

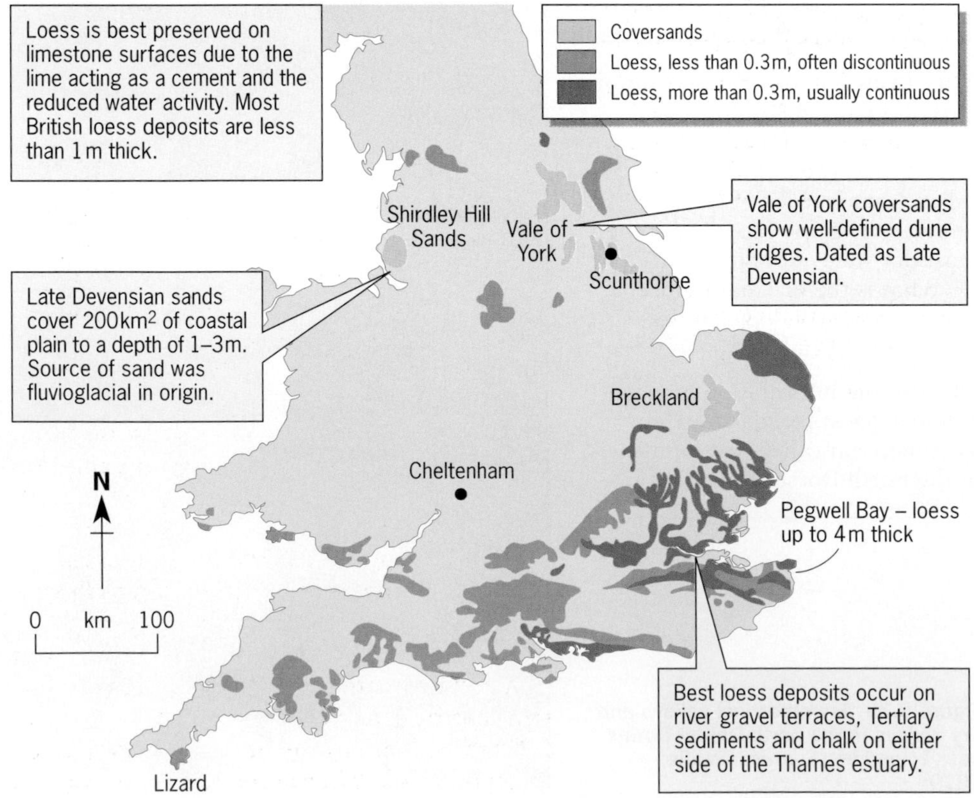

Figure 8.40 Loess deposits and aeolian coversands in England and Wales (*Source:* Goudie, 1990)

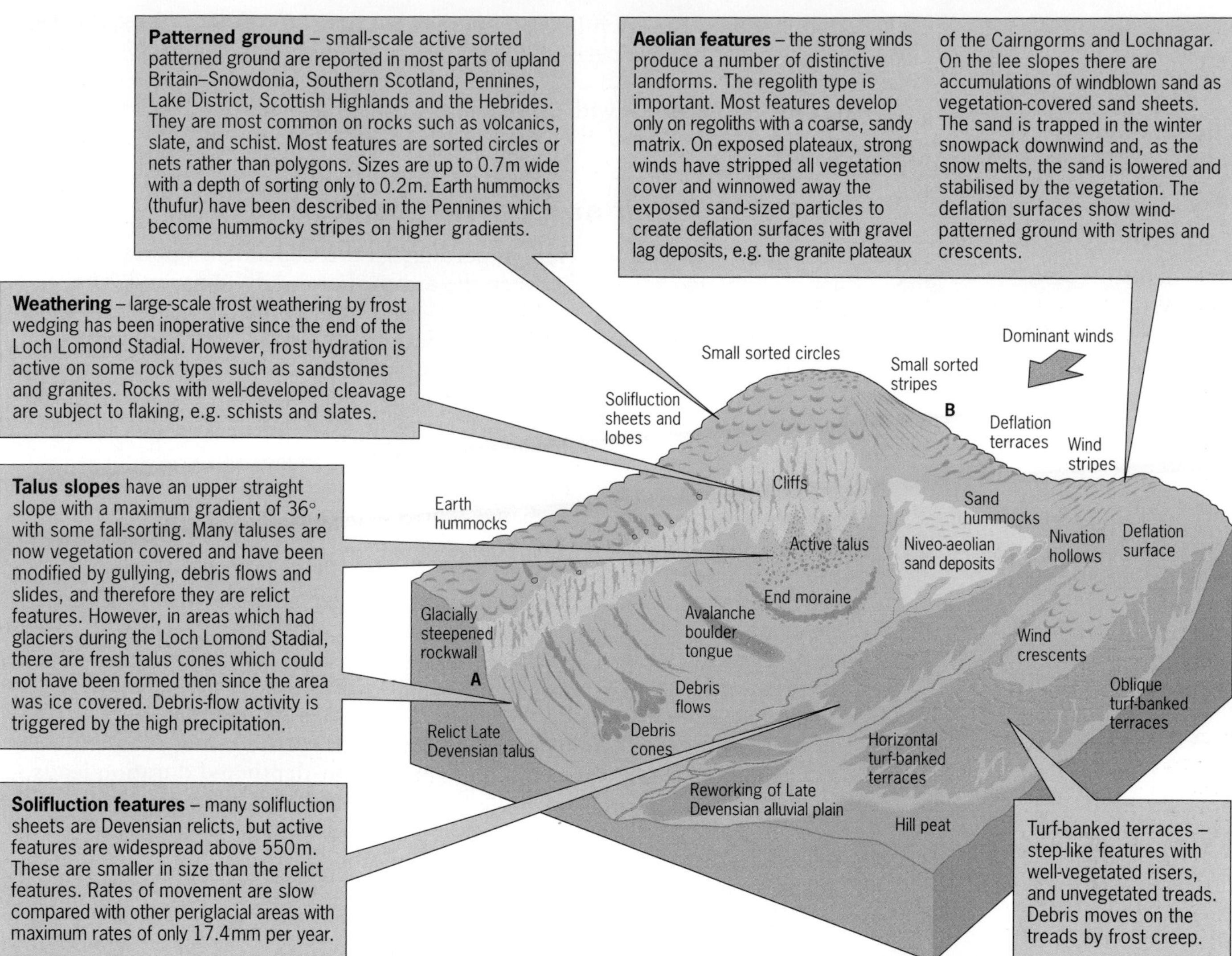

Patterned ground – small-scale active sorted patterned ground are reported in most parts of upland Britain–Snowdonia, Southern Scotland, Pennines, Lake District, Scottish Highlands and the Hebrides. They are most common on rocks such as volcanics, slate, and schist. Most features are sorted circles or nets rather than polygons. Sizes are up to 0.7m wide with a depth of sorting only to 0.2m. Earth hummocks (thufur) have been described in the Pennines which become hummocky stripes on higher gradients.
Aeolian features – the strong winds produce a number of distinctive landforms. The regolith type is important. Most features develop only on regoliths with a coarse, sandy matrix. On exposed plateaux, strong winds have stripped all vegetation cover and winnowed away the exposed sand-sized particles to create deflation surfaces with gravel lag deposits, e.g. the granite plateaux of the Cairngorms and Lochnagar. On the lee slopes there are accumulations of windblown sand as vegetation-covered sand sheets. The sand is trapped in the winter snowpack downwind and, as the snow melts, the sand is lowered and stabilised by the vegetation. The deflation surfaces show wind-patterned ground with stripes and crescents.
Weathering – large-scale frost weathering by frost wedging has been inoperative since the end of the Loch Lomond Stadial. However, frost hydration is active on some rock types such as sandstones and granites. Rocks with well-developed cleavage are subject to flaking, e.g. schists and slates.
Talus slopes have an upper straight slope with a maximum gradient of 36°, with some fall-sorting. Many taluses are now vegetation covered and have been modified by gullying, debris flows and slides, and therefore they are relict features. However, in areas which had glaciers during the Loch Lomond Stadial, there are fresh talus cones which could not have been formed then since the area was ice covered. Debris-flow activity is triggered by the high precipitation.
Solifluction features – many solifluction sheets are Devensian relicts, but active features are widespread above 550m. These are smaller in size than the relict features. Rates of movement are slow compared with other periglacial areas with maximum rates of only 17.4mm per year.
Turf-banked terraces – step-like features with well-vegetated risers, and unvegetated treads. Debris moves on the treads by frost creep.
Small sorted circles
Small sorted stripes
Dominant winds
Solifluction sheets and lobes
B
Deflation terraces
Wind stripes
Cliffs
Earth hummocks
Sand hummocks
Active talus
Niveo-aeolian sand deposits
Nivation hollows
Deflation surface
End moraine
Glacially steepened rockwall
Avalanche boulder tongue
Wind crescents
A
Debris flows
Oblique turf-banked terraces
Relict Late Devensian talus
Debris cones
Horizontal turf-banked terraces
Reworking of Late Devensian alluvial plain
Hill peat

Figure 8.41 Schematic representation of the range of active periglacial features on UK mountains. Those to the left of an imaginary line linking A and B are characteristic of frost-susceptible regolith rich in silt and fine sand. This is formed from the weathering of rocks such as schist, slate and lavas. Those to the right of imaginary line A–B are characteristic of sandy regolith produced by the weathering of sandstones and granites. (*Source:* Ballantyne, 1987)

The present-day periglaciation of upland Britain

Periglacial environments are those which are characterised by cold non-glacial processes, especially ground freezing. Britain has no permafrost, but in the higher mountains there are annual freezing cycles of 2–4 months with ground freezing to a depth of about 0.5 m. This results in active periglacial features even if they are not as well developed as those which occur in high-latitude areas or higher-altitude mountains.

Temperatures are relatively mild due to Britain's Maritime climate. Ben Nevis (1343 m), Britain's highest mountain, has an average annual temperature just above 0 °C, but mean monthly temperatures are below freezing for 2–4 months of the year. The lowest temperature recorded in the last 20 years was –17.3 °C. Precipitation totals are high, with most upland areas receiving over 2000 mm and some western mountains over 4000 mm per year. In the Scottish Highlands, the average snowlie (with over 50 per cent cover) exceeds 100 days per year at 600 m. The highest peaks have over 220 days per year. The strength of the wind is particularly significant. On Ben Nevis during 13 years of observations, there was an average 261 gales per year of over 82 km per hr.

Britain's mountains have active periglacial features which are summarised in Figure 8.41. There are also many relict features which date from the Late

Figure 8.42 Periglacial patterned ground, Cairngorms, Scotland

Devensian and the Loch Lomond Stadial (11 000–10 000 BP) (Fig. 8.42). The distribution of active periglacial features is controlled by lithology, altitude, vegetation cover, aspect and gradient. Lithology is an important control on the range of features, with sandstones, granites and slates forming the best-developed features.

8.9 Permafrost and environmental change

The landforms which develop on permafrost are dependent upon the heat energy exchange balance between the atmosphere and the ground. The permafrost layer is very sensitive to changes in this budget. In particular, the seasonally active layer within which many landforming processes work thickens rapidly with additional heat inputs. These additions may be natural (e.g. climatic fluctuations) or anthropogenic, i.e. caused by human activity (e.g. structures).

The energy exchange buffer zone

How the air–ground heat exchange works depends upon the nature of a surface store consisting of the vegetation cover, the organic surface layer (litter and peat) and the snow cover (Fig. 8.43). The ability of this buffer zone to store and transmit the heat energy controls the behaviour of the active and permafrost layers. For instance, heat inputs from the atmosphere may remain unchanged, yet if vegetation is reduced, e.g. deforestation, subsurface heat transfer will alter (Fig. 8.44). At a study site near Inuvik, northern Canada, the depth of the active layer increased from 38 cm to 183 cm in eight years, following stripping of the spruce/birch woodland.

Snow acts as a 'blanket'. Thus, increased snow depth and duration leads to higher ground temperatures and a thicker active layer. For example, in 1971, snow fences were erected to keep snow from a highway near Schefferville, the iron ore town of northern Quebec, Canada. They did their job, but the thicker, longer-lasting snow banks collected by the fences allowed the ground

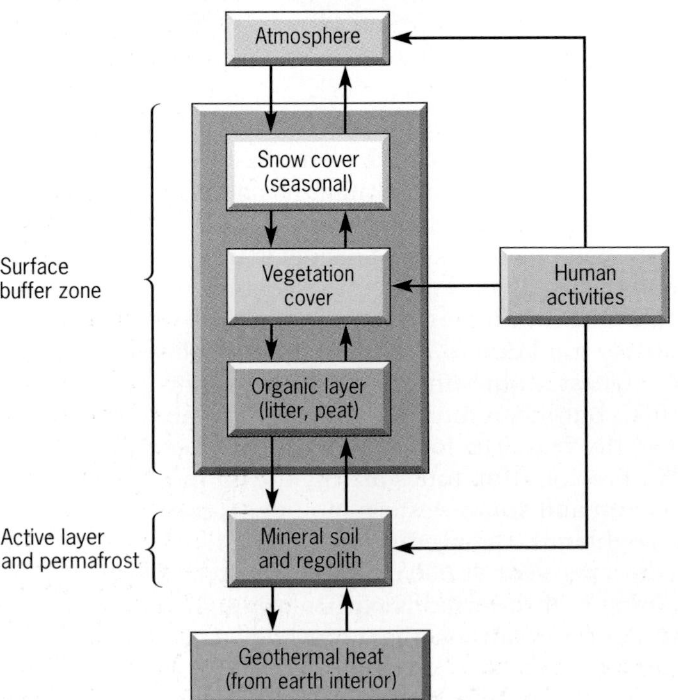

Figure 8.43 Heat energy transfers and permafrost

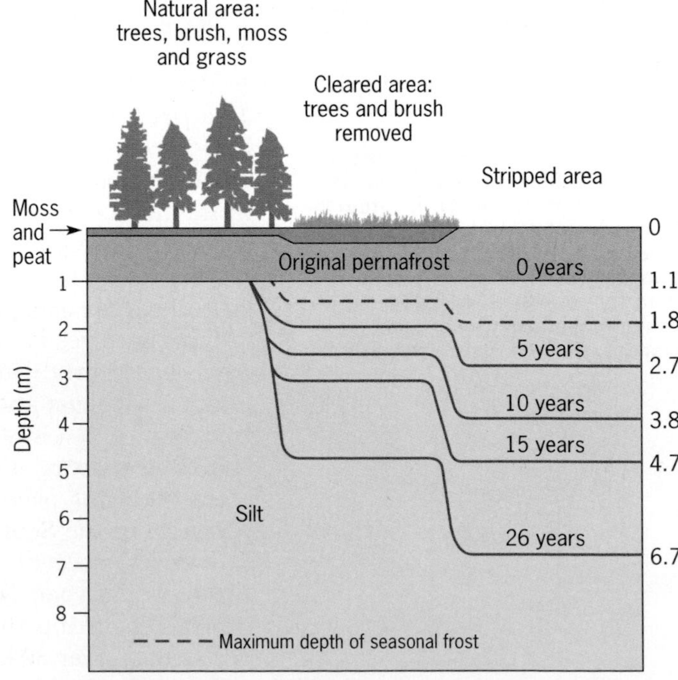

Figure 8.44 Permafrost degradation under surface disturbances in central Alaska over a 26-year period

beneath to become warmer. By the late 1970s, mean ground temperatures at depths of 10 m had risen by 1–2 °C. The thicker active layer, and thinner permafrost, caused ground subsidence and deformation of the nearby road bed.

Impacts of human activity

Deforestation and snow fences are only two of a wide range of human impacts on these sensitive periglacial landforms. They illustrate the general understanding that human activities modify the buffer zone and tend to increase heat energy inputs to the permafrost. The weight of roads, railbeds, airstrips, buildings, etc., exert downward and lateral forces which, along with heat transmitted from these structures, transfer heat energy into the permafrost. As we have seen earlier in this chapter, degradation of the permafrost caused by the melting of the ice and the thickening of the active layer result in ground subsidence and slumping. The impacts include building subsidence and collapse, railbed, airstrip and road buckling, and pipe fracturing. Around settlements and industrial sites, lateral spread of permafrost degradation may produce an extensive zone of chaotic hummocks and waterlogged hollows.

Engineers use three basic techniques to minimise this heating of the permafrost:

1 raise the main structures above the ground;

2 spread the load by setting the structures on broad 'rafts';

3 insulate structures thoroughly to cut down heat loss.

These techniques add significantly to the cost of settlement and economic activity in periglacial regions.

Tourists seeking 'the wilderness experience' are arriving in ever greater numbers, yet many permafrost landforms are fragile and have low carrying capacities. Even small numbers of visitors can trigger erosion, and conservationists are increasingly concerned. For example, intense lobbying by environmental groups has led to the Tuktoyaktuk pingos of the Mackenzie delta being the first area given Canadian Landmark status ('Landmark' designation is to give protection to high-quality features too limited in size to become a National Park).

Climatic change

Current concern about global warming focuses upon the greenhouse effect of human activities releasing carbon dioxide into the atmosphere. Yet as the Pleistocene glaciation illustrates, natural mechanisms cause significant shifts in global temperatures over differing timescales. For example, a minor cooling from around AD 1550 to 1850 across the northern hemisphere caused the so-called 'Little Ice Age', and an expansion of periglacial conditions. From that time, however, the permafrost zone has been in retreat, e.g. the permafrost layer beneath the peat plateaus of northern Manitoba, Canada, was degrading at 0.5–1.0 m/yr by the 1930s. Long-term studies at Mezen, north-east of Archangel, Russia, show that from 1850 to 1950 the permafrost boundary retreated northwards at an average rate of 400 m/yr.

If the current global warming forecasts are accurate, the permafrost region retreat will accelerate. Warming is likely to be greatest in higher latitudes, rising as much as 7 °C by the mid-twenty-first century. Winter temperatures in particular will rise, and snowfall will increase by up to 50 per cent. The end result of such changes is illustrated for North America in Figure 8.45.

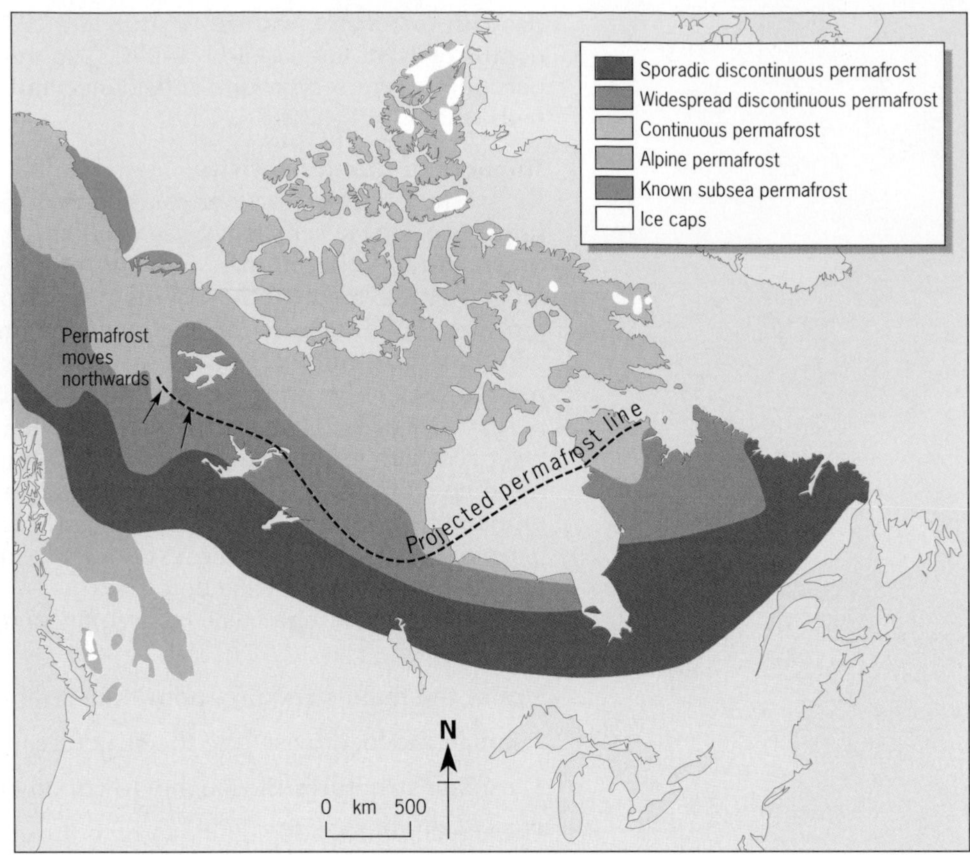

Figure 8.45 Permafrost distribution in Canada, and projected changes following climate warming

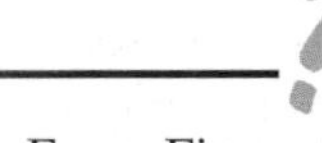

35 From Figure 8.45:

a Measure the average northward shift in the permafrost zone, if the global warming forecast is accurate. (Take measurements along a minimum of five well-spaced transects.)

b Suggest reasons for the variations in the northward retreat forecast for different parts of Canada.

36 From your understandings of permafrost processes and landforms gained in this chapter, describe and explain the main landform changes likely across the zone between the existing and projected permafrost limits.

37 What will be the main advantages and disadvantages for human activities of this northward shift of the permafrost limit?

Summary

- Periglacial processes take place in cold climates beyond the margins of ice masses.
- Permafrost is the key environmental characteristic, and means permanently frozen ground.
- Landforms are the result of processes at work in the seasonally active surface layer.
- On areas of low relief, the main periglacial landform types are a variety of mounds, hollows and patterned ground.
- On exposed rock surfaces, repeated freeze–thaw cycles produce a cover of angular debris which form as block fields or, on slopes, collect as rock glaciers and talus ridges.
- Valley shapes are modified by solifluction and other mass-movement processes, to produce valley cambering, stepped cross-profiles, and slope-foot and valley-floor deposits.
- During the brief summer months, vigorous stream action creates unstable valley and channel features, caused most commonly by incision and braiding.
- Extensive loess or limon plains result from aeolian transport of dry, loose silt particles from periglacial surfaces. The materials may be carried over great distances.
- A variety of periglacial landforms created during the later stages of the Pleistocene glaciation are widespread across the UK.
- Permafrost surfaces and many periglacial landforms are fragile and sensitive to environmental change, specially temperature fluctuations, which may be caused by human activities, e.g. building structures, global warming.

9 Arid and semi-arid environments

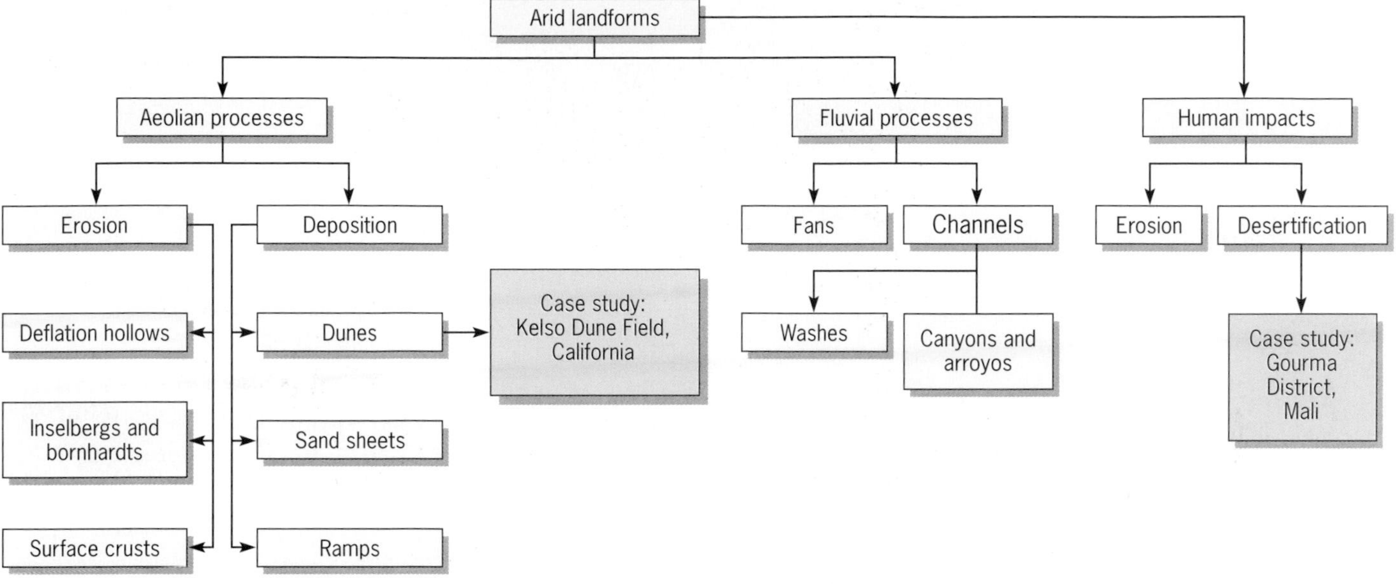

Table 9.1 Two examples of an aridity index

Precipitation effectiveness index

$$\frac{P}{E} = \frac{Pm \times 10}{1\ Em}$$

Pm = Mean monthly precipitation
Em = Mean monthly evaporation rate
P = Precipitation
E = Evaporation

Arid/semi-arid boundary: $\frac{P}{E} = 16$

Semi-arid/humid boundary: $\frac{P}{E} = 31$

Potential evapotranspiration index *(Khosla)*

$$Lm = \frac{Tm - 32}{9.5}$$

Lm = Mean monthly water loss in inches
Tm = Mean monthly temperature

Arid/semi-arid boundary: Lm = 20

Semi-arid/humid boundary: Lm = 40

9.1 Introduction

Arid lands may be defined as those regions where mean annual precipitation is less than 250 mm, and where evaporation rates (*outputs*) persistently exceed precipitation rates (*inputs*). It is this permanent moisture budget deficit that is the crucial characteristic. In formal terms: potential evapotranspiration exceeds precipitation (PET > P). Thus, semi-arid lands are commonly defined as receiving 250–500 mm mean annual precipitation, but the vital feature is a moisture budget deficit for at least eight months a year. Beware, however, as threshold figures given in books and atlases do vary. A useful device which measures the precipitation–evaporation relationship is the *aridity index* (Table 9.1 and Fig. 9.2).

Using these criteria, we can identify two main types of desert: hot deserts (10–33° latitudes), and continental interior temperate deserts (hot summers, severe winters). Some classifications include a third category of polar deserts, with low year-round precipitation and temperature figures.

Figure 9.1 The Gobi Desert, Mongolia

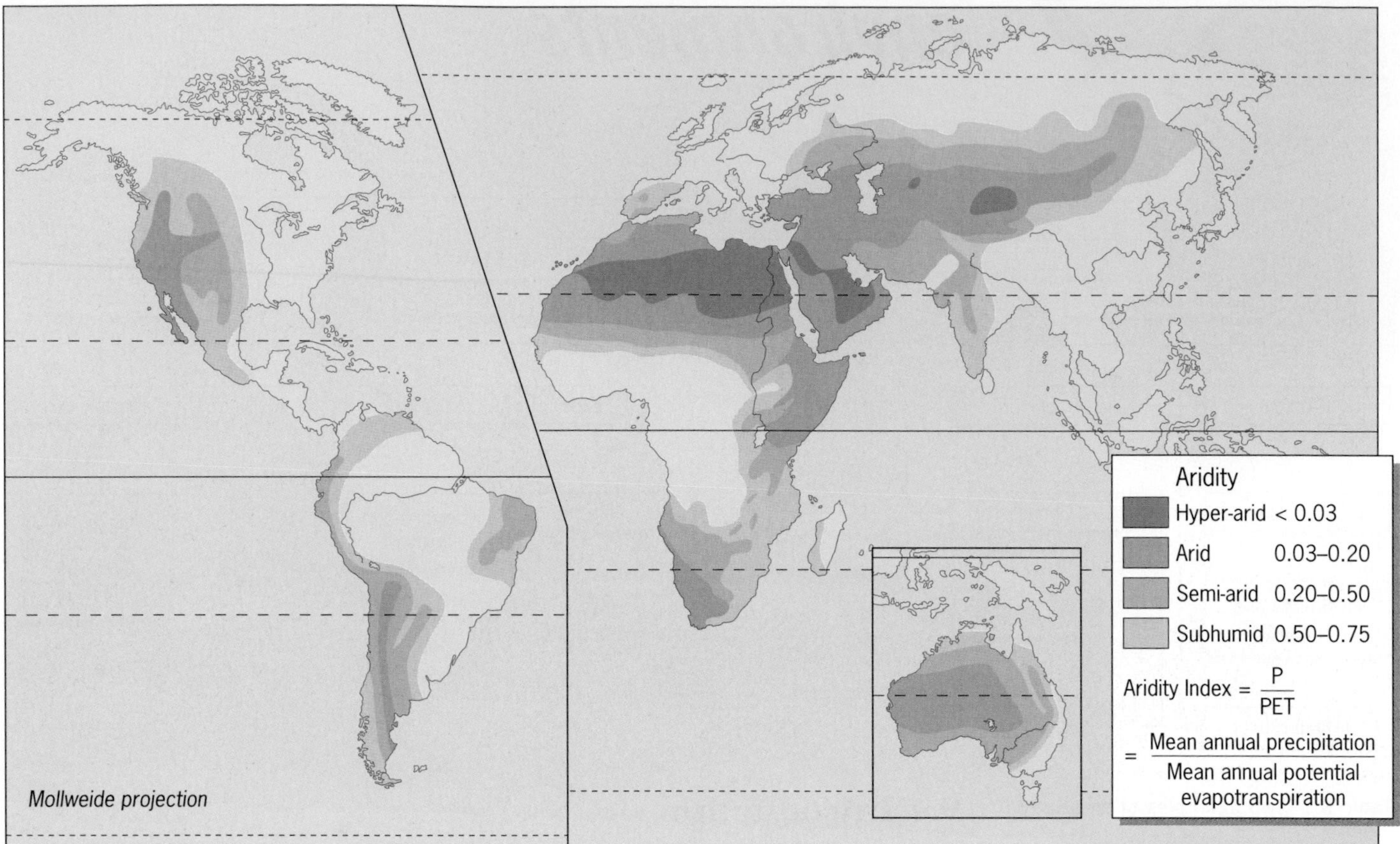

Figure 9.2 World map of arid lands (*Source:* Cooke et al., 1993)

The popular image of a desert as a 'sand sea' or *erg* is misleading. An erg (a region of shifting sand) has a minimum size of 125 km^2, and the largest, the Rub al Khali in Arabia, covers 560 000 km^2. There are indeed extensive sand-covered areas, but they make up only approximately one-fifth of the arid surface area (Fig. 9.3). Sand covers 40 per cent of Australia, and one-third of Arabia is covered by dunes. In the Sahara the proportion is one-ninth.

The remainder consists of a wide diversity of rock and stone landscapes and landforms (Table 9.2). One striking feature of arid lands is that, because of the absence or sparseness of vegetation cover, the landforms are highly visible (Fig. 9.1).

9.2 Agents and processes at work in arid lands

Wind and water are the agents of erosion which shape arid landforms. Of the two, water is the more important. This may seem surprising at first, but running water does more work (i.e. erosion–transportation–deposition) than wind, except in the very driest of deserts, e.g. parts of the Namib of south-west Africa. Of course, there are many localities where wind is the primary agent for a wide range of erosional and depositional landforms. A sand-dune is the best-known example.

Most erosional work is done by mechanical processes. Chemical, biochemical and biological weathering activity is restricted by the low levels of soil moisture, small biomass and slow decomposition rates. However, many landforms are the result of the work of both wind and water, and of mechanical and chemical weathering. Chemical processes associated with moisture and solutes can loosen the surface of a rock and ease the task of mechanical processes, while repeated temperature changes cause expansion

Table 9.2 A geological classification of deserts

Type	Example
Mountain and basins	South-western USA
Plateaux and wadis	Iran/Afghanistan; Libya
Ancient shields	Western Australia
Sand seas (ergs)	C & NW Australia; Central Arabia
Stony deserts (reg; hamada)	Southern Iran; Peru

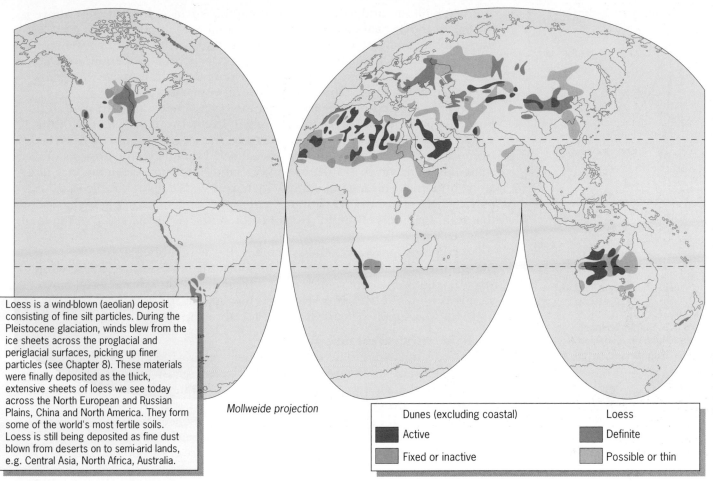

Loess is a wind-blown (aeolian) deposit consisting of fine silt particles. During the Pleistocene glaciation, winds blew from the ice sheets across the proglacial and periglacial surfaces, picking up finer particles (see Chapter 8). These materials were finally deposited as the thick, extensive sheets of loess we see today across the North European and Russian Plains, China and North America. They form some of the world's most fertile soils. Loess is still being deposited as fine dust blown from deserts on to semi-arid lands, e.g. Central Asia, North Africa, Australia.

Mollweide projection

Dunes (excluding coastal)
Active
Fixed or inactive

Loess
Definite
Possible or thin

Figure 9.3 The distribution of windblown (aeolian) landforms (*Source:* Chorley et al., 1984)

Figure 9.4 Exfoliation of rocks: mechanical weathering, Namibia

and contraction of a surface rock layer, which in time becomes loosened (Fig. 9.4). Moisture can then penetrate behind this weathered 'skin'. As the water evaporates, salts crystallise and expand, causing further detachment of the surface layer (exfoliation). Mechanical processes continue the disintegration, and expose a fresh surface to the weathering cycles. Overall, weathering rates are slow in arid environments.

?

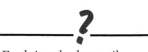

1 Explain why low soil moisture, sparse vegetation cover and slow organic decomposition limit the activity of chemical and biochemical weathering processes in arid environments (see Chapter 1, pp.7–12).

9.3 The wind at work (aeolian processes)

Wind is a stream of moving air and, like water, contains energy to carry out erosion, transportation and deposition. Thus we can identify both erosional and depositional landforms.

Erosion by wind

Wind creates erosional landforms by two mechanical processes:

- *Abrasion* The disintegration of rock surfaces as a result of physical impacts by wind-borne particles;
- *Deflation* The picking up and removal of loose particles by the wind, resulting in a lowering of the land surface.

Wind velocity and abrasion

If you are lying on a sandy beach and a breeze springs up, you may become irritated by sharp stinging sensations, and by sand grains sticking on your skin. Look along the surface of the beach (not facing directly into the wind!)

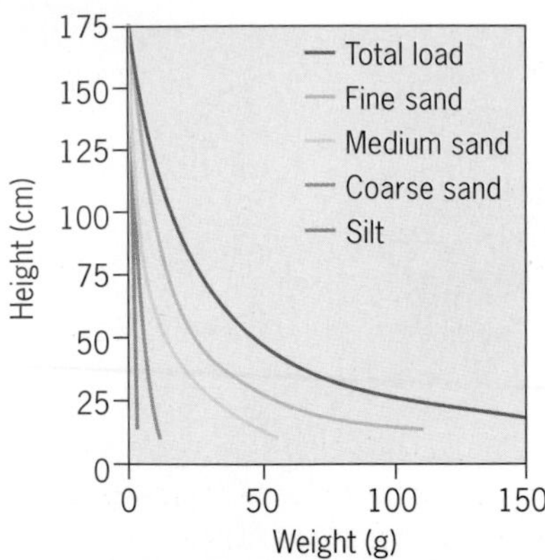

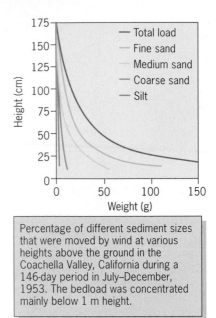

Percentage of different sediment sizes that were moved by wind at various heights above the ground in the Coachella Valley, California during a 146-day period in July–December, 1953. The bedload was concentrated mainly below 1 m height.

Figure 9.5 Sand transport: weight/height relationships (*Source:* Skinner and Porter, 1987)

?

2 From Figure 9.5:
a Describe the distribution of wind-borne sediment load;
b Measure the percentage of total load made up by each particle category at 25, 50 and 100 cm height.

and you will see sand grains rolling along this surface. Immediately above is a 'mist' of moving sand grains – it is these that are 'abrading' your skin! Look more closely and you will see that an individual particle is not permanently suspended, but is bouncing along. This process is called *saltation*. Now stand up: the stinging continues, perhaps up to your knees, but above this there are few impacts, just a refreshing breeze on your face.

This simple (if painful) example gives us some important information. First, the particle load is controlled by wind velocity and the nature of the surface material, e.g. when the sand is wet, the grains will not be detached from the surface. Second, abrasion is achieved mainly by the impacts of sand-calibre particles (0.1–1.0 mm), e.g. 'sandblasting' to clean the face of a stone building. Third, wind velocity increases with height, but the sand particles are concentrated within 50 cm of the land surface. Figure 9.5 shows that total load falls off rapidly with height, and that the make-up of the load changes with height.

We can thus make a general statement that wind abrasion is at a maximum in a narrow zone 15–50 cm above the land surface, where wind velocity and particle movement are both moderate. At ground level, particle supply is high but transport velocity is low. Above 50 cm, wind velocity is high, but the supply of sand available for abrasion is small.

Particle entrainment and transport

There are similarities between the ways in which wind and running water pick up (entrain) and move their particle loads (see Chapter 2). Thus, the key variables of the wind we need to know about are its *velocity* and the amount of *turbulence*. Crucial land surface variables are *surface roughness*, and the *cohesion* and *grain size* of the materials. The presence or absence of *water* and *vegetation* also affect how the wind and the surface interact.

The entrainment and movement of particles are controlled by the interactions of six kinds of force (Figure 9.6).

Shear is the major driving force applied by the wind to surface materials. Shear works by producing a small pressure difference between the upwind and downwind sides of a particle. The reduced pressure on the downstream side encourages forward motion, i.e. in the direction of the wind. Surfaces of coarse materials cause a shallow zone of increased eddies and turbulence, which also assists lift. Once the initial grains begin to saltate (move in a bouncing motion), their ballistic impacts (bombardment) on surface grains become the main force for further entrainment. The zone of movement broadens and fans out downwind, in a process known as *avalanching*.

There are two important relationships between wind speed and particle size carried. First, tiny clay and silt particles need a higher wind velocity than larger sand grains before they are detached from the surface. This is because clay and silt particles adhere tightly together and also create a smoother surface. If the cohesion and smoothness are disturbed, however, then wind turbulence quickly entrains the tiny particles. For example, a breeze may lift little material from a clay or silt surface, but if the wheels of a vehicle churn up this surface, then a cloud of 'dust' rises.

Second, the difference between the fluid and impact threshold lines indicates the role of impact forces. When there is surface motion of the particles (Fig. 9.7), the impacts apply some force to help the 'lift' of the wind. So, entrainment of particular particle sizes needs a lower wind speed than if the wind flow is working alone as a fluid.

Once in motion, dust-sized particles can remain in suspension for long periods and rise to considerable heights (Fig. 9.8). Studies of glaciers in the Swiss Alps reveal thin reddish layers, made of dust carried north from the Sahara, deposited on the snow and subsequently incorporated into the ice.

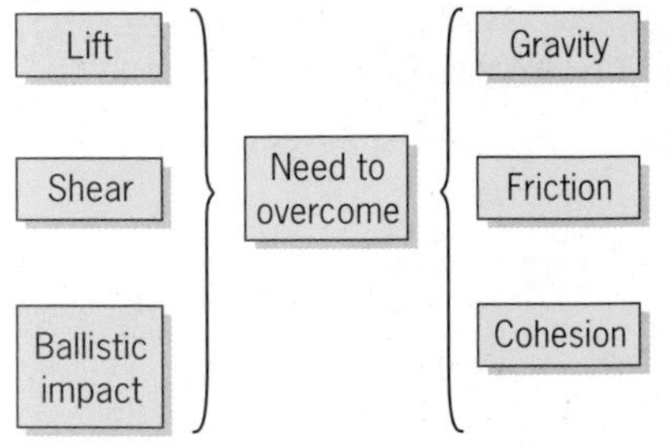

Figure 9.6 Forces influencing the entrainment and movement of particles

Impacted sand grains bounce into the air stream and are carried along by the wind as gravity pulls them back to the land surface where they impact other grains, repeating the process. After reaching its maximum height, each follows a ballistic trajectory, like that of a bullet.

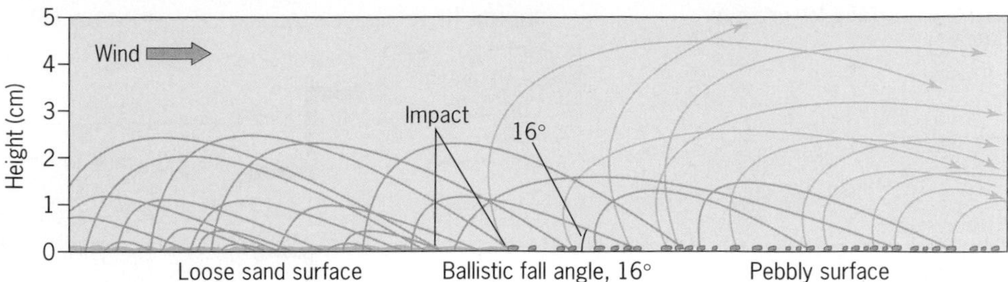

Figure 9.7 The transport of sand grains by saltation (*Source:* Skinner and Porter, 1987)

Only exceptionally strong winds can carry sand grains larger than 1 mm in suspension. Thus, true 'sandstorms' are rare. As our simple 'beach abrasion' example has shown, most of the sand-calibre load is entrained and transported near the ground surface by repeated bouncing motions known as *saltation* (Fig. 9.7). Grains too coarse to be lifted at a particular wind strength may be rolled along the surface (cf. the bedload of a stream, section 2.7). This process is known as *creep*. Impacts with other surface particles and from saltating grains slowly abrade the coarse particles, reducing their mass until they too achieve 'lift-off'.

?

3 Using Figure 9.8a and b, construct a graph to show the relationship between distance from source and dust deposition for the five sites. Describe this relationship briefly, and suggest factors which influence the distribution.

4a From Figure 9.8, what is the maximum distance from source that dust haze was recorded on at least 15 per cent of observation days?

b At an average wind speed of 15 km/hr, how long would this transportation take?

Deflation and selective erosion

Continued removal of weathered particles by *deflation* will result in a lowering of the land surface. In systems terms, when deflation outputs from the surface store exceed the material inputs, then the land surface will be progressively lowered. We are concerned, therefore, with the net balance of the surface store (Fig. 9.9).

The rate and extent of deflation vary widely. Where strong winds move across loose, dry, erodible surfaces, deflation may reach 1 m a year and affect thousands of square kilometres, e.g. the desertification along the southern edges of the Sahara in the African Sahel, during the severe droughts of the 1960s and 1970s. Under less extreme and exceptional conditions, surface deflation is commonly 1–5 mm a year. Long-continued removal can produce huge *deflation hollows*, such as the Qattara Depression of the northern Sahara. This is 320 km long, up to 160 km wide, and has been lowered to 134 m below sea level. As the Bodelé Depression example of Figure 9.8 shows, these are major sources of wind-blown dust and sand.

a Harmattan dust haze recorded over the Atlantic Ocean (% of total observations)

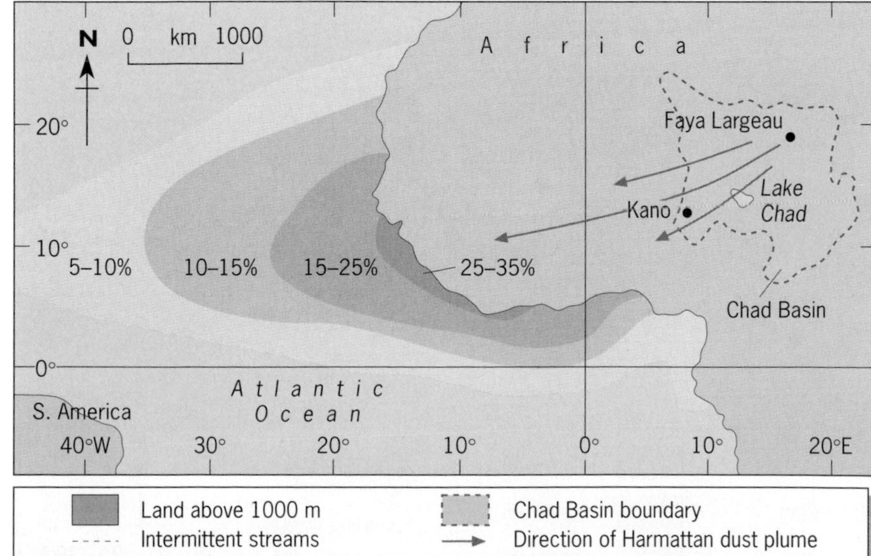

b The Chad Basin. Dust deposition at five collection sites in northern Nigeria over a period of 82 days during the 1978 Harmattan season

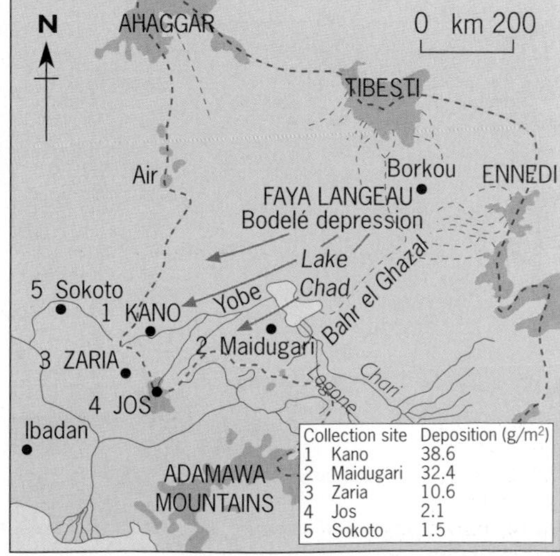

Collection site	Deposition (g/m²)
1 Kano	38.6
2 Maidugari	32.4
3 Zaria	10.6
4 Jos	2.1
5 Sokoto	1.5

Figure 9.8 The transport of Saharan dust (*Source:* McTainsh and Walker, 1982)

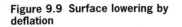

Figure 9.9 Surface lowering by deflation

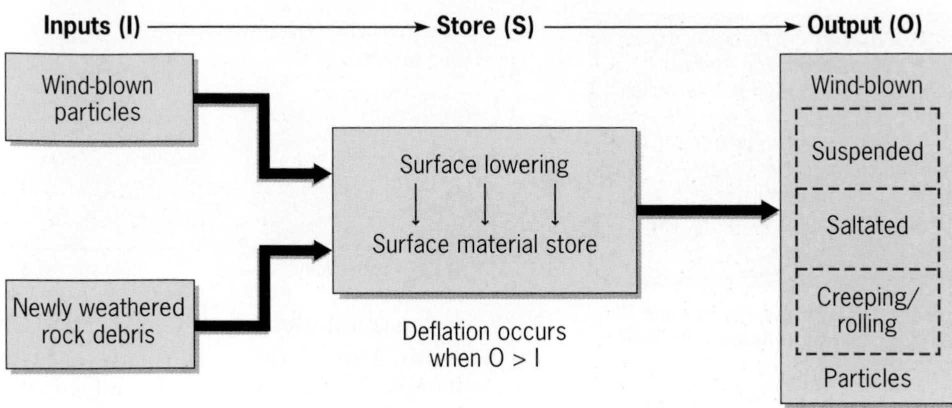

5 Explain why the following variables are given as controlling factors on the final depth of deflation in a locality: • the thickness of the regolith; • the rate of bedrock weathering; • the depth of the water table.

6 From Figure 9.10:

a Calculate • the average length of the yardang ridges in this part of northern Chad; • the mean distance between the ridge crests. (Use a minimum sample of five measurements.)

b Draw a cross-section along line A–B, and label the landform types.

c Use a map and your responses to questions a and b to describe the landscape of the region.

Although deflation may affect extensive surfaces (cf. sheet erosion by running water), erosional impacts are commonly selective and localised. This is because of local variations in wind and surface character (Fig. 9.9). For example, where the surface geology consists of an alternating series of resistant and less resistant rocks, aligned roughly parallel with the prevailing wind direction, differential erosion will occur. Deflation creates hollows across the more readily weathered beds, leaving the harder rocks as ridges. Abrasion then 'streamlines' these ridges to *yardangs,* elongated and streamlined wind-eroded ridges (Fig. 9.10).

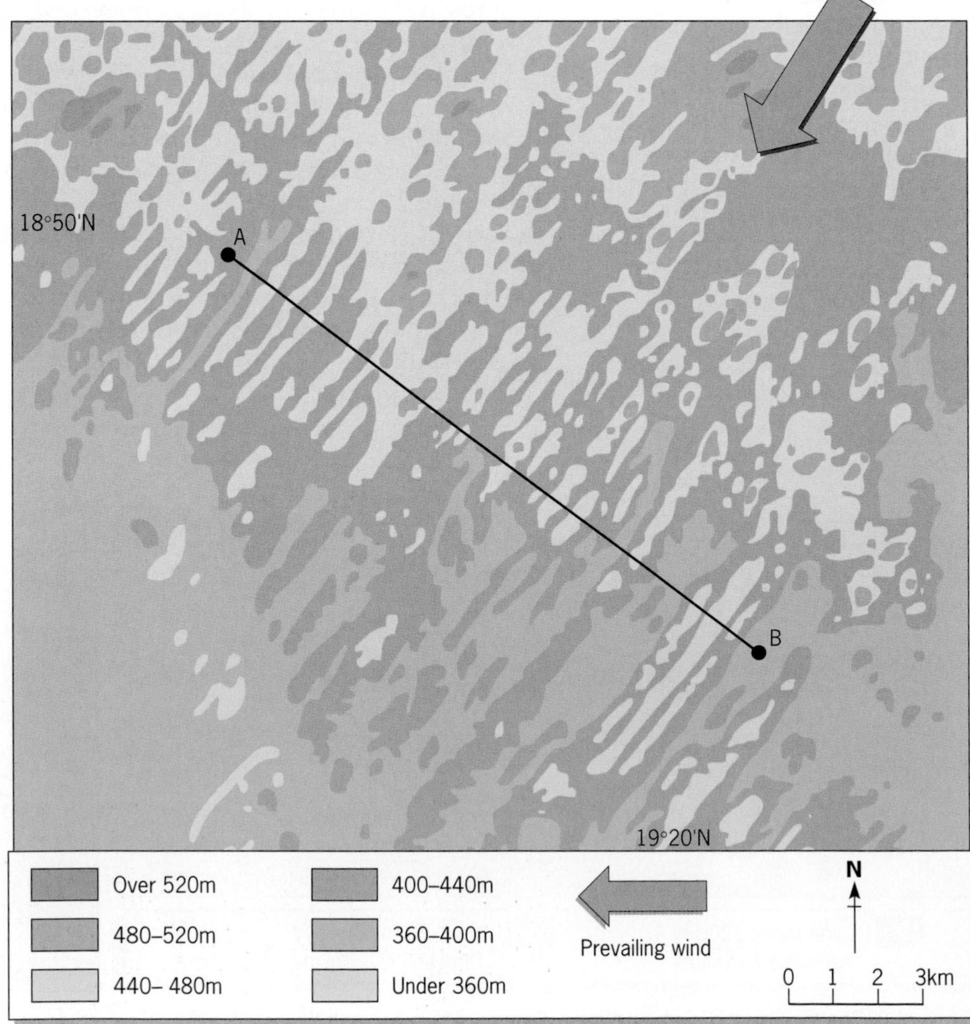

Figure 9.10 'Ridge-and-groove' yardang landscape, northern Chad (*Source:* Collard, 1990)

Figure 9.11 Buttes: 'The Teapots', Monument Valley, Utah, USA

Inselbergs and bornhardts

Driving across rock and stone deserts, it sometimes seems that the vast, flat landscape is endless. Then, shimmering in the heat haze, a strange tower-shape appears, far ahead. As its outline becomes clearer, it emerges as a lonely hill mass projecting abruptly from the desert surface. This is an *inselberg*, a steep-sided isolated hill, mountain or ridge, rising abruptly from an adjoining monotonously flat plain. They are not confined to deserts, but are particularly distinctive features of arid and semi-arid environments. The *buttes* and *mesas* of the high plateaux of the western USA are spectacular examples and, like Ayers Rock in Australia, are major tourist attractions (Fig. 9.11).

There are two key variables in the morphology (form) of inselbergs. First, some inselbergs rise straight from the desert surface, while others are mounted on a surrounding platform of intermediate slope. Second, inselbergs may be angular or rounded. Well-rounded examples, especially those developed on crystalline rocks such as granites, are known as *bornhardts*. Both types may persist for tens of millions of years, because weathering processes in arid regions work very slowly on such strong rocks.

There are two main theories about how inselbergs and bornhardts are formed. Both involve *differential erosion*, where the isolated mass consists of particularly resistant rock. The first, proposed by King, suggests that the resistant mass has been isolated as a result of *slope retreat* and is related to pediment formation (Fig. 9.12a). The second theory is that the form develops initially as a subsurface feature and is progressively exposed by *surface downwasting* (Fig. 9.12b). This is similar to the theory proposed by Linton

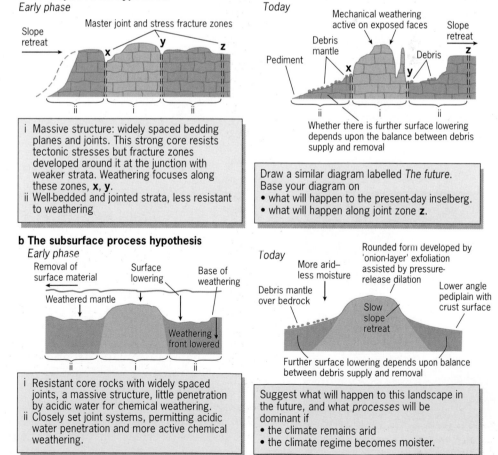

a The slope retreat hypothesis

Early phase

Slope retreat
Master joint and stress fracture zones
x y z
ii i ii

i Massive structure: widely spaced bedding planes and joints. This strong core resists tectonic stresses but fracture zones developed around it at the junction with weaker strata. Weathering focuses along these zones, **x**, **y**.
ii Well-bedded and jointed strata, less resistant to weathering

Today

Mechanical weathering active on exposed faces
Slope retreat
Debris mantle
Debris
Pediment
x y z
ii i ii
Whether there is further surface lowering depends upon the balance between debris supply and removal

Draw a similar diagram labelled *The future*. Base your diagram on
- what will happen to the present-day inselberg.
- what will happen along joint zone **z**.

b The subsurface process hypothesis

Early phase

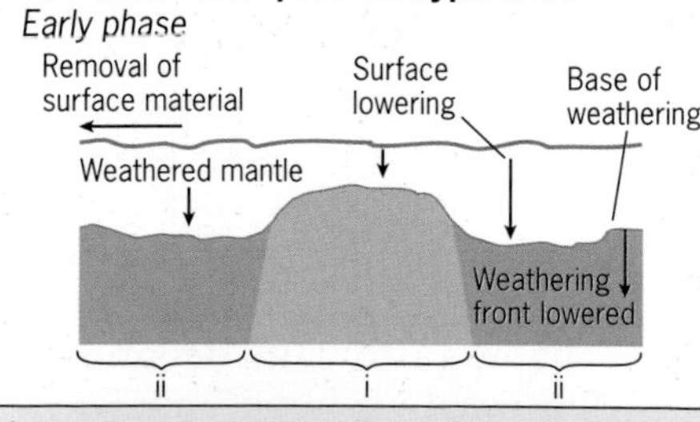

i Resistant core rocks with widely spaced joints, a massive structure, little penetration by acidic water for chemical weathering.
ii Closely set joint systems, permitting acidic water penetration and more active chemical weathering.

Today

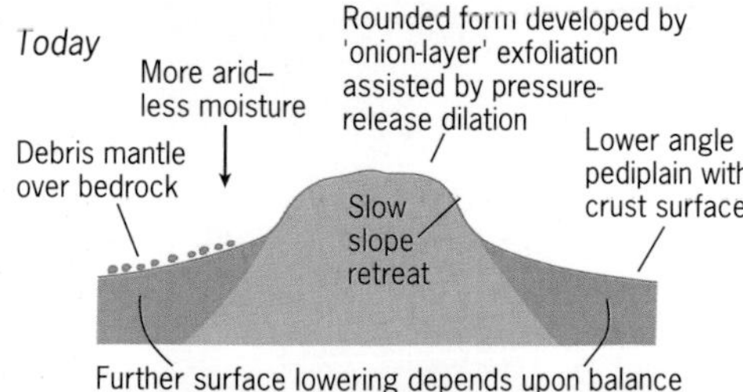

Suggest what will happen to this landscape in the future, and what *processes* will be dominant if
- the climate remains arid
- the climate regime becomes moister.

Figure 9.12 Inselbergs and bornhardts

Figure 9.13 Desert pavement and salt flat, Owens Valley, Sierra Nevada California, USA

for the development of tors in Britain. This explanation involves subsurface chemical weathering, and so today's desert inselbergs may have begun life when the regional climate was moister.

The slope retreat hypothesis seems to fit the butte and mesa form, while the surface downwasting hypothesis is a more convincing explanation of the rounded bornhardt type. It is likely, too, that both sets of processes have been involved in the development of some inselbergs, e.g. slope retreat increases in importance as the rock tower is increasingly exposed by surface lowering.

Surface crusts

Extensive areas of flat or gently sloping desert surfaces are covered with a mantle of weathered debris (Fig. 9.13). This mantle often consists of a coarse crust capping a layer of fine material. If undisturbed by extreme floods or human activities, these are stable surfaces with low rates of change. They are the end-product of two long-continued processes – surface deflation by wind, and selective stone sorting caused by the presence of subsurface moisture (Fig. 9.14). The progressive removal of the finer particles may ultimately create a *desert armour*: 'A surface layer of coarse particles concentrated chiefly by deflation' (Skinner and Porter, 1987). The clasts which make up the surface often fit together so tightly that they form a resistant *desert pavement* protecting the finer matrix below. Pavements vary in clast size from gravel crusts, known as *reg*, to large stone surfaces known as *hamada*.

Figure 9.14 The development of a stone crust, or pavement

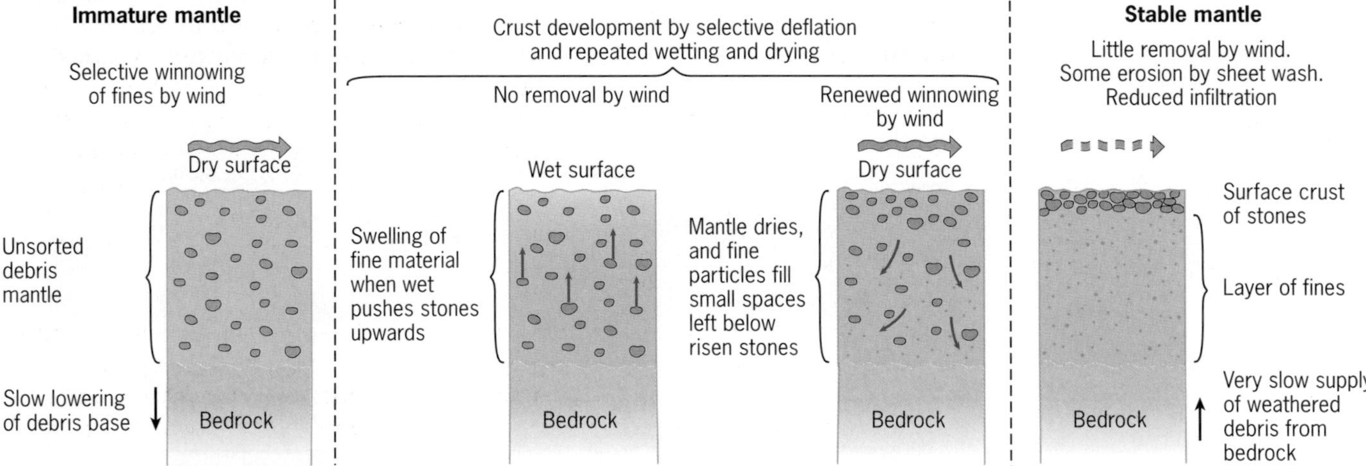

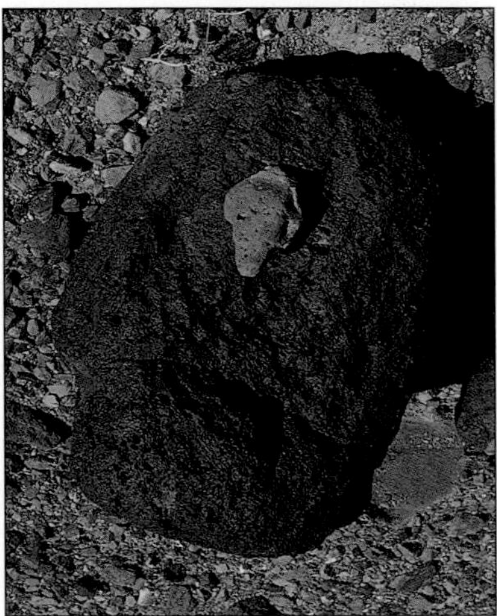

Figure 9.15 Desert varnish, Panamint Mountains, California, USA

Further surface erosion relies on slow abrasion of the pavement clasts. This abrasion cuts angular facets on the stones and polishes them. These distinctive clasts are known as *ventifacts*: surface stones (clasts) which have been abraded and shaped by wind-blown sediment.

When surfaces become stabilised and individual stones remain exposed in situ, a dark, shiny patina or 'skin' develops, called *desert varnish* (Fig. 9.15): 'a thin, dark, shiny coating, consisting of manganese and iron oxides, formed on the surfaces of stones and rock outcrops in desert regions after long exposure' (Skinner and Porter, 1987). The definition tells us that some chemical reactions have taken place, and that water has played its part.

It is clear that abrasion and deflation are both at work in a given landscape. Nor must we forget that water is involved, e.g. the occasional storms and flash floods assist the selective erosion of a debris surface, washing out the finer particles, re-sorting coarser materials, etc. (see sections 9.5, 9.6). Repeated wetting and drying also increases the cohesiveness of clay and silt surfaces or crusts, encourages the development of duricrusts, and reduces erodibility.

7 Identify three changes in environmental conditions which would • increase and • reduce the rate of deflation, and explain how each would work. (You may find annotated diagrams useful in your answers.)

8 Suggest changes in conditions which might cause renewed deflation across an 'armoured' surface or pavement.

Duricrusts

The surface armour may be further strengthened by an evaporite cement (comparable to when you build a crazy paving area in your garden and set the blocks in a cement base). Soluble minerals brought towards the surface in solution by capillary action, or present in standing water, are precipitated as the water evaporates. Where these evaporites dominate, a *duricrust* develops. This may occur directly as a result of surface evaporation and precipitation, or indirectly, where a subsurface horizon of the 'concrete' is precipitated and is subsequently exposed by surface lowering.

Duricrusts are named according to their chemical composition (Table 9.3). They develop best on *playas*, the flat beds of non-permanent lakes, and are seen in their extreme form in *salt flats*.

Table 9.3 Main types of duricrust

Type	Composition	Example
Caliche or calcrete	Calcium carbonate ($CaCo_3$)	W. Australia
Silcrete	Silicon oxide (SiO_2)	Lake Eyre Basin, Australia
Gypcrete	Gypsum ($CaS0_4 2H_2 0$) Hydrous calcium sulphate	Namib Desert, Namibia
Sodium chloride	Sodium chloride (Na_2Cl_3)	Death Valley, USA

9.4 Depositional landforms

Depositional landforms are found where, today or in the past, the supply (input) of wind-borne particles into a locality has exceeded their removal (output) over long periods. Such landforms fall into three main categories: relatively flat surfaces, perhaps patterned with ripples, known as *sand sheets*; hills of sand, known as *dunes* (the world's highest dunes reach 400 m) (Fig. 9.16); and sloping masses of sand piled against the windward flanks of mountains, known as *sand ramps* or *climbing dunes*. A less common type of ramp is the *falling dune*, deposited from wind eddies in the lee of a ridge.

Sand-dunes

Dunes are the commonest aeolian depositional landforms. They are found in hot and temperate deserts, and along coasts in many non-arid regions (Chapter 5, pp.99–101). Dunes in arid environments occur in a wide variety of forms, but seven main categories have been identified (Fig. 9.17). Their

Figure 9.16 Transverse dunes, Namibia

Figure 9.17 Principal dune types (*Source:* Skinner and Porter, 1987)

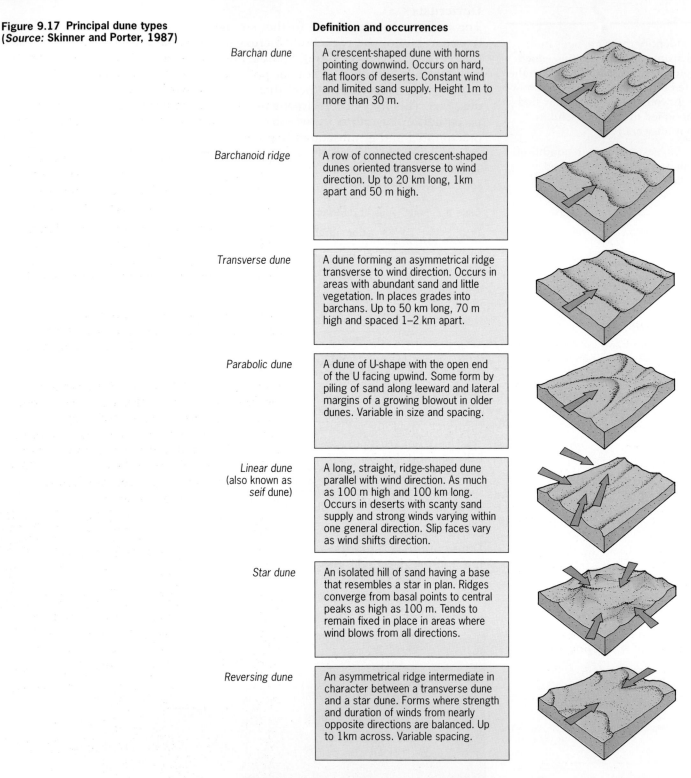

	Definition and occurrences	
Barchan dune	A crescent-shaped dune with horns pointing downwind. Occurs on hard, flat floors of deserts. Constant wind and limited sand supply. Height 1m to more than 30 m.	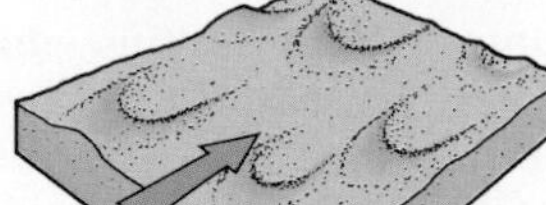
Barchanoid ridge	A row of connected crescent-shaped dunes oriented transverse to wind direction. Up to 20 km long, 1km apart and 50 m high.	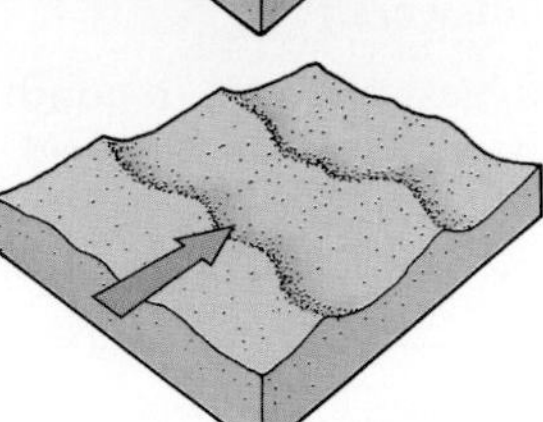
Transverse dune	A dune forming an asymmetrical ridge transverse to wind direction. Occurs in areas with abundant sand and little vegetation. In places grades into barchans. Up to 50 km long, 70 m high and spaced 1–2 km apart.	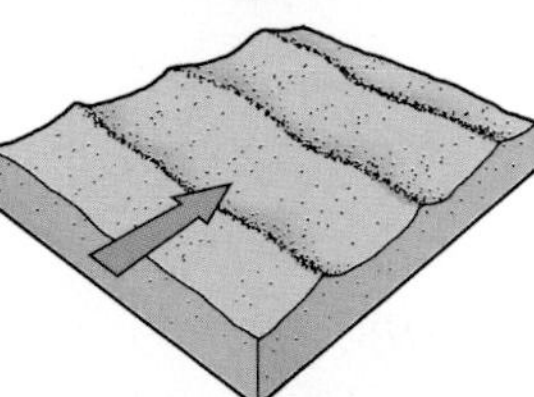
Parabolic dune	A dune of U-shape with the open end of the U facing upwind. Some form by piling of sand along leeward and lateral margins of a growing blowout in older dunes. Variable in size and spacing.	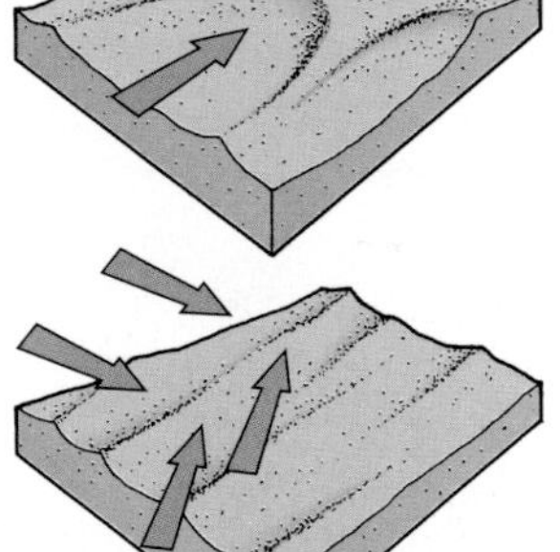
Linear dune (also known as *seif* dune)	A long, straight, ridge-shaped dune parallel with wind direction. As much as 100 m high and 100 km long. Occurs in deserts with scanty sand supply and strong winds varying within one general direction. Slip faces vary as wind shifts direction.	
Star dune	An isolated hill of sand having a base that resembles a star in plan. Ridges converge from basal points to central peaks as high as 100 m. Tends to remain fixed in place in areas where wind blows from all directions.	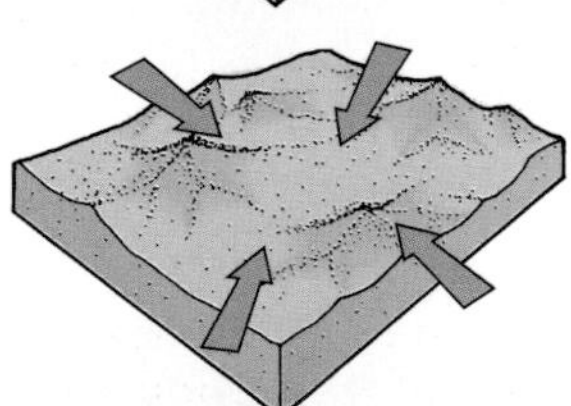
Reversing dune	An asymmetrical ridge intermediate in character between a transverse dune and a star dune. Forms where strength and duration of winds from nearly opposite directions are balanced. Up to 1km across. Variable spacing.	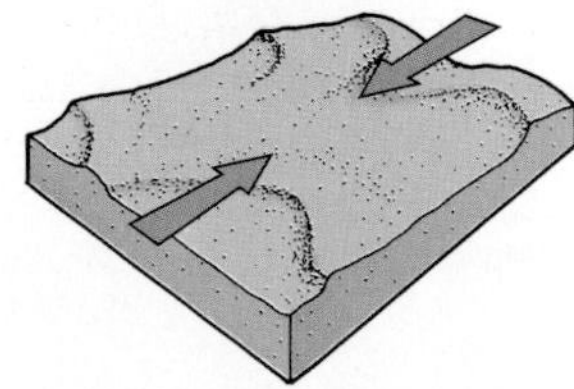

location, form, size and permanence depend upon the relationships between *wind* (direction, strength, turbulence/airflow, regularity), the *supply* of sand-calibre material, and the *environmental conditions* needed for continued sand accumulation. A study in the Wahiba erg (sand sea) of Oman recorded that sand movement was greatest during summer. The persistent south-west monsoon winds moved a whole dune mass approximately 3 m in 10 days. Wind directions fluctuate in winter and spring, and only the dune crests are reshaped, moving at up to 1 m a day.

9 From Figure 9.17, draw up a summary table to show • dune type; • wind relationships; • size/form characteristics.

Dune dynamics

Because sand-dunes are made of unconsolidated particles they are susceptible to movement and change, and may seem to lack permanence. Yet large dune systems take thousands of years to accumulate, and may last 50 000 years or more, although considerable reworking and reshaping of the materials may take place. When studying a particular set of dunes, we need to find out which of the three following conditions they are in (sand sheets and ramps fall into the same categories):

- *Active* Landforms on which surface sand transportation and deposition are currently taking place (see the Kelso Dunes Case Study, p.190). Active dunes may be migrating in the net transport direction (transverse or barchan dunes), extending (linear dunes), or growing in height (star dunes).
- *Dormant* Landforms on which surface sand transport and deposition are currently absent or at a low level (e.g. low wind energy regime; insufficient sand supply; well-developed vegetation cover). Yet such dunes are capable of becoming active again as a result of minor changes in environmental conditions.
- *Relict* Landforms produced in past times when climatic and/or depositional conditions were different, and have been stabilised for a period of at least 1000 years. They can revert to an active state only as a result of major environmental changes.

The Simpson and Strzelecki Deserts of central Australia illustrate the scale, age and persistence of major dune fields (Fig. 9.18). This vast system consists of longitudinal linear dunes which were extended NNW by persistent SSE winds over thousands of years. The more detailed view of the Birdsville area (Fig. 9.19) shows that individual dunes are 100 km or more long. Their average height is 12 m, but may reach 50 m; they are up to 200 m wide at the base and their crests are 1–2 km apart. Between the dunes, spreads of coarse pebbles form desert pavements, or *gibber* plains.

?

10 Draw a sketch map of the area shown on Figure 9.18 to show the main physical units, e.g. dune field, playa area, river system.

11 Use Figures 9.18, 9.19 and your sketch map to describe the landscape of this region.

12 To what extent does the landscape of Figures 9.18 and 9.19 support the view that many arid landforms are the result of the work of water as well as wind?

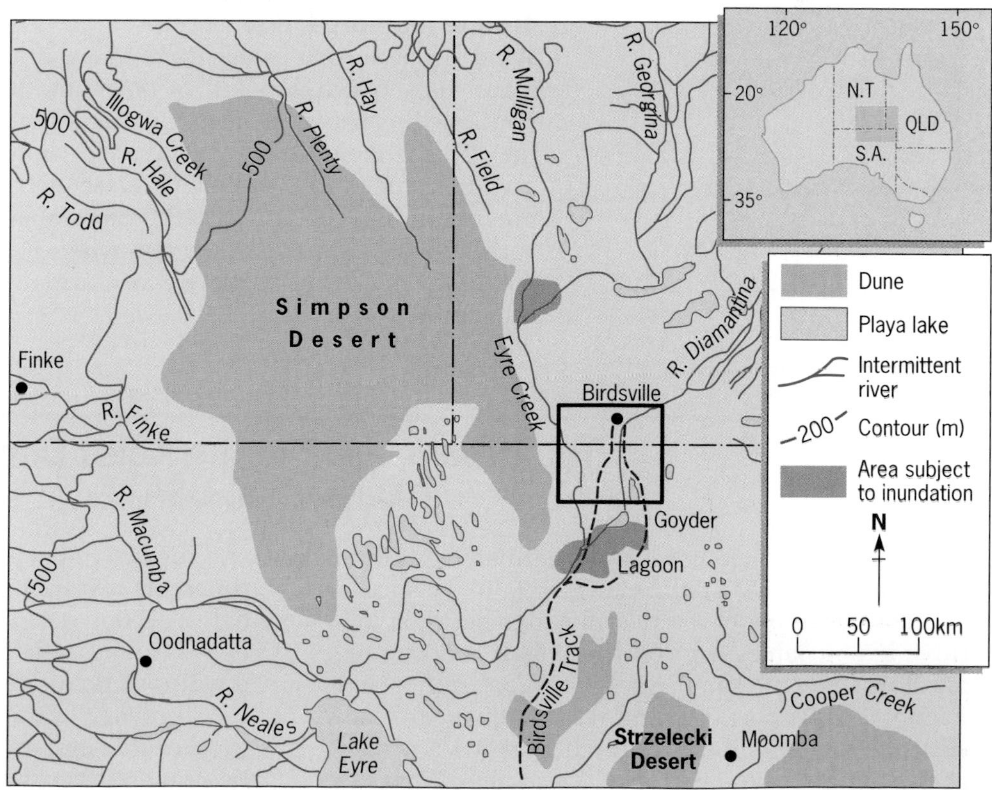

Figure 9.18 The Simpson Desert, Australia. Numerous intermittent watercourses flow either through the desert to Lake Eyre or disappear within the desert (*Source:* Nanson et al., 1992)

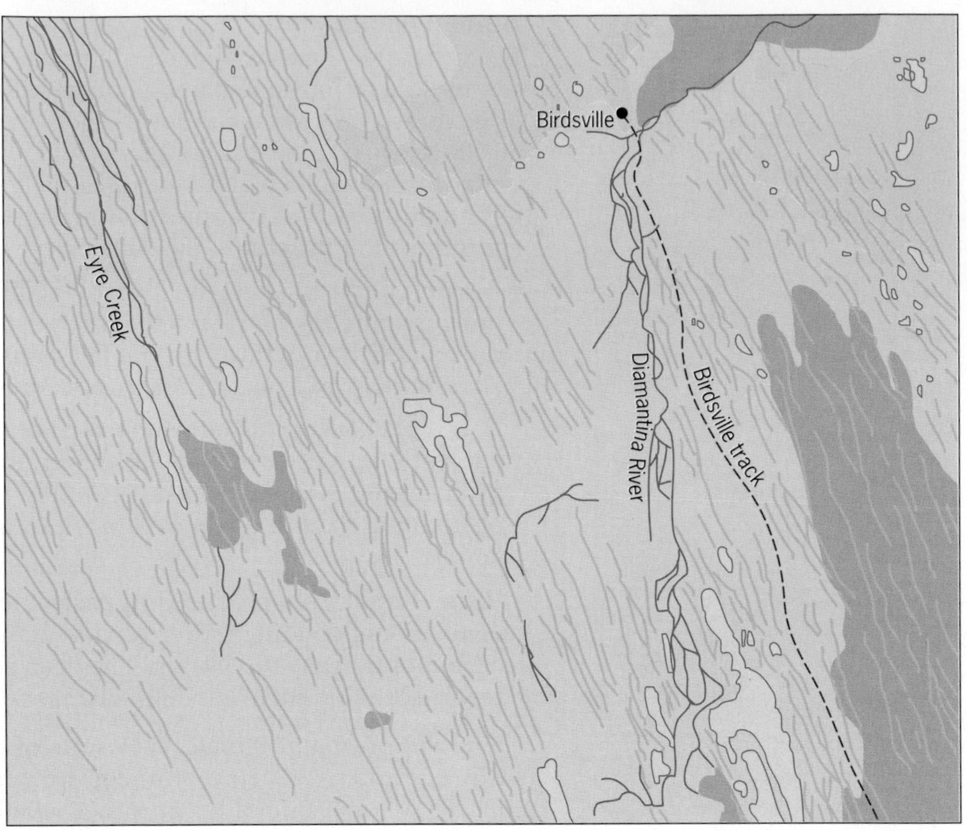

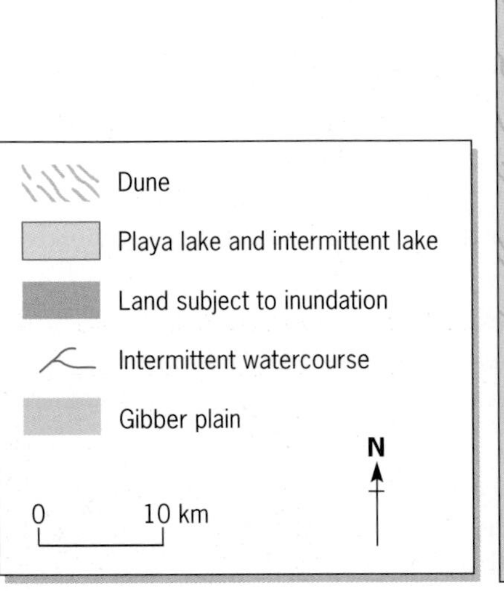

Figure 9.19 Study site near Birdsville, Simpson Desert (*Source:* Nanson et al., 1992)

Today this longitudinal dune system is largely dormant, and has been formed within the past 10 000 years. These dunes consist of reworked sand from an earlier system built around 80 000 BP. Relict remains of this original system are found in the cores of today's dormant dunes.

A dune mobility index

Wind strength and surface dryness are crucial variables influencing whether a dune is active and mobile or dormant. Scientists have calculated threshold relationships between the two variables and developed a *dune mobility index* (M). This is the ratio between the percentage of time wind speed exceeds the threshold for sand transport (5 m/sec at 1 m above the ground) and effective precipitation (Precipitation (P)/Potential evaporation (PE)). Dunes are fully active when M>200 and dormant when M<50. Between these two figures, local factors determine activity or dormancy, e.g. character of materials, vegetation character.

The development of the Kelso Dunes, California, USA

The Kelso dune field covers approximately 100 km^2 of the Mojave Desert, California (Fig. 9.20). It is an excellent example of the relationships between the three key variables controlling aeolian depositional landforms: (i) the sand supply source; (ii) the wind energy to transport the sand via transport pathways; (iii) conditions suitable for the deposition and accumulation of the particles. Kelso also illustrates two further important understandings: first, the timescale involved in the evolution of large dune fields; second, the variation in supply, transport and deposition activity over time, as environmental conditions fluctuate.

Figure 9.21 summarises the Kelso Dunes' geomorphic system. The Mojave River is a non-permanent river with its headwater catchment in the

San Bernadino Mountains east of Los Angeles. Irregular winter storms and seasonal snow melt generate short-lived pulses of discharge and sediment load eastwards to the Mojave Desert. This desert is a series of basins rimmed by lines of mountains (Fig. 9.20). The Mojave River cuts the Afton Canyon through one of the mountain ridges, and disperses on to a basin floor. Here it loses velocity and has built up an extensive alluvial fan (see section 9.5). During periods of moister climate, a lake has formed in the basin. At these times, the Mojave River becomes a delta. The landform is thus called a *delta fan*.

From this 'sink' or sediment store, the dominant westerly and north-westerly winds move the sand along the transportation corridor of the basin known as the 'Devil's Playground'. This corridor, 25 km long and 10 km wide, is a complex mixture of bare rock surfaces, dried-out lake beds (*playas*), sand sheets and dunes.

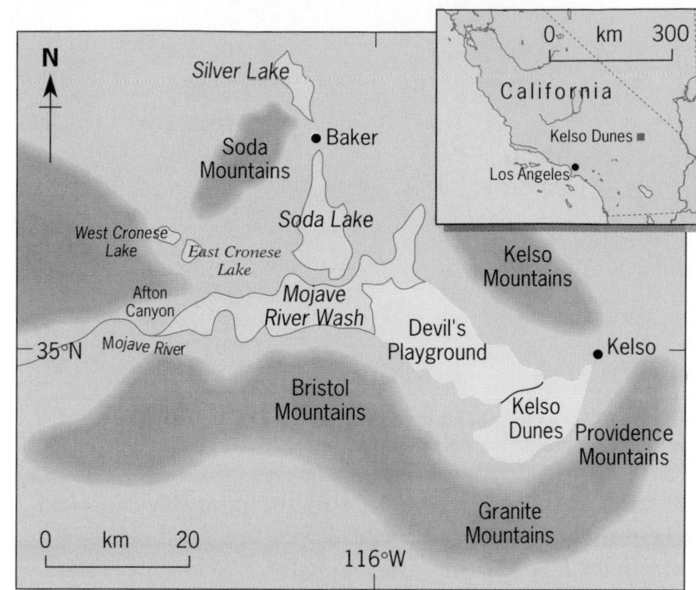

Figure 9.20 The Kelso dune field, Mojave Desert, California, USA

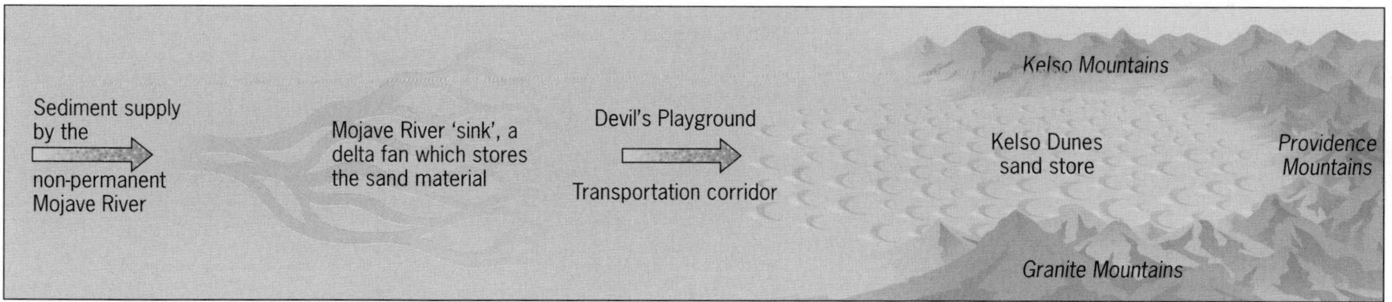

Figure 9.21 The Mojave River–Kelso Dunes geomorphic system

At the south-eastern end of Devil's Playground, the wind-borne sands collect in an embayment surrounded on the north, east and south by mountain ridges. The accumulation of sand in this embayment has produced the Kelso Dunes sand store (Fig. 9.22).

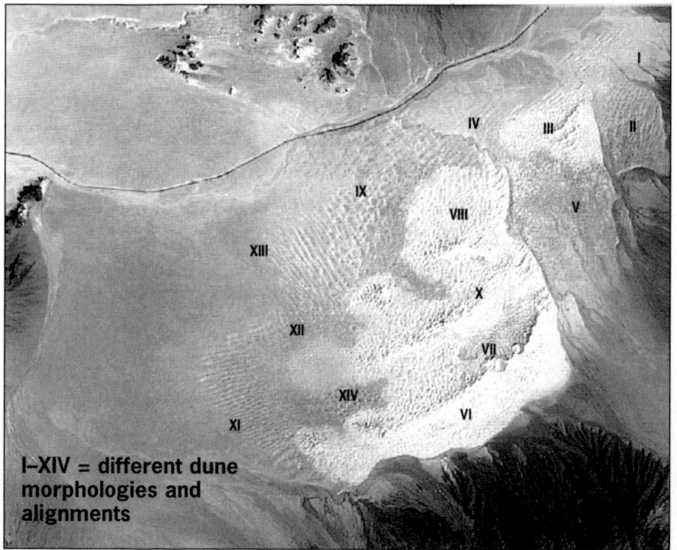

Figure 9.22 Landsat image of the Kelso dune field

This dune field is built of a complex mixture of transverse, longitudinal, crescentic dunes and sand sheets. Individual dunes are rarely more than 15 m high. Today, only 40 per cent of the total dune field consists of active dunes. The other landforms are dormant or relict features, often degraded to less than 5 m high, and with extensive vegetation cover.

The estimate that 60 per cent of the dune field is currently dormant suggests that present-day conditions are marginal for mobile, active dune formation. The M ratios (see p.190) for the Kelso environment are 100–130, which explains why only the higher, more exposed dune areas with greater wind speeds are active today. The mean annual rainfall is 76 mm, and for the whole dune field to become active once more, scientists estimate that this would need to drop to around 40 mm a year.

Phases of evolution

The cross-section of Figure 9.23 shows the general landform sequence. This sequence can be explained by variations in climatic conditions and sediment supply. Each depositional landform type is associated with a period of increased sediment supply, transport and deposition (Table 9.4). During drier periods,

Kelso Dunes
California, USA

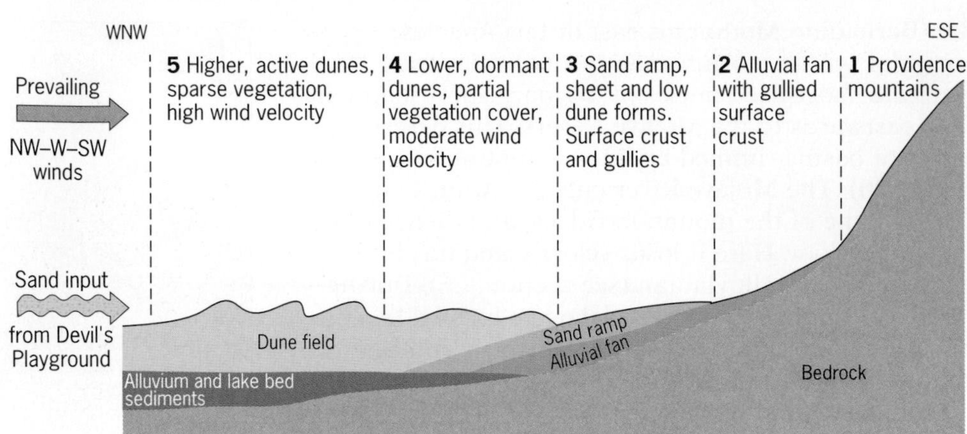

Figure 9.23 Generalised landform sequence across the Kelso Dunes

basin lakes dry up, alluvial fan surfaces dry out, and vegetation cover is reduced. Thus the sand stores of the delta fan and the alluvial fans provide increased supplies for entrainment and transport by the wind. The result is that sand accumulation and active dune formation occur mainly during these more arid episodes. Table 9.4 shows the alternation of moist and dry episodes and their geomorphic results.

?

13 From Table 9.4, list the periods of active dune formation.

14 Why is dune formation most rapid in periods of increased aridity?

15 Explain why periods of relatively high rainfall are important in the development of the Kelso Dunes.

16 At present, only 40 per cent of the Kelso dune field is 'active'. Suggest what changes in environmental conditions could cause the whole dune field to become (**a**) active, (**b**) dormant.

17 Use Table 9.4 to help you describe and explain the set of landforms shown on the cross-section of Figure 9.23.

Table 9.4 A generalised chronology of the Kelso Dune Field, California

Years BP	Environmental conditions	Geomorphic activity
18 400–16 000	Moist conditions	Lake Mojave levels high; sedimentation of delta fan and lake bed. Alluvial fan formation along basin fringes
16 000–15 000	Increasing aridity	Lake levels fall, alluvial fans dry out. Sand supply increases and sand ramps are built up along the mountain flanks
15 000–13 700	Less dry conditions	Reduced aeolian activity
13 700–12 000	Moist conditions	Lake Mojave levels high; renewed basin sedimentation
12 000–9000	Increasing aridity	Lakes and alluvial fans dry out. Widespread dune formation in the Kelso basin and the building of sand ramps against the mountain flanks
9000–7000	Decreasing aridity	Reduced aeolian activity but a gravel/boulder crust develops on the sand ramps
7000–5000	Increased aridity	Increased aeolian activity. Little new sand accumulation but a reworking and redistribution of dunes which spread further to the south-east
5000–4000	Fluctuating aridity	Extensive crescentic dunes form in the east of the dune field, largely from a reworking of existing sand
4000–2000	Moderately moist	Little aeolian activity, with most dunes well vegetated and dormant
2000–1500	Increasing aridity	Increasing sand supply and aeolian activity produce crescentic dune accumulation across the west of the dune field
1500–850	Moister conditions	Lakes fill irregularly, increased cover, reduced aeolian activity
850–450	Increasing aridity	Lake beds and delta fan dry; increased sand supply and aeolian activity. Widespread active dune formation
450–350	Moist conditions	Reduced aeolian activity. Most dunes dormant
350–100	Fluctuating	Little aeolian activity
100–0	Mostly drier	Active dune formation in central areas

NB The dates are approximate, and the timing and character of the sequence varied from place to place across the Kelso Dune Field

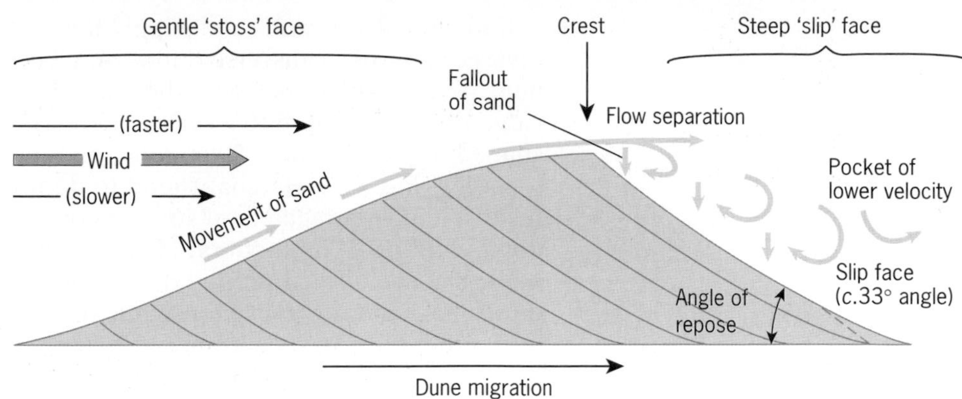

Figure 9.24 Basic wind/dune form relationships (*After:* Skinner and Porter, 1987)

Airflow and dune form

Wind strength, direction and persistence are important variables in dune formation. A fourth factor is *airflow*. Drag or friction between the land surface and the moving air creates *turbulence*: the rougher and more irregular the surface, the more turbulent the airstream. The simplest example of the land/wind relationship is seen in the form of the basic asymmetrical mobile dune (Fig. 9.24). The two-way nature of this relationship is illustrated by ripple patterns on sand surfaces. They are formed by small-scale waves in the airflow, and once formed help to perpetuate the waves, which in turn sustain the ripples.

In regions with a single prevailing wind direction, and a store of sand to transport and deposit, dunes form either transverse to or parallel to this dominant airstream. The controlling factor is the type of airflow. Where the turbulence has a *wave* form, dunes tend to form *transverse* to the wind direction (Fig. 9.25). Notice the importance of the eddies of wind which rotate in the lee of the dune crests.

The second type of turbulence is *helical* or *spiral* airflow (Fig. 9.26). In these conditions, longitudinal *linear*, also known as *seif*, dunes form parallel to the wind direction, e.g. the Simpson and Strzelecki dunes of central Australia (Figs 9.18, 9.19).

a Wave-like pattern in the wind erodes and deposits alternate ridges and hollows

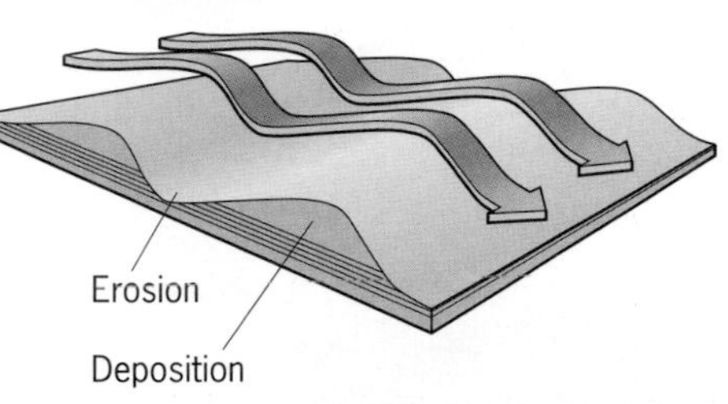

b Slip faces develop

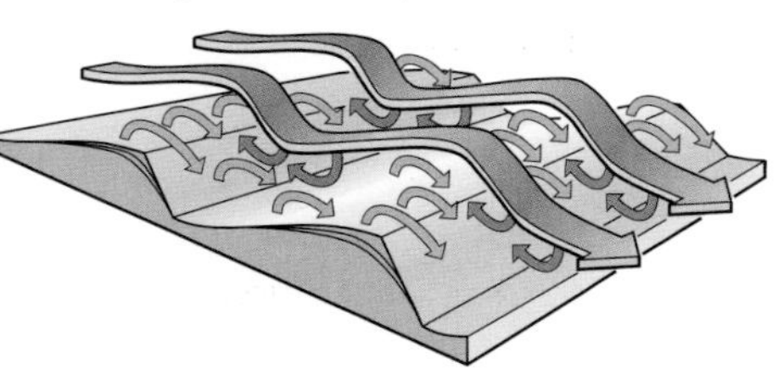

Figure 9.25 Two stages of transverse dune formation

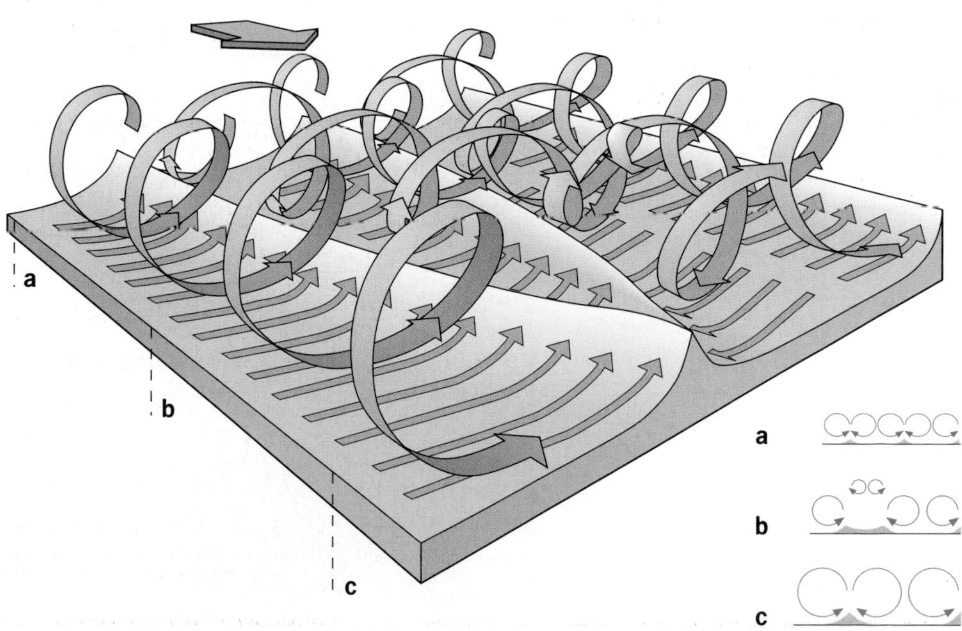

Figure 9.26 Helical airflow and linear dunes (*Source:* Embleton and Thornes, 1979)

If airflow waves and sand supply fluctuate locally, and take convergent and divergent paths, transverse dunes may become curved. This may generate helical flow, which increases the curvature and ultimately the single sand ridge may break up into a series of crescentic *barchan* dunes or tongue-shaped *linguoid* dunes. Thus, seif ridges and barchan crescents may occur together (Fig. 9.1). Barchans tend to dominate where sand supply is insufficient to provide total surface cover. Active barchans migrate downwind by a combination of wave flow and helical flow (Fig. 9.27).

We can see, therefore, that dune landscapes are very complex. First, some dunes have great persistence, while others evolve from one form to another over time. For instance, transverse ridges may become curved, break into barchans, elongate to linguoids and finally emerge as linear seif ridges. Second, any sand-built landscape may contain several types of dune. Third, any particular dune-field may contain, at a moment in time, dunes which are active, dormant and relict.

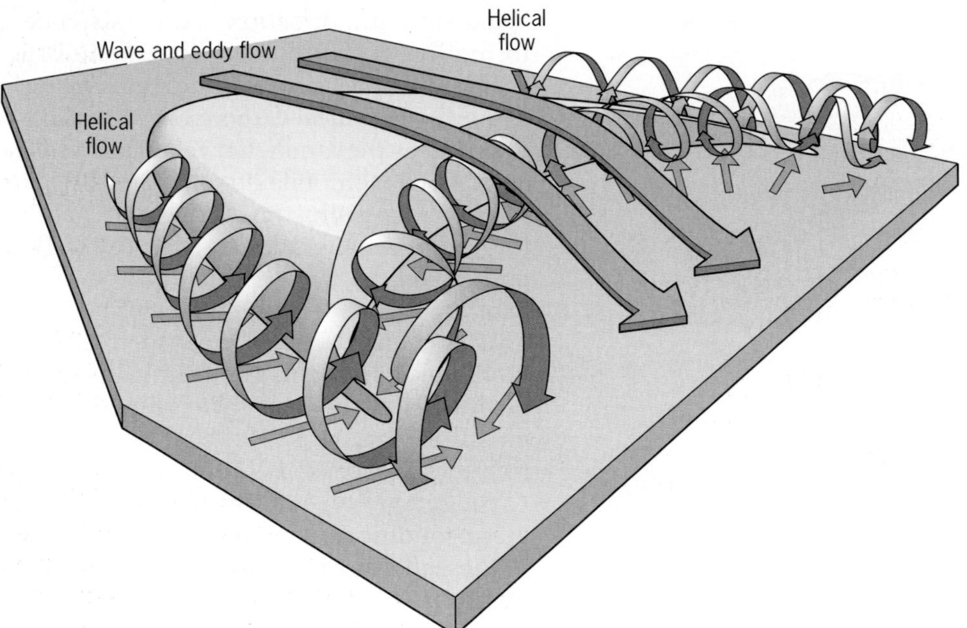

Figure 9.27 Airflow and barchan development (*After:* Embleton and Thornes, 1979)

9.5 Water at work

Permanent surface water is rare under natural conditions in arid environments. Water does its geomorphic work in short-lived, intense and often localised episodes. Precipitation inputs occur mainly in irregular, intense storms, e.g. intensities of 25 mm an hour are not unusual. Thus, for perhaps a few hours or days a year, high-energy conditions exist, e.g. the flash flood. This concentrated energy can be applied to the vigorous erosion and transportation of material, and channel networks are etched upon the landscape. Once the rain ceases, however, the surface water flow and its associated energy decline rapidly, causing deposition of sediment. When the surfaces dry out, aeolian processes take over once more.

The influence of surface character

How the water behaves when it arrives at the land surface depends upon the nature of that surface. So, unconsolidated active dunes and sand sheets are easily erodible and susceptible to sheetwash and gullying, but they are highly permeable and infiltration rapidly reduces surface flow and energy available for erosion and transportation. In contrast, dormant dunes and

The Harquahala Basin is fringed by bare mountain ridges which rise up to 450 m above the basin floor. The mean annual rainfall is less than 300 mm and arrives in irregular, sudden storms. There are no permanent streams, and the main axis of the basin is crossed by Centennial Wash, a broad, flat ephemeral stream channel which has cut short narrow entrance and exit canyons through the surrounding mountains at Harrisburg Narrows and Mullens Cut.

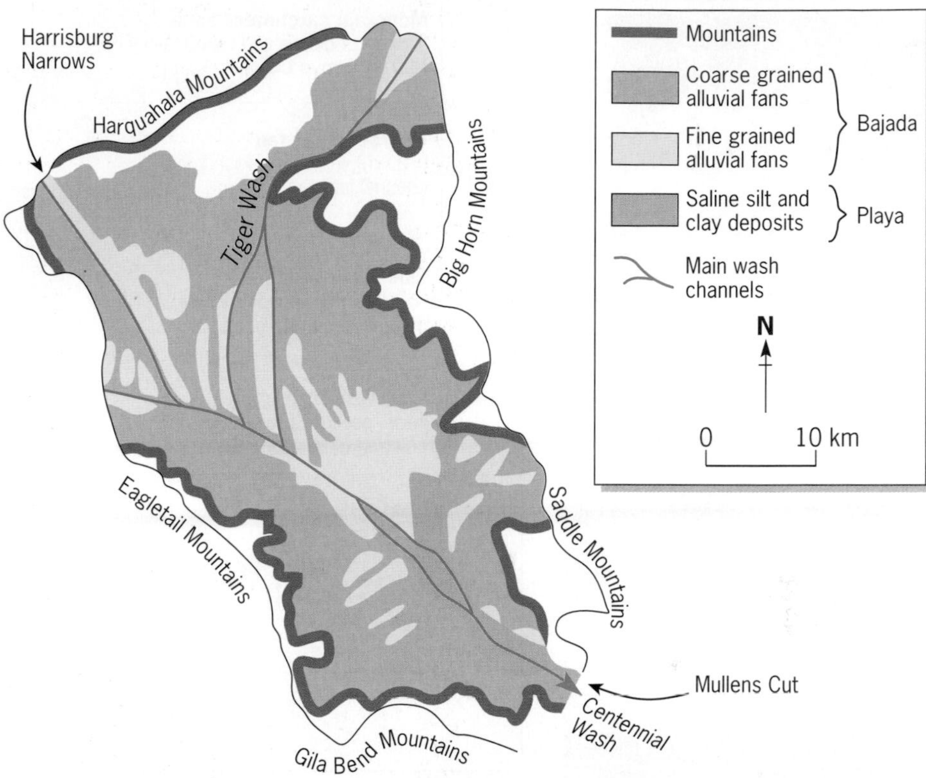

Figure 9.28 Geomorphology of the Harquahala Valley, Arizona, USA

sheets with surface crusts, e.g. reg or gibber plains, are initially more resistant. Yet once the protective crust is broken, erosion of the underlying sand accelerates and long-lasting gully systems are incised.

Mountain and basin landform sequences

The Harquahala Basin in Arizona, USA (Fig. 9.28), illustrates the landform sequence which results from the influence of the precipitation input/land surface interaction upon the erosion–transportation–deposition system. Mechanical weathering in the fringing mountains provides ample debris supply. The impermeable rocks, steep slopes and sparse vegetation cause rapid overland flow into gullies during the infrequent storms. The high-energy discharge entrains and transports debris through narrow canyons or *arroyos*.

At the exit from the mountains, channel gradients suddenly decrease, causing deposition as alluvial fans. Repeated flash floods over several thousand years have extended these fans to fill the basin floor. Along each flank the individual fans have merged to form a depositional zone known as a *bajada*. The 'toes' of the bajadas are periodically truncated by flood discharges along Centennial Wash. The sedimentary sequence indicates the progressive down-basin loss of energy and load transport capacity. Coarse material is deposited first, near the mouths of the mountain arroyos. The lower components of the bajada consist of finer materials. Finally, along the bed of Centennial Wash, near the basin exit, a *playa* has formed. This is a flat surface consisting of fine silts, clays and evaporite deposits, where standing water, ponded back by rock bars at Mullens Cut, has evaporated.

Alluvial fans

A simple fan receives its water and sediment inputs from a single canyon or arroyo. On the fan surface, unless entrenchment is vigorous, channels become choked with debris, causing the water to spill outwards across the

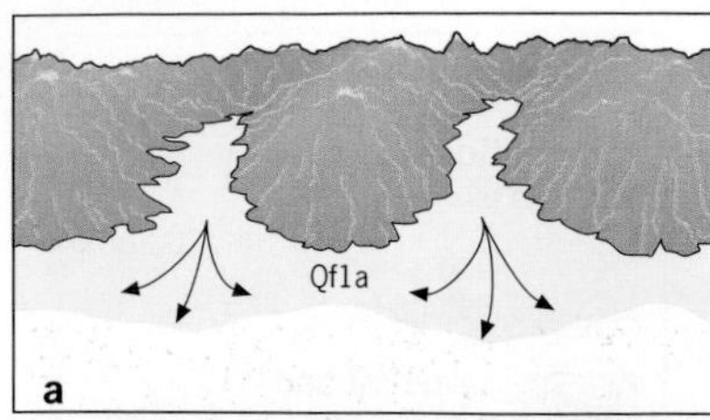

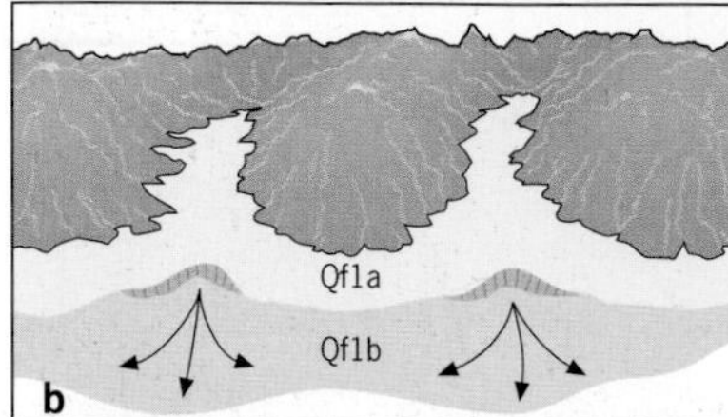

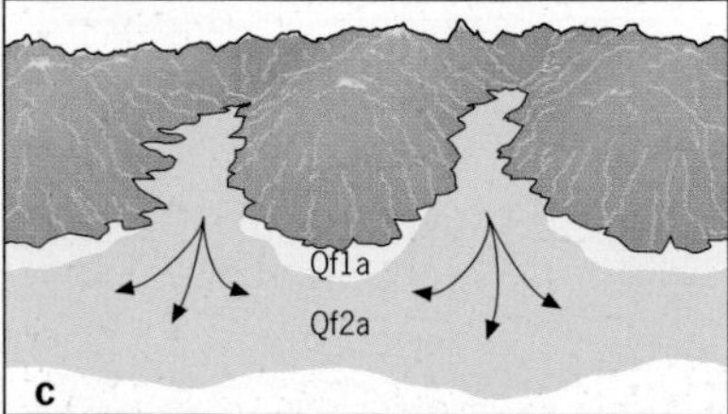

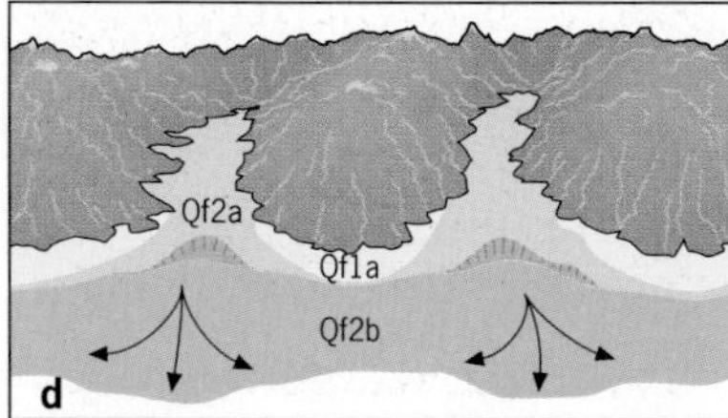

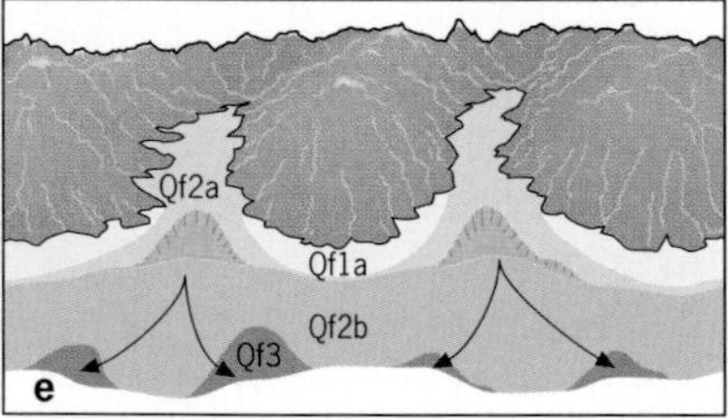

a Aggradation at canyon mouths, then coalescence of fans into bajadas during the early to middle Pleistocene (Qf1a).
b Subsequent increase in stream power relative to sediment load, causing fanhead entrenchment and a shift of deposition to lower parts of the fan (Qf1b).
c Deposition of Qf2 over much of Qf1 during the late Pleistocene.
d Reduction in sediment load relative to stream power during the early Holocene, causing fanhead entrenchment.
e Erosion of soil from interfluves of Qf2 by sheetwash and deposition on the lower fan during the late Holocene, forming Qf3.

Figure 9.30 Evolution of the Ajo bajada, Arizona, USA (*Source:* Parker, 1995)

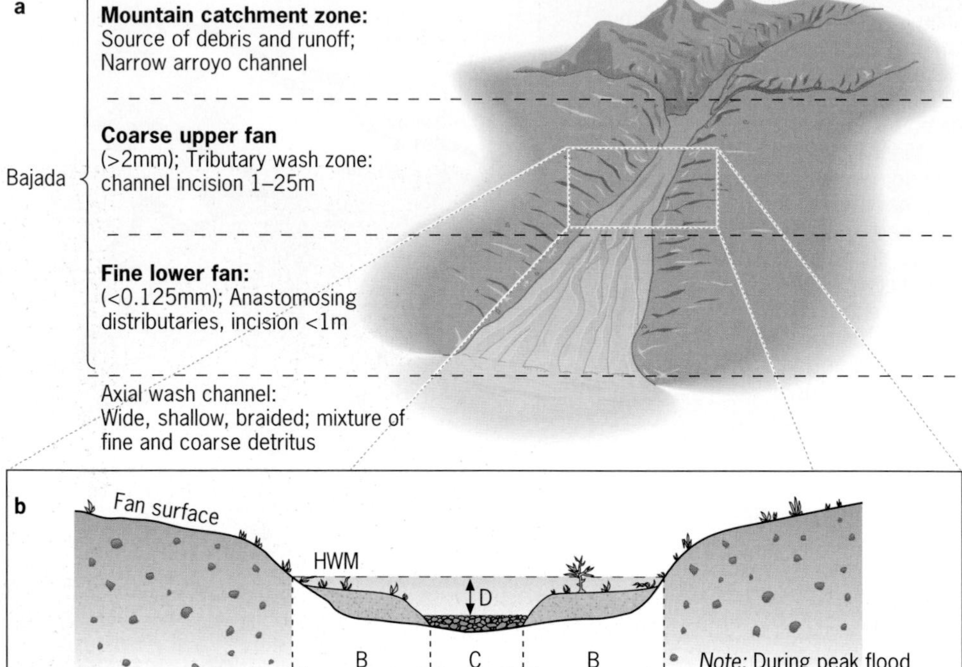

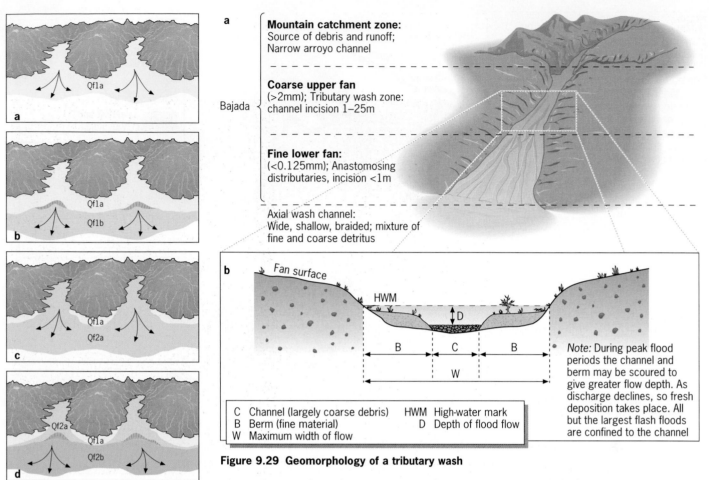

Figure 9.29 Geomorphology of a tributary wash

upper fan, seeking lower routes. This creates the fan or cone shape with sets of **anastomosing** channels criss-crossing the surface. The longitudinal surface profile is commonly gently concave, although some have straight or even convex profiles. Gradients are usually 3–6° and rarely exceed 10°. The fan cross-profile has a broad convex shape. The channel network extends across the whole fan surface, but at any given time certain channel directions are dominant. (Fan development has similarities with delta processes, discussed in Chapter 5.) Hence the aggradation pattern shifts around the fan over time. Figure 9.29 summarises these characteristics and the significance of the main channel or *wash* in the functioning of the fan system.

Large fans may be 15 km long and wide, and have been built over more than a million years. They are complex landforms consisting of several *segments*. Each segment has been created under a distinctive set of environmental conditions. When these conditions change, e.g. moister climate or reduced debris supply, so a new segment begins to build. The Ajo bajada of the Organ Pipe Cactus National Monument, Arizona, has evolved from two adjacent fans since the early Pleistocene period (1.5 million years BP) and has five segments (Fig. 9.30). The huge Hanaupah fan, in Death Valley, has three well-defined segments, the earliest accumulating by 800 000 BP (Fig. 9.31).

?

18 Describe and explain the distribution of particle sizes across a fan.

19 Why is an alluvial fan most likely to have a gently concave longitudinal profile?

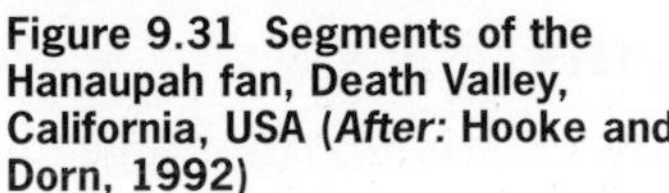

Figure 9.31 **Segments of the Hanaupah fan, Death Valley, California, USA (*After:* Hooke and Dorn, 1992)**

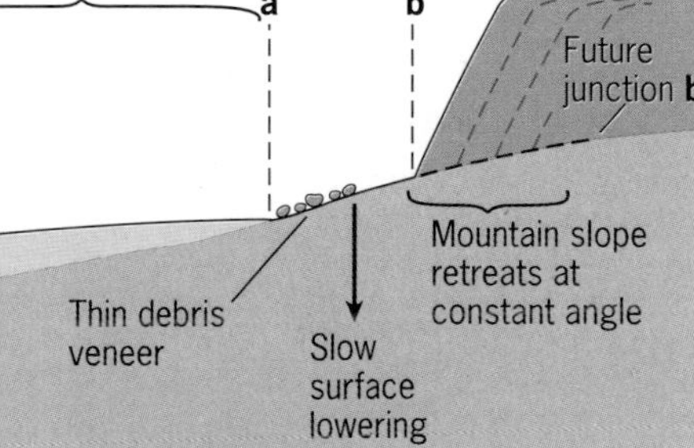

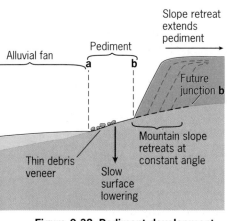

Figure 9.32 Pediment development

?

20 Construct a fully annotated systems diagram to summarise the process of alluvial fan development in an arid environment.

21 Essay: By the use of examples, discuss the factors which influence the location, evolution and morphology of fans in arid environments.

These examples possess environmental conditions favourable to vigorous fan formation:

- A debris source in unvegetated mountains with steep slopes;
- Long dry spells when weathered debris accumulates, and provides ample supply during the infrequent heavy storms;
- Extensive steep mountain catchment with an abrupt edge against a gently sloping surface, e.g. plain or basin;
- A mountain channel system focusing on one exit.

Pediments

The landform and slope system we have studied so far has three elements: mountain front; alluvial fan/bajada; wash or playa. The boundary of each is marked by an angular break of slope. (This angularity, common in arid landscapes, contrasts with the more gradual, curved slope boundaries frequent in moist environments.) Some landscapes, however, include a fourth element, a gently sloping bedrock surface with a discontinuous debris mantle between the mountain front and the alluvial fan. This additional landform is a *pediment*: a gently sloping surface, cut across bedrock and thinly or discontinuously veneered with weathered debris, that slopes away from the base of a mountain front in an arid or semi-arid environment.

Pediments are produced by the retreat of the mountain front and the gradual lowering of the pediment surface (Fig. 9.32). Pediments slope at up to 10°, and eroded debris delivered from the mountain front is transported across the pediment by surface waters during storms and enters the alluvial fill at the pediment toe. The width of the pediment is controlled by rock type, and the balance between debris supply and removal and, of course, time.

9.6 Stream channels and valley forms

In any environment there is a close relationship between stream discharge and channel form (see Chapter 2). Streamflow in arid and semi-arid environments is, with a few major exceptions, e.g. the River Nile, non-permanent (seasonal or ephemeral). Channel form is adapted to spasmodic surges of water and sediment. Two main types can be identified: the canyon, with near-vertical walls and a flat floor, with its smaller relative, the arroyo, and the wash, a shallow, poorly defined valley criss-crossed by braided channels. The term *wadi* can be used for gorge-like landforms of all scales. Large wadis and canyons have long histories and probably owe their development to periods of moister climate.

Canyons and arroyos

The Canyon de Chelly, Arizona (Fig. 9.33), is an excellent example of the power of seasonal and ephemeral streams to erode vertically into strong bedrock. It is a deep *box canyon*, so-named because of the box-like shape of vertical sides and flat floor. It has been incised more than 200 m into thick sandstones which are strong enough to stand as cliffs. The high energy discharges apply their energy mainly to vertical erosion, but the width of the canyon tells us that some has been applied laterally. The steepness of the canyon walls illustrates the slow rates of mechanical weathering and erosion in this dry climate, for there has been very little slope retreat. The flat floor is aggradational, built of up to 100 m of alluvial fill.

The key understanding is that the work of canyon formation is concentrated in brief, high-energy pulses, separated by long spells of very low activity. We can follow the story of an individual flood surge by the study of an arroyo (Fig. 9.34). Figure 9.35a represents the dry, inactive phase of the photograph. During a rainstorm, overland flow delivers runoff and weathered debris rapidly to the arroyo channel, creating a flash flood. Figure 9.35b summarises the situation at maximum discharge. Once the rain ceases stream flow and available energy decline rapidly. Sediment load is dumped and the alluvial fill is built up once more. The final surface flow is in braided channels across the gravel and sand arroyo floor. When surface flow ceases, the arroyo returns to a condition similar to that of Figure 9.35a.

Arroyo networks can spread quickly in semi-arid regions if environmental conditions change. For example, there has been widespread entrenchment of arroyos across the dry plains of the western USA over the past century. This is the result of the reduction of vegetation cover caused by increased cattle grazing as ranching has spread westwards.

?

22 What differences would you expect to find in the cross-section of Shell Canyon (Fig. 9.34) after a flood episode, and what factors would influence these changes from before the flood? (Remember, the arroyo has four elements: walls; rock bed; alluvial infill; alluvial arroyo surface.)

Figure 9.33 Canyon landscape: Canyon de Chelly, Arizona, USA

Figure 9.34 Arroyo, Shell Canyon, California, USA

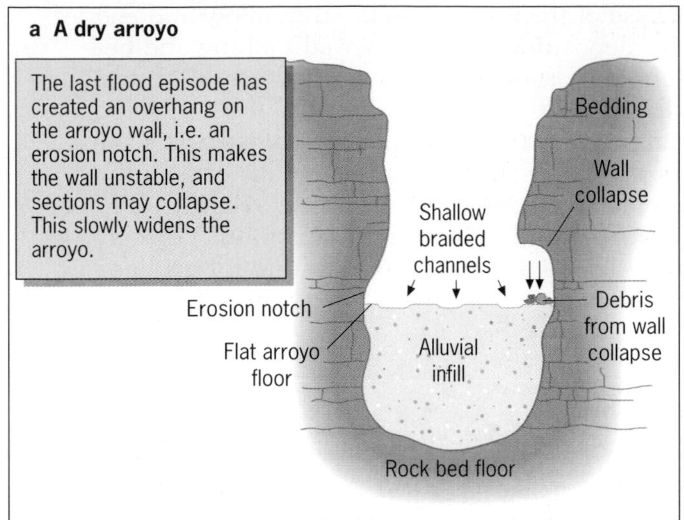
a A dry arroyo
The last flood episode has created an overhang on the arroyo wall, i.e. an erosion notch. This makes the wall unstable, and sections may collapse. This slowly widens the arroyo.
Bedding
Wall collapse
Shallow braided channels
Debris from wall collapse
Erosion notch
Flat arroyo floor
Alluvial infill
Rock bed floor

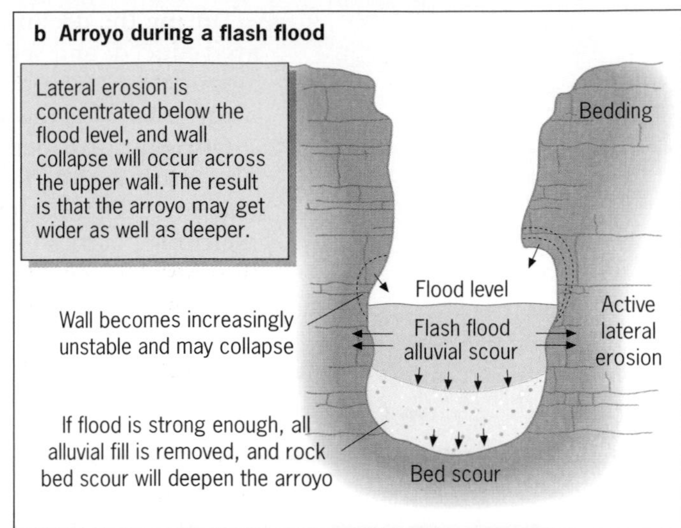
b Arroyo during a flash flood
Lateral erosion is concentrated below the flood level, and wall collapse will occur across the upper wall. The result is that the arroyo may get wider as well as deeper.
Bedding
Flood level
Wall becomes increasingly unstable and may collapse
Flash flood alluvial scour
Active lateral erosion
If flood is strong enough, all alluvial fill is removed, and rock bed scour will deepen the arroyo
Bed scour

Figure 9.35 Canyon and arroyo development: sequence through a flood episode

Figure 9.36 Bridge reconstruction across the Salt River bed, Phoenix, Arizona, USA. Bed scour during a severe flood undermined the bridge, causing collapse. The final remnants of standing water, the braided channels and the broad, shallow form of the valley floor can be seen. The alluvial sediments are up to 15 m thick.

Washes: channel and valley forms on low-relief landscapes

Across basins and plains where streams are not bounded by strong rock conditions, channels and valley forms tend to be broad and shallow (Fig. 9.36). The combination of low long-profile gradients, permeable bed materials and abundant sediment supply minimises vertical incision, and encourages lateral dispersal of streamflow. In many cases, the bank materials are unconsolidated sediments, erodible when lateral forces are applied by a stream.

As with canyons and arroyos, channel form is adapted to the irregular inputs of water and sediment. Figure 9.37a shows that the broad wash is built of alluvial infill (an aggradational landform) set within the surrounding desert surface. It is the equivalent to the floodplain of a permanent stream, but its width is controlled not by meanders (Chapter 3, pp.45–7), but by lateral shifts of braided channels (Fig. 9.37b). As water levels rise during a flood episode, streamflow develops a braided pattern. There is some vertical incision and bed scour (channels 2, 3 on Fig. 9.37b), but the key characteristic is that even high-energy flash floods struggle to cope with the volume of sediment supply. This energy/load imbalance explains the tendency to braided flow (Chapter 3, section 3.4). Lateral erosion causes the channels to shift across the wash surface, perhaps causing widening of the valley floor (channel 1 on Fig. 9.37b).

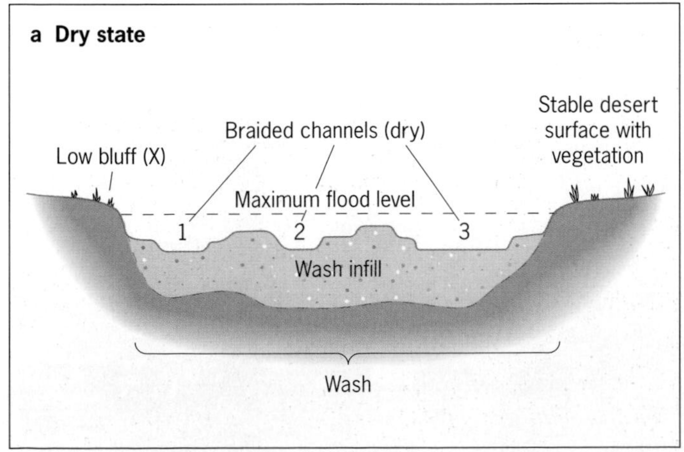
a Dry state
Stable desert surface with vegetation
Braided channels (dry)
Low bluff (X)
Maximum flood level
1
2
3
Wash infill
Wash

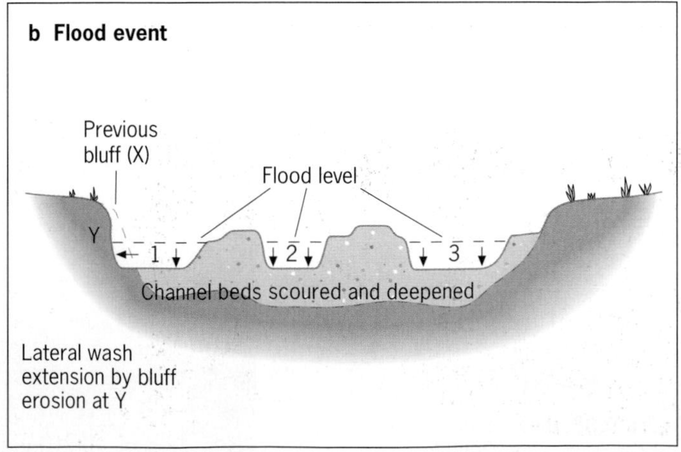
b Flood event
Previous bluff (X)
Flood level
Y
1
2
3
Channel beds scoured and deepened
Lateral wash extension by bluff erosion at Y

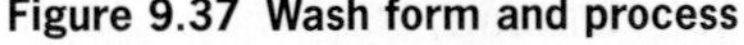
Figure 9.37 Wash form and process

During the declining phase of the flood episode, streamflow and energy are reduced, and sediment deposition causes renewed infilling and bed aggradation. Finally, discharge ceases, leaving a new pattern of braided dry channels across the wash surface.

Only during exceptional floods is the wash fully inundated and scoured. The result is that these broad valley and channel forms are constantly changing. This can create management problems, as the reconstruction of the bridge across the Salt River bed in Figure 9.36 illustrates. Furthermore, the fresh, unvegetated deposits are a ready sediment store for aeolian processes.

9.7 Human impacts

People have lived in arid and semi-arid environments for thousands of years, without causing major disruptions to the natural processes and landforms. Over the past few decades, however, human intrusion has accelerated. Population growth has caused intensified pressures on areas already settled, and encouraged or forced movement into increasingly arid margins. The recreation and tourism explosion has drawn millions of people into severe and beautiful environments.

Surfaces and landforms vary in their stability and in their *resilience* or *fragility*, that is, their ability to withstand changed inputs and impacts. Yet all have critical thresholds beyond which their form and stability will begin to change rapidly. In systems terms, positive feedback mechanisms take over from the negative feedback mechanisms and endanger the dynamic equilibrium of the landform. For instance, active sand dunes appear to have little stability, but if they are unvegetated, they may be appropriate sites for off-road recreational vehicle (ORV) activities (Fig. 9.38). They provide challenging terrain, yet they are resilient as the combination of wind and gravity soon repairs the impacts. In contrast, desert pavements and crusts are stable forms, but they are dangerously fragile. Once vehicle tyres break through the surface armour, the finer materials are easily eroded by wind and water, and gullied 'badlands' can quickly develop. There is even concern that Ayers Rock in Australia, which has resisted millions of years of natural processes, is being damaged by the hundreds of thousands of tourist feet which tramp across it each year.

Recreational impacts on landforms are much less serious than those created by increasing human and domestic animal pressures. These pressures disrupt natural processes and reduce landscape stability. One result is the widespread *desertification* occurring around the fringes of many

Figure 9.38 Blown Sand Mountain and dune buggies, Ocotillo Wells, Anza Borrego Desert, California, USA

hot and temperate deserts. Across the Sahel zone of the southern edge of the Sahara, a combination of increased population and grazing animals, and repeated droughts, has triggered accelerated erosion. This is a semi-arid region with expanses of dormant sand sheets and dunes. Such 'dormant' landforms are stable, but require only minor changes in environmental conditions to return to the 'active' state (see section 9.4). The reduction of vegetation cover and disruption of the surface crusts have caused this critical threshold to be crossed. The sand deposits are becoming active and mobile once more (see Case Study, Mali). Notice that, in this case, desertification does not mean the spread of desert-like conditions to *new* areas, but a *return* of such conditions to landforms which had been developed by arid processes in the past, and subsequently stabilised.

When wind, water, humans and animals combine: an example from Mali, Africa

The Gourma region of Mali lies within the northern arc of the River Niger (Fig. 9.39). Approximately 50 per cent of the surface consists of dormant and relict dune systems which accumulated at least 10 000 years ago. They are linear dunes aligned parallel to the prevailing easterly winds. They are typically 15–40 m high, with gently rounded cross-profiles and steeper northern slopes. The dune ridges are 50–1500 m apart, and individual ridges may be 10 km long. Figure 9.40 illustrates the landforms in the Singama district. Between the dunes, bedrock and stone crusts make up the surfaces, criss-crossed by sand-choked ephemeral stream channels.

The mean annual rainfall ranges from 400 mm in the south of Gourma to 200 mm in the north. Rainfall occurs on about 25 days during June–September. Over the past several thousand years there has been slow soil development on the dunes and a savanna grassland, with scattered trees, has evolved. The soil profiles and the vegetation have stabilised the dunes to a dormant and in places a relict state.

For centuries the Gourma dune ecosystem has been used by nomadic peoples for grazing and fuelwood, at sustainable levels. From the 1950s, however, rapid population growth has increased the grazing pressures and the fuelwood demands. When the droughts of the 1960s and 1970s struck, therefore, the vegetation was already under stress, and the surface trampling impacts of animals widespread. In 1980, a survey concluded that large areas of the Gourma dunefield had become 'a treeless grassland with large areas of bare soil surfaces ... and the destruction of the original plant cover has led to increasing soil erosion' (Barth, 1982).

Figure 9.41 shows the impact of one heavy rainstorm (intensity 25 mm/hr) in August 1977 on a

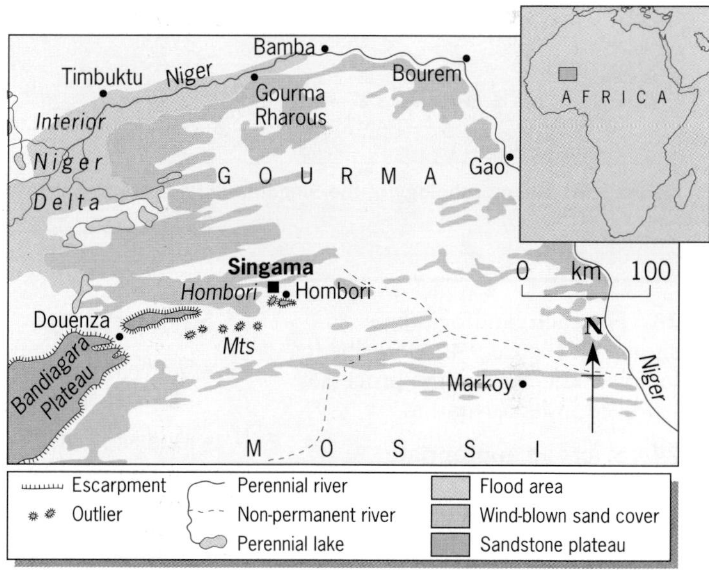

Figure 9.39 Distribution of wind-blown sand accumulations in the Gourma (*Source:* Barth, 1982)

dune largely stripped of vegetation. A series of gullies up to 2 m deep and 3 m wide were cut into the south-facing slope as far as the dune crest. The runoff and load fanned out at the dune foot, before flash flood flow removed the sediment westwards, until surface flow ceased. The fresh channel infill and the exposed dune face became a source of material for aeolian processes. Surveys in 1978 and 1979 showed significant sand movement in the Singama area whenever wind speeds exceeded 10 m/sec. Dust clouds obscured visibility at a local airport on 28 days in the dry season.

It is clear that short-term climate fluctuations, coinciding with intensified human pressures, have reactivated parts of the Gourma dunefield. In this region, desertification means the shift in dune condition from dormant or relict to active.

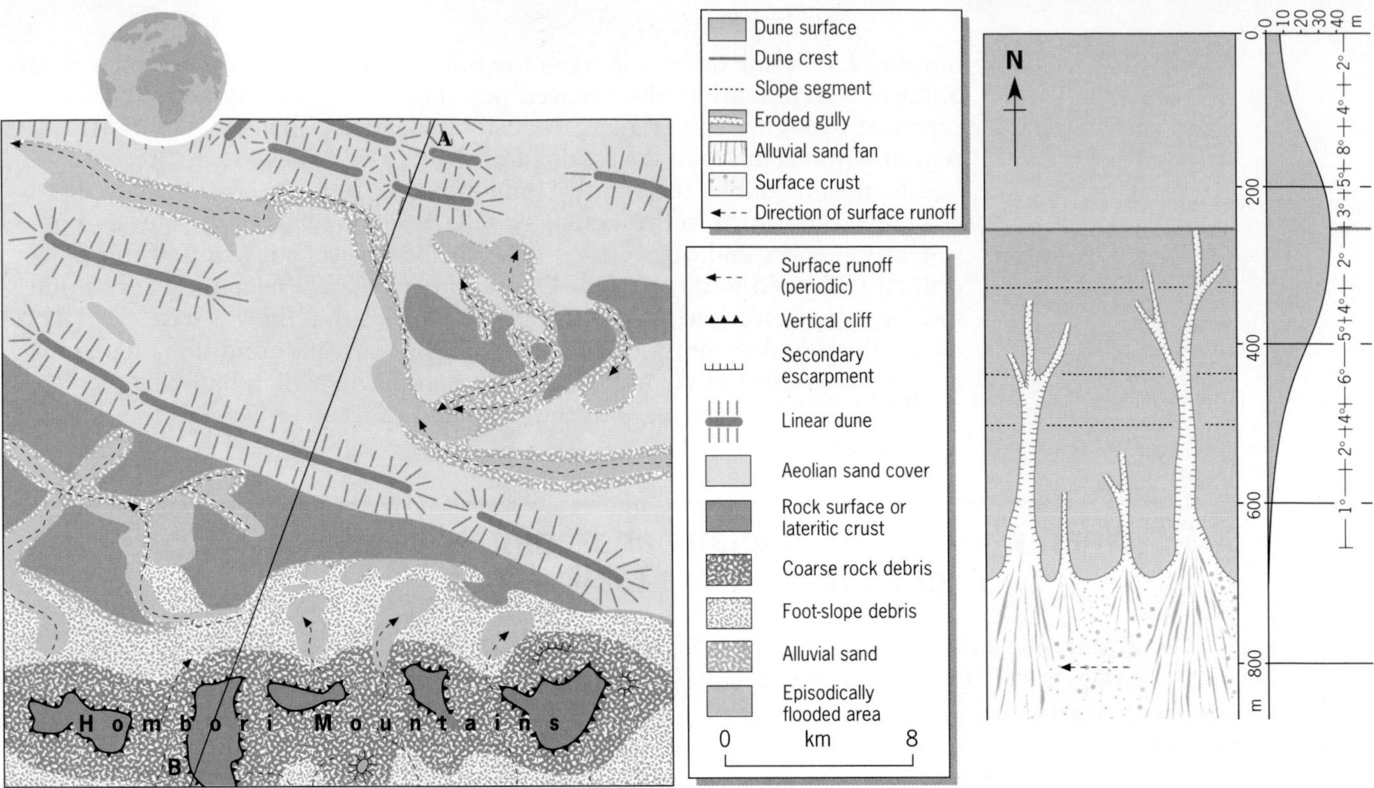

Figure 9.40 Geomorphology of the Singama dune system in the Gourma

Figure 9.41 Dune erosion at Singama

23 For each landform type shown on Figure 9.40, describe its location and outline the processes involved in its formation.

24 Select an appropriate sampling technique, and calculate the proportion of the total area shown on Figure 9.40 covered by sand.

25 Draw a sketch cross-section along line A–B on Figure 9.40 and label the landforms.

Summary

- Arid environments are characterised by climates with permanent moisture budget deficits, and incomplete or non-existent vegetation cover.
- Arid and semi-arid landscapes consist of collections of erosional and depositional landforms produced by the work of wind and water.
- The main aeolian processes of erosion are abrasion and deflation.
- Wind energy transports particles, mainly of silt and sand, by creep, saltation and suspension motion.
- Landforms created mainly by wind erosion processes vary from isolated hills (inselbergs, bornhardts, buttes, mesas), through surface crusts, pavements and pediments, to yardangs and deflation hollows.
- The main landform types produced by aeolian deposition are sandsheets, dunes and ramps.
- Sand-dunes may be active, dormant or relict, and individual dune fields may contain examples of each.
- Irregular and intense inputs of surface water play an important role in landform formation in all but the driest of deserts.
- The erosional landform types produced mainly by the work of seasonal or ephemeral streams are sharply incised canyons, arroyos, wadis, or broad shallow washes, criss-crossed by braided channels.
- The main depositional landforms from stream action are alluvial fans, bajadas and playas.
- Increased human activities are causing both localised erosion, e.g. recreational impacts and extensive desertification, e.g. from overgrazing, and there is a growing realisation that arid and semi-arid environments need careful management.

References

Ballantyne, C K (1987), 'The present-day periglaciation of upland Britain', in Boardman, J (ed.), *Periglacial Processes and Landforms*, Cambridge University Press.

Barth, H K (1982), 'Accelerated erosion of fossil dunes in the Gourma region (Mali) as a manifestation of desertification', *Aridic Soils and Geomorphic Processes, Catena Supplement*, Brunswick.

Barton, M E, Coles, B J and Tiller, G R (1983), 'A statistical study of the cliff-top slopes in part of the Christchurch Bay coastal cliffs', *Earth Surface Processes and Landforms*, 8.

Benn, D I (1993), 'Moraines in Coire na Creiche, Isle of Skye', *Scottish Geographical Magazine*, 109 (3).

Beven, K and Carling, P (1992), 'Velocities, roughness and dispersion in the lowland River Severn', in Carling, P and Petts, G E (eds), *Lowland Floodplain Rivers*, Wiley.

Bird, E C F (1985), *Coastline Changes: a global review*, Wiley.

Carling, P A (1986), 'The Noon Hill flash floods: 17 July 1983', *Transactions of the Institute of British Geographers*, NS 11.

Carr, A P, Blackley, M W L and King, H L (1982) 'Spacial and seasonal aspects of beach stability', *Earth Surface processes and Landforms*, 7.

Carter, R W G (1988), *Coastal Environments: an introduction to the physical, geological and cultural system of coastlines*, Academic Press.

Chorley, R J, Schumm, S A and Sugden, D E (1984), *Geomorphology*, Methuen.

Clark, M J, Rickets, P J and Small, R J (1976), 'Barton does not rule the waves', *Geographical Magazine*, 48 (10).

Clowes, A and Comfort, P (1987), *Process and Landform*, Oliver & Boyd.

Collard, R (1990), *The Physical Geography of Landscape*, Unwin Hyman.

Cooke, R U and Doornkamp, J C (1990), *Geomorphology in Environmental Management*, Oxford University Press.

Cooke, R U, Warren, A and Goudie, A (1993), *Desert Geomorphology*, UCL Press.

Drage, B, Gillman, J F, Hoch, D and Griffiths, L (1983), 'Hydrology of North Slope coastal plain streams', *Permafrost: Fourth International Conference Proceedings*, National Academy Press.

Embleton, C and King, C A M (1975), *Periglacial Geomorphology*, Arnold.

Embleton, C and Thornes, J (1979), *Process in Geomorphology*, Arnold.

Gilvear, D J and Harrison, D J (1991), 'Channel change and the significance of floodplain stratigraphy: 1990 flood event, lower River Tay, Scotland', *Earth Surface Processes and Landforms*, 16.

Goudie, A (1990), *The Landforms of England and Wales*, Blackwell.

Gupta, A and Dutt, A (1989), 'The Auranga: description of a tropical monsoon river', *Zeitschrift für Geomorphologie*, 33 (1), March.

Hall, A M (1983), 'Weathering and landscape evolution in NE Scotland' (PhD thesis, Aberdeen).

Hambrey, M (1994), *Glacial Environments*, UCL Press.

Hanson, J D (1988), *Coasts*, Cambridge University Press.

Harrison, S (1993), 'Solifluction sheets in the Bowmont Valley, Cheviot Hills', *Scottish Geographical Magazine*, 109 (2).

Harvey, A M (1991), 'The influence of sediment supply on the channel morphology of upland streams: Howgill Fells, NW England', *Earth Surface Processes and Landforms*, 16 (7), November.

Hooke, R and Dorn, R L (1992), 'Segmentation of alluvial fans in Death Valley, California: new insights from surface exposure dating and laboratory modelling', *Earth Surface Processes and Landforms*, 17.

Jones, R L and Keen, D H (1993), *Pleistocene Environments in the British Isles*, Chapman & Hall.

Kesel, R H, Yodis, E G and McCraw, D J (1992), 'An approximation of the sediment budget of the lower Mississippi River prior to human modification', *Earth Surface Processes and Landforms*, 17 (7), November.

Klimaszewski, M (1993), 'Corries and glaciation in Glacier National Park, Montana, USA', *Zeitschrift für Geomorphologie*, 37 (1), March.

Krüger, J (1994), 'Sorted polygons on recently deglaciated terrain in the highland of Maelifellssandur, South Iceland', *Geografiska Annaler*, 76 (1–2).

Lancaster, N (1994), 'Controls on aeolian activity: some new perspectives from the Kelso Dunes, Mojave Desert, California', *Journal of Arid Environments*, 27.

Lewkowicz, A G (1992), 'Factors influencing the distribution and limitation of active-layer detatchment slides on Ellesmere Island, Arctic Canada', in Dixon, J C and Abrahams, A D (eds), *Periglacial Geomorphology*, Wiley.

MAFF (1993), *Coastal Defence and the Environment*, HMSO.

MAFF (1995), *Shoreline Management Plans: a guide for coastal defence authorities*, HMSO.

McKirdy, A (1992), *Coastal Engineering*, English Nature.

McTainsh, G H and Walker, P H (1982), 'The nature and distribution of Harmattan dust', *Zeitschrift für Geomorphologie*, 26 (4), December.

Nanson, G, Chen, X Y and Price, D M (1992), 'Lateral migration, thermo-luminescence chronology and colour variation of longitudinal dunes near Birdsville in the Simpson Desert (central Australia)', *Earth Surface Processes and Landforms*, 17.

New Forest District Council, *Barton on Sea Coastal Protection* (information leaflet).

Parker, K C (1995), 'Effects of geomorphic history on soil and vegetation patterns on arid alluvial fans', *Journal of Arid Environments*, 30.

Pethick, J (1984), *An Introduction to Coastal Geomorphology*, Arnold.

Pérez, F L (1992), 'Miniature sorted stripes in the Páramo de Piedras Blancas (Venezuelan Andes)', in Dixon, J C and Abrahams, A D (eds), *Periglacial Geomorphology*, Wiley.

Pringle, A W (1985), 'Holderness coastal erosion and the significance of ords', *Earth Surface Processes and Landforms*, 10.

Prosser, R (1977), *Geology Explained in the Lake District*, David & Charles.

Ritchie, W and Walker, H J (1974), 'River in the frozen north', *Geographical Magazine*, March.

Robinson, L A (1977), 'Erosive processes on the shore platform of NE Yorkshire', *Marine Geology*, 23.

Selby, M J (1987), 'Slope and weathering', in Gregory, K J and Walling, D E (eds), *Human Activity and Environmental Processes*, Wiley.

Skinner, B J and Porter, S C (1987), *Physical Geology*, Wiley.

Small, R J (1989), *Geomorphology and Hydrology*, Longman.

Strahler, A H (1966), *The Earth Sciences*, Harper & Row.

Strunt, H (1983) 'Pleistocene drapiric upturnings of lignites and clayey sediments as periglacial phenomena in Central Europe', in *Permafrost: Fourth International Conference Proceedings*, July, National Academy Press.

Summerfield, M A (1991), *Global Geomorphology*, Longman.

Sunamura, T (1978), 'Mechanisms of shore platform formation on the south-eastern coast of the Izu Peninsula, Japan', *Journal of Geology*, vol. 86.

Thornes, C R (1992), 'River meanders: nature's answer to the straight line', inaugural lecture, Nottingham University, May.

Trudgill, S (1988), 'Integrated geomorphological and ecological studies on rocky shores in southern Britain', *Field Studies*, 7.

Vines, H and Spencer, T (1995), *Coastal Problems: geomorphology, ecology and society at the coast*, Arnold.

Whittow, J (1984), *Dictionary of Physical Geography*, Penguin.

Glossary

Ablation The wastage or removal of surface snow or ice by melting.

Accretion A process of accumulation or incremental growth.

Agents Wind, ice, running water, oceans: those agencies which energise erosion, transportation and deposition.

Aggradation The building up of a land surface by deposition.

Alluvial fan A fan-shaped mass of material deposited by streams when transportation energy declines, e.g. at the foot of a mountain range.

Anastomosis The branching and rejoining of distributary streams across a channel, delta or fan surface.

Base level The lowest level to which a stream can erode its valley.

Bed armour A layer of clasts of similar size in a channel bed which fit together to resist entrainment.

Berm A ridge of coarse sand and shingle on an upper beach, deposited by spring tides and storm waves.

Bifurcation ratio The quantitative relationship between the different orders of magnitude of streams in a drainage basin.

Closed-system pingo An ice-cored hill formed from local groundwater expelled under pressure by permafrost formation.

Col A lower section or gap in a mountain ridge.

Critical threshold The level of energy and/or mass at which crucial changes of process occur.

Critical tractive force The level of applied energy at which particles of specific mass and calibre begin to move.

Cuspate foreland A triangular accumulation of shingle and sand jutting into the sea.

Drainage basin The total area that contributes water to a river.

Drainage reversal Reversal in the direction of stream flow.

Driving variable The key variable influencing a variable.

Energy The capacity to produce activity.

Entrainment The process by which clasts are drawn into a moving body of water, ice or air.

Erratic A glacially transported and deposited rock whose composition differs from that of the bedrock on which it sits.

Firn Granular snow, at least one year old and intermediate in density between snow and ice.

Floodplain The part of a river valley that is inundated during floods.

Fluvial landforms Landforms created by the work of rivers.

Force The amount of energy applied.

Gravity The force of the earth on an object at rest.

Helical flow A spiral or corkscrew-like motion of water as it moves along a river channel, or an airstream which moves in a spiral form.

Ice budget The annual balance between accumulation and ablation in a glacier.

Ice-wedge pseudomorph A caste of a former ice wedge that has been infilled by sediment.

Incised meander A river meander cut deeply into bedrock as a result of rejuvenation.

Interlocking spur Protrusions of higher land characteristic of the upper sections of a river valley resulting from the river's winding course.

Karst Topography formed by limestone solution, characterised by a general lack of surface drainage.

Lag gravel The coarsest materials remaining on a desert surface after selective removal by wind action.

Lithology The character of rock – its structure, composition, texture and hardness.

Littoral zone The environment between the highest and lowest levels of spring tides.

Long profile The longitudinal section of a stream or glacier, drawn from source to mouth.

Mass balance A comparison of the inputs and outputs of processes, e.g. a measure of the change in total mass of a glacier during a year. The concept can be applied to sedimentary landforms, e.g. a beach.

Negative feedback The process by which a change is counteracted to return a system to a state of equilibrium.

Nivation The wearing down of rocks beneath and adjacent to a snow mass.

Open system A system whose boundaries are open to inputs and outputs of energy and matter.

Open-system pingo An ice-cored hill formed through the freezing of water that has moved from a distant higher source.

Ord A low section of beach where cliff erosion rates are increased. Characteristic of the Holderness coastline, where they are irregularly spaced and move southwards with the direction of longshore drift.

Plane A surface separating two beds of rock. It can also be a line of failure in a landslip.

Pool See **Riffle and pool**.

Positive feedback The process by which an initial change in the input of energy and matter through a system triggers progressive changes which may destabilise the system.

Process–response system A system in which the components of structure and the processes of function, and the relationships between them, are identified.

Prograding The process by which a shoreline advances seawards by sediment accumulation.

Ramp A change of gradient in a solid rock or sediment accumulation.

Recovery/relaxation time The period of time taken by a system to adjust to new environmental conditions after disruption by changed inputs of energy and matter.

Recurrence interval The average period of time between two events of similar magnitude.

Regelation The refreezing of water into ice within a glacier, usually caused by a reduction in pressure.

Regolith The surface mantle of rock debris and soil overlying bedrock.

Relict A feature which was formed by processes which are no longer operative in an environment but which still exists as an anomaly in the changed present-day climatic conditions.

Riffle and pool An alternating sequence of shallower reaches and gravel bars (riffles) and deeper reaches (pools) along a stretch of a river channel.

River capture The diversion of the headwaters of a river into a neighbouring drainage basin, leaving a beheaded stream in the original basin.

Sinuosity The degree to which the planform of a stream channel meanders.

Sinuosity ratio The amount of meandering exhibited by a stream, measured as the ratio between channel length and valley length.

Spit A narrow, elongated accumulation of sand and shingle projecting into the sea.

Storm beach A linear mass of coarse, rounded clasts deposited on an upper beach during a storm.

Streamflow The flow of surface water in a well-defined channel.

Subaerial A feature which occurs, or a process which operates, on the Earth's surface, i.e. not subterranean or submarine.

Sweep zone The mobile sediment on a beach deposited during summer aggradation and removed offshore during winter erosion.

Talik A layer of unfrozen ground beneath the seasonally frozen surface zone and above or within the permafrost.

Tectonic activity Internal forces of the Earth's crust at work, e.g. earthquakes, volcanic eruptions.

Thalweg The longitudinal profile of a stream bed.

Till Unsorted sediments deposited directly by glacier ice.

Toe armouring The temporary protection of a cliff foot by collapsed cliff material.

Tor A stack of well-jointed blocks projecting from a platform of solid rock. Often developed on granite.

Wave-cut notch An area at a cliff foot indented by direct wave erosion – most well developed in solid rock cliffs.

Index

Published by Collins Educational
77–85 Fulham Palace Road
London W6 8JB

An imprint of HarperCollins*Publishers*

First published 1997

ISBN 0 00 326686 9

Edited by Ron Hawkins

Designed by Jacky Wedgwood

Design production by Adrienne Lee

Picture research by Caroline Thompson

Computer artwork by Jerry Fowler, Hardlines and Barking Dog

Typeset in 10 on 12 point New Aster

Printed and bound in Hong Kong

Acknowledgements

Every effort has been made to contact the holders of copyright material, but if any have been inadvertently overlooked the publisher will be pleased to make the necessary arrangements at the first opportunity.

Photographs
The publisher would like to thank the following for permission to reproduce photographs:
Des Bowden, Fig. 9.16
British Geological Survey/National Environment Research Council, Figs 8.1, 8.10
Professor Ronald I Dorn, Arizona State University, Fig. 9.31
Tony Waltham Geophotos, Figs 1.11, 3.11, 3.15, 8.6, 8.9, 8.12, 8.14, 8.20, 8.23, 8.26, 8.35;
Reprinted from: *The Genus and Significance of 'hummocky moraine': evidence from the Isle of Skye, Scotland, 1992.* Article by D I Benn, published in the *Quaternary Science Review* 11: 781–99 (1992), with kind permission from Elsevier Science Ltd, The Boulevard, Langford Lane, Kidlington OX5 1GB, UK, Fig. 7.25;
Dr Nicholas Lancaster, Quaternary Sciences Center, Desert Research Institute, University of Nevada, Fig. 9.22;
London Aerial Photo Library, 5.38, 6.7, 8.2;
Ordnance Survey, Fig. 7.27;
Science Photo Library, Fig. 5.47;
© The Telegraph plc, London, 1996, Fig. 4.1;
Woodfall Wild Images, Figs 5.24, 5.25, 5.39, 5.41.

All other photographs in the book have been supplied by the authors, Victoria Bishop and Robert Prosser

Cover picture
Farm at Holmpton, Humberside – coastal landslide. *Source:* Tony Waltham Geophotos